The Properties of Petroleum Fluids

Second Edition

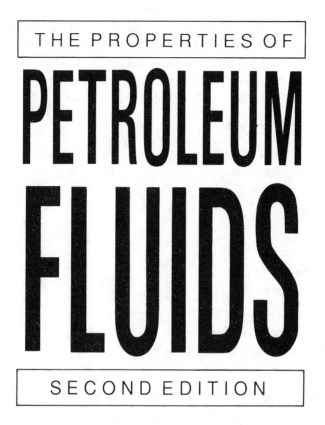

THE PROPERTIES OF PETROLEUM FLUIDS

SECOND EDITION

WILLIAM D. McCAIN, Jr.

PennWell Books
PennWell Publishing Company
Tulsa, Oklahoma

Copyright© 1990 by
PennWell Corporation
1421 South Sheridan Road
Tulsa, Oklahoma 74112-6600 USA

800.752.9764
+1.918.831.9421
sales@pennwell.com
www.pennwellbooks.com
www.pennwell.com

Marketing Manager: Julie Simmons
National Account Executive: Barbara McGee

Director: Mary McGee
Production/Operations Manager: Traci Huntsman

Library of Congress Cataloging-in-Publication Data Available on Request

McCain, William D., Jr. 1933-
   The properties of petroleum fluids, 2nd ed.
   ISBN13 978-0-87814-335-1
   P. cm.
   Includes bibliographies and index.
   1. Petroleum. 2. Gas. 3. Oil reservoir engineering.   I. Title.
TN870.5.M386 1989
665.5dc20
89-35421 CIP

Printed in the United States of America

9 10 11 12 13   13 12 11 10 09

The effort that went into the creation of this book is dedicated to my parents.

Contents

Nomenclature

English

a, b, c	constants in various equations of state
a, b, c, A_o, B_o	constants in Beattie-Bridgeman equation of state
a, b, c, A_o, B_o, C_o, α, γ	constants in Benedict-Webb-Rubin equation of state
a_c	constant in Soave-Redlich-Kwong and Peng-Robinson equations of state
a_T	temperature-dependent coefficient in Soave-Redlich-Kwong and Peng-Robinson equations of state
a_{Ti}	temperature-dependent coefficient of component i
a_{Tj}	temperature-dependent coefficient of component j
A	Avogadro's number of molecules per molecular weight
A	cross-sectional area
A	sum of mole fraction of carbon dioxide and hydrogen sulfide in Equation B–17
A, B, A′, B′	coefficients in various equations of state
A'_j, B'_j	coefficients for component j
A, B, E	coefficients in Beattie modification of Beattie-Bridgeman equation of state
AGP	additional gas produced, a parameter in reservoir-gas specific gravity equation, defined by Equation 7–11
API, °API	liquid gravity in °API, defined by Equation 8–2

$$6.29 \times 10^{-6} \frac{bbl}{cc}$$

$$\rho_w = 62.37 \, lb/cuft$$

xv

b_j coefficient for component j in various equations of state

b_j plotting factor defined by equation 14–4

b_o oil shrinkage factor, defined by Equation 8–4

B mole fraction of hydrogen sulfide in Equation B–17

$B, C, ...$ virial coefficients

B_g gas formation volume factor, specific to dry gas, defined by Equation 6–1

B_o oil formation volume factor, defined by Equation 8–3

B_{ob} oil formation volume factor at bubble-point

B_{oD} see Table 10–3

B_{oDb} see Table 10–3

B_{oSb} see Table 10–3 *formation Volume factor of oil*
P 9.276

B_t total (two-phase) formation volume factor, defined by Equation 8–6

B_{tD} see Table 10–3

B_w water formation volume factor, defined in Chapter 16

B_{wb} water formation volume factor at the bubble point of water

B_{wg} wet-gas formation volume factor, defined by Equation 7–12

c_g gas coefficient of isothermal compressibility, defined by Equations 6–4

c_o liquid coefficient of isothermal compressibility, defined by Equations 8–7 or Equation 8–24

c_{pr}	gas pseudoreduced coefficient of isothermal compressibility, defined by Equation 6–14
c_w	water coefficient of isothermal compressibility, defined by Equations 16–2 or Equation 16–3
C	constant of integration or constant of proportionality
$(CN)_{pb}$	defined by Equation B–38
$(CN)_{Bob}$	defined by Equation B–48
C_p	gas heat capacity at constant pressure
d	molar density $1/V_M$
$EXP(x)$	equivalent to e^x
f	fugacity, defined by Equation 15–3
f_j	fugacity of component j, defined by Equation 15–18
f_g	fugacity of gas in equilibrium with liquid
f_{gj}	fugacity of component j of gas in equilibrium with liquid
f_L	fugacity of liquid in equilibrium with gas
f_{Lj}	fugacity of component j of liquid in equilibrium with gas
F_J	factor in Stewart-Burkhardt-Voo equation, defined by Equation B–9
G	chemical potential
G_j	chemical potential of component j
GPM	liquids content of gas, gallons per Mscf
GPM_j	liquids content of component j in gas, gallons per Mscf of gas

J	factor in Stewart-Burkhardt-Voo equation, defined by Equation B-5
J'	factor in Stewardt-Burkhart-Voo equation, defined by Equation B-3
K	equilibrium ratio, defined by Equation 12-14
K	factor in Stewart-Burkhardt-Voo equation, defined by Equation B-6
K'	factor in Stewart-Burkhardt-Voo equation, defined by Equation B-4
K_j	equilibirum ratio of component j
K^C_j	calculated value of equilibrium ratio of component j
K^T_j	trial value of equilibrium ratio of component j
ln	natural logarithm, base e
log	common logarithm, base 10
L	length
L_c	heating value (heat of combustion)
L_{cj}	heating value of component j
L_v	latent heat of vaporization
m	mass
m	constant in Soave-Redlich-Kwong and Peng-Robinson equations of state
m'	mass of one molecule
m_R	mass of reservoir gas
M	molecular weight

M_a	apparent molecular weight, defined by Equation 3–35
M_{air}	apparent molecular weight of air
M_{C7+}	apparent molecular weight of heptanes plus fraction
M_g	apparent molecular weight of gas
M_j	molecular weight of component j
M_L	apparent molecular weight of liquid in equilibrium with gas
M_o	apparent molecular weight of liquid
M_{oR}	apparent molecular weight of reservoir liquid
M_{STO}	apparent molecular weight of stock-tank liquid
n	number of moles, total moles
n'	number of molecules
n_g	number of moles of gas in equilibrium with liquid
n_j	number of moles of component j
n_L	number of moles of liquid in equilibrium with gas
n_{Lf}	moles of liquid remaining at end of differential vaporization
n_{Li}	moles of liquid at start of differential vaporization
n_R	number of moles of reservoir gas
$\bar{n}_g$	fractional moles of gas, n_g/n
$\bar{n}_{g1}\ \bar{n}_{g2},\ ...$	fractional moles of gas formed in stages 1, 2, ... of stage separation process
$\bar{n}_L$	fractional moles of liquid, n_L/n

$\bar{n}_{L1}, \bar{n}_{L2}, \ldots$	fraction moles of liquid formed in stages 1, 2, ... of stage separator process
p	pressure
Δp	pressure change
$\Delta \text{pressure}$	$p_{b\ corr} - p_{b\ field}$, Equation 11–8
p_b	bubble-point pressure
$p_{b\ corr}$	bubble-point pressure obtained from correlation
$p_{b\ field}$	bubble-point pressure obtained from production data
p_c	critical pressure
p_{cj}	critical pressure of component j
p_d	dew-point pressure
p_j	partial pressure of component j in gas mixture, defined in Chapter 3
p_k	convergence pressure
p_{pc}	pseudocritical pressure, defined by Equation 3–42
p'_{pc}	pseudocritical pressure adjusted for nonhydrocarbon components, see Equation 3–45
p_{pr}	pseudoreduced pressure, defined by Equation 3–43
p_r	reduced pressure, defined by Equation 3–41
p_{rj}	reduced pressure of component j
p_R	reservoir pressure
p_{sc}	pressure, standard conditions, see Table 6–1
p_{SP}	separator pressure

$p_{SP1}, p_{SP2}, \ldots$	pressure of stages 1, 2, … of stage separation process
p_v	vapor pressure
p_{vj}	vapor pressure of component j
p_{vr}	reduced vapor pressure, p_v/p_c
P	parachor in interfacial tension equations
P_j	parachor for component j
r	electrical resistance
R	producing gas-oil ratio
R	universal gas constant
R_s	solution gas-oil ratio (gas solubility in oil), defined by Equation 8–5 *ie gas solubility*
R_{sb}	solution gas-oil ratio at bubble point
R_{sD}	see Table 10–3 *Pg.281*
R_{sDb}	see Table 10–3 *Solution GOR*
R_{sSb}	see Table 10–3
R_{sw}	solution gas-water ratio (gas solubility in water)
R_w	water resistivity, defined by Equation 16–5
R_{SP}	separator producing gas-oil ratio
$R_{SP1}, R_{SP2}, \ldots$	producing gas-oil ratio from stages 1, 2, … of stage separation process
R_{ST}	stock-tank producing gas-oil ratio
S	salinity of brine

T	temperature
ΔT	temperature change
T_B	temperature at normal boiling point
T_{Bj}	temperature at normal boiling point of component j
T_c	critical temperature
T_{cj}	critical temperature of component j
ΔT_h	reduction in hydrate-forming temperature
T_{pc}	pseudocritical temperature, defined by Equation 3–42
T'_{pc}	pseudocritical temperature adjusted for nonhydrocarbon components, see Equation 3–44
T_{pr}	pseudoreduced temperature, defined by Equation 3–43
T_r	reduced temperature, defined by Equation 3–41
T_{rj}	reduced temperature of component j
T_R	reservoir temperature
T_{sc}	temperature, standard conditions
T_{SP}	separator temperature
$T_{SP1}, T_{SP2}, \ldots$	temperature of stages 1, 2, ... of stage separation process
v	molecular velocity
v	specific volume
V	volume

V_b	volume of liquid at the bubble point
VEQ	equivalent gas volume of stock-tank liquid, a parameter in reservoir gas specific gravity equation, defined by Equation 7– 10
V_j	partial volume of component j, defined in Chapter 3
V_M	molar volume, i.e., volume of one mole
V_{Mc}	molar volume at the critical point
V_{Mg}	gas molar volume
V_{ML}	liquid molar volume
V_o	liquid volume
V_R	gas volume calculated at reservoir conditions
V_{sc}	gas volume calculated at standard conditions
V_t	total volume
$\left(\dfrac{V_t}{V_b}\right)_F$	see Table 10–3
V_w	water volume
ΔV_{wp}	change in water volume during pressure reduction in B_w equation, Equation 16–1
ΔV_{wT}	change in water volume during temperature reduction in B_w equation, Equation 16–1
w_{C1}	lb methane/lb mole mix (mole fraction times molecular weight of component)
w_{C2}	lb ethane/lb mole mix (mole fraction times molecular weight of component)
w_{CO2}	lb carbon dioxide/lb mole mix (mole fraction times molecular weight of component)

W_{H2S}	lb hydrogen sulfide/lb mole mix (mole fraction times molecular weight of component)
w_{mix}	lb mix/lb mole mix (molecular weight of total mixture)
w_{N2}	lb nitrogen/lb mole mix (mole fraction times molecular weight of component)
w_j	weight fraction of component j
W	moisture content, Equation B–76
W	weight percent solute, Equation B–89
W_1	weight fraction of methane in mixture, defined by Equation 11–1, used as percent in Figure 11–6
W_2	weight fraction of ethane in ethane and heavier, defined by Equation 11–2, used as percent in Figure 11–6
x_j	mole fraction of component j in liquid
x_{ji}	mole fraction of component j in liquid at initial conditions
x_{jf}	mole fraction of component j in liquid at final conditions
y_j	mole fraction of component j in gas
Y	condensate volume
z	compressibility factor ($z = pV/nRT$), defined by Equation 3–40
z_c	compressibility factor at critical point
z_g	compressibility factor of gas ($z_g = pV_{Mg}/RT$)
z_j	mole fraction of component j in total mixture

| z_L | compressibility factor of liquid ($z_L = pV_{ML}/RT$) |
| z_{sc} | gas compressibility factor at standard conditions |

Greek

α	temperature-dependent coefficient in Soave-Relich-Kwong and Peng-Robinson equations of state
α_j	temperature-dependent coefficient of component j
B	coefficient of isobaric thermal expansion, defined by Equation 8–25
γ_{C7+}	specific gravity of heptanes-plus fraction, defined as for liquid
γ_g	gas specific gravity, defined by Equation 3–36
γ_{gR}	reservoir gas specific gravity
γ_{gSP}	separator gas specific gravity
γ_{gSP1}	first-stage separator gas specific gravity
γ_{gSP2}	second-stage separator gas specific gravity
γ_{gST}	stock-tank gas specific gravity
γ_o	liquid specific gravity, defined by Equation 8–1
γ_{oj}	specific gravity of component j of a liquid
γ_{STO}	stock-tank oil specific gravity
γ_w	water (brine) specific gravity, defined in Chapter 16
δ_{ij}	binary interaction coefficient in Soave-Redlich-Kwong and Peng-Robinson equations of state
Δ	indicates difference

ϵ	pseudocritical temperature adjustment factor for nonhydrocarbon components, Equations 3–44 and 3–45
ϵ_j	error function of component j, Chapter 15
ϵ_J	factor in Stewart-Burkhardt-Voo equation, defined by Equation B–7
ϵ_K	factor in Stewart-Burkhardt-Voo equation, defined by Equation B–8
μ	dynamic viscosity
μ_g	gas viscosity (dynamic)
μ_{gj}	gas viscosity of component j
μ_{g1}	gas viscosity (dynamic) at atmospheric pressure
μ_o	oil viscosity (dynamic)
μ_{ob}	oil viscosity (dynamic) at bubble-point pressure
μ_{oD}	oil viscosity (dynamic) at atmospheric pressure (dead oil)
μ_w	water viscosity (dynamic)
μ_{w1}	water viscosity (dynamic) at atmospheric pressure
ν	kinematic viscosity, defined by Equation 6–15
ρ_a	apparent liquid density of solution gas
$\rho_{a,C1}$	apparent liquid density of methane
$\rho_{a,C2}$	apparent liquid density of ethane
ρ_{bs}	liquid density at reservoir pressure and 60°F
ρ_{C3+}	density of propane and heavier part of mixture

ρ_g	density of gas
ρ_{gc}	density of gas at its critical point
$\Delta\rho_{H2S}$	adjustment to liquid density due to hydrogen sulfide content
ρ_L	density of liquid in equilibrium with gas
ρ_o	density of liquid
ρ_{oj}	density of component j as a liquid at standard conditions
ρ_{ob}	density of liquid at bubble-point conditions
ρ_{oR}	density of reservoir liquid at reservoir conditions
$\Delta\rho_p$	adjustment to liquid density due to pressure
ρ_{po}	density of pseudoliquid
ρ_{pr}	pseudoreduced gas density, defined by Equation B−18, B−11
ρ_{STO}	density of stock-tank oil, at standard conditions
$\Delta\rho_T$	adjustment to liquid density due to temperature
ρ_w	density of water (brine)
σ	interfacial tension
σ_{gw}	gas-water interfacial tension
ϕ_j	fugacity coefficient of component j, defined by Equation 15−21
ϕ_{gj}	fugacity coefficient of component j in gas
ϕ_{Lj}	fugacity coefficient of component j in liquid

ω	acentric factor
ω_j	acentric factor of component j
Ω_a	constant in equation of state, 0.45724 in Peng-Robinson equation, 0.42747 in Soave-Redlich-Kwong
Ω_b	constant in equation of state, 0.07780 in Peng-Robinson equation, 0.08664 in Soave-Redlich-Kwong equation

Subscripts

a	apparent, also used with Ω_a to indicate constant in a_c equation
actual	actual (as in real gas)
air	air
A, B, C	different chemical species, different components of a mixture
A	argon
b	bubble point, also used with Ω_b to indicate constant in b equation
B	boiling point
c	critical, also used with L_c to represent combustion
corr	value of property obtained from correlation
C1, C2, C3, ...	methane, ethane, propane, ...
C3+	property of propane-plus fraction of the petroleum mixture
C7+	property of the heptanes-plus fraction of the petroleum mixture

CO2	carbon dioxide
C+	property of the plus fraction of the petroleum mixture
d	dew point
dry	dry, with L_c to indicate no water vapor prior to combustion
D	property measured in a differential vaporization (see Table 10–3) also used for dead oil
f	final value or final conditions
field	value of property obtained from production history
F	property measured in a flash vaporization, see Table 10–3
g	gas
g1	gas at atmospheric pressure
H2S	hydrogen sulfide
i	initial value or initial conditions or different components of a mixture
ideal	property of an ideal gas or ideal gas mixture
ij	i and j represent different components of a mixture
ijk	i, j, and k represent different components of a mixture
j	different components of a mixture
jf	j represents different components of a mixture, f represents final value or final conditions
ji	j represents different components of a mixture, i represents initial value or initial conditions

L	liquid, usually in the context of liquid in equilibrium with a gas
mix	total mixture
M	molar
N2	nitrogen
o	oil or liquid
O2	oxygen
p	pressure (in the context of constant pressure in C_p) or pressure dependent
pc	pseudocritical
po	pseudoliquid
pr	pseudoreduced
r	reduced
R	reservoir (used only when necessary to distinguish between quantities in same calculation, such as ρ_{oR} and ρ_{STO})
s	solution (gas in oil)
sc	standard conditions
sw	solution (gas in water)
S	property measured in a separator test, see Table 10–3
SP	separator
SP1	first-stage separator
SP2	second-stage separator

ST	stock tank
STO	stock-tank oil (used only when necessary to distinguish between quantities in same calculation, such as ρ_{oR} and ρ_{STO})
t	total or two phase
T	temperature or temperature dependent
v	used in p_v to indicate vapor pressure and L_v to indicate vaporization
w	water (oilfield brine)
wet	wet, with L_c to indicate saturated with water vapor prior to combustion
wg	wet gas
1, 2, 3, ...	intended primarily to indicate different conditions of pressure and temperature, also used to indicate stages of separation

Superscripts

C	calculated value
T	trial value

Abbreviations

A	angstrom unit (10^{-8} cm)
AGP	additional gas produced, a parameter in reservoir-gas specific gravity equation
°API	degree (American Petroleum Institute)
atm	atmosphere
bbl	barrel
bbl/d	barrels per day

BTU	British thermal unit
cc	cubic centimeter
cm	centimeter
cp	centipoise
cu	cubic
cu ft	cubic foot
cu ft/d	cubic feet per day
cu ft/lb	cubic feet per pound
cu ft res gas	cubic feet of reservoir gas reported at reservoir conditions
cu ft STO	cubic feet of stock-tank liquid reported at standard conditions
cu m	cubic meter
C	indicates critical point on diagram
C_1, C_2, ...	methane, ethane, ... subscript indicates number of carbon atoms
C_{2+}	ethane and heavier fraction of mixture
C_{3+}	propane and heavier fraction of mixture
C_{7+}	heptanes and heavier fraction of mixture
d	day
diff vap	differential vaporization
eq wt	equivalent weight
ft	foot

ft-lb	foot-pound
°F	degrees Fahrenheit
g	gram
g/cc	grams per cubic centimeter
gal	gallons
g mole	gram mole
gr	grain
GOR	gas-oil ratio
GPM	gallons per Mscf
hp-hr	horsepower hour
Hg	mercury
in	inch
in Hg	inches of mercury
kg/sq cm	kilograms per square centimeter
k mole	kilogram mole
k Pa	kilo pascal
K	Kelvins
kw-hr	kilowatt hour
l	liter
lb	pound
lb/cu ft	pounds per cubic foot

lb/gal	pounds per gallon
lb mole	pound mole
lb/sq in	pounds per square inch
lb/sq ft	pounds per square foot
lim	limit
liq	liquid
ln	natural logarithm (base e)
log	common logarithm (base 10)
m	meter
meq	milliequivalent weight
mg	milligram
microsip	$psi^{-1} \times 10^6$
ml	milliliter
mm	millimeter
mm Hg	millimeters of mercury
mol wt	molecular weight
MMscf	million standard cubic feet
MMSTB	million stock-tank barrels
Mscf	thousand standard cubic feet
Mscf/d	thousand standard cubic feet per day
MSTB	thousand stock-tank barrels

MSTB/d	thousand stock-tank barrels per day
oz/sq in	ounce per square inch
ppm	parts per million
psi	pounds per square inch
psia	pounds per square inch absolute
psig	pounds per square inch gauge
res	reservoir
res cu ft	cubic feet of reservoir gas reported at reservoir conditions
res bbl	barrels of reservoir liquid or gas (or both) reported at reservoir conditions
°R	degrees Rankin
scf	standard cubic feet, volume of gas reported at standard conditions
scf/STB	standard cubic feet of gas per barrel of stock-tank liquid
sec	second
sip	psi^{-1}
sp. gr.	specific gravity
sq	square
sq cm	square centimeter
sq ft	square foot
sq in	square inch

SP	separator
SP bbl	barrels of separator liquid
ST	stock tank
STB	barrels of stock-tank oil (stock-tank barrel) reported at standard conditions, also used for barrels of water at standard conditions
STB/d	stock-tank barrels per day
STO	stock-tank oil
vs	versus
VEQ	equivalent gas volume of stock-tank liquid, a parameter in reservoir-gas specific gravity equations
wt	weight

Preface

■Substantial progress in our knowledge of the physical properties of reservoir fluids has occurred during the last fifteen years. Correlations have been improved. Observations of production behavior and laboratory results have led to better rules of thumb for use in determining the type of fluid in the reservoir. Tabular and graphical correlations have been reduced to equation form for use in computers. Laboratory procedures for analyzing reservoir fluids have been standardized. And techniques for converting laboratory results into useable fluid properties have been clarified. New information with regard to the chemical nature of petroleum has become available. Development in use of equations of state to calculate gas-liquid equilibria has been rapid.

For these reasons a rewrite of the first edition of this book became necessary. Original text and figures have been used where possible; however, some rearrangement was indicated based on experience in classroom use of the book. Revisions and additions to bring the book up to date were made as needed.

Each correlation presented in this book has been compared with other correlations of the same property and with available experimental data; only the most accurate correlations are presented. Verifications of the best correlations were based on a large set of fluid property data provided to Texas A&M University by Core Laboratores, Inc. Essential identifying features of the data (such as company, geographical or geological location, date of sampling and analysis, well or field name) were removed; however, the data represent most areas of the free world in which petroleum exploration and production were active during 1980-1986. Recognition is due to Phil Moses of Core Laboratories for providing these data.

Most of the chapters of the book are intended to be complete, that is, to provide all of the required information, equations, and correlations available for that particular topic. However, chapters 4, 14, and 15 do not fit this pattern.

Chapter 4 does not address all available equations of state. It merely shows the history of the art of developing equations of state. Chapter 14 is a very simple introduction into K-factor correlations. Chapter 15 only introduces the topic of equation-of-state gas-liquid equilibria calculations. The underlying theory of the calculations is illustrated; however, none of the difficulties of application to petroleums nor any of the sophisticated techniques of solving the equations are addressed.

This book was written for use as the textbook for a three-semester-hour sophomore- or junior-level course in a petroleum engineering curriculum. Since a secondary objective was to provide a reference book for practicing petroleum engineers, some of the material is unnecessary to prepare the undergraduate student for further petroleum engineering coursework.

Chapter 1 can be omitted in its entirety for those users with a good understanding of organic chemistry. However, I recommend that it be covered in the undergraduate petroleum engineering course because it emphasizes those parts of earlier chemistry courses which are important to petroleum engineering. Chapters 4 and 15 should not be studied in a first course in fluid properties. They provide an introduction into equation-of-state gas-liquid equilibria calculations. Normally, such calculations are not required of undergraduates or most practicing petroleum engineers. Chapter 7 contains material not usually needed by the undergraduate. Most of Chapter 14 can also be omitted, as it is reference material for those with an interest in K-factor correlations.

The textbook should be presented in order as the chapters are numbered. Chapters 1 through 3 give reference materials, mostly covered in earlier chemistry and physics courses, which underlie all of the remaining chapters. Chapter 2 must precede Chapter 5. Chapter 3 must precede Chapter 4. Chapter 5 is the most important one in the book: it sets the stage for everything that follows. Chapters 6 and 7 are a two-chapter sequence of gas properties which must be preceded by chapters 3 and 5. Chapters 8 through 11 are a four-chapter sequence of black-oil properties, which uses material from Chapter 5 and, to some extent, Chapter 6. Chapters 12 through 15 are a four-chapter sequence of gas-liquid equilibria calculations which uses material from chapters 2 through 8, 10, and 11. Chapters 16 and 17 are a two-chapter sequence of water properties which more or less stands alone. However, the definitions in these chapters will be more easily understood after study of earlier chapters.

The book can be covered in one semester with the following schedule: Chapter 1: 4 class days; Chapter 2: 4 days; Chapter 3 (omit kinetic theory): 3 days; Chapter 5: 2 days; Chapter 6 (omit heating value and Joule-Thomson effect): 2 days; Chapter 8: 2 days; Chapter 9: 2 days; Chapter 10: 3 days; Chapter 11: 6 days; Chapter 12 (omit differential vaporization): 3 days; Chapter 13: 3 days; Chapter 14 (convergence pressure only): 1 day; Chapter 16: 2 days; Chapter 17: 2 days. This schedule leaves several class meetings for examinations or other topics desired by the instructor.

A brief discussion of the definitions of certain words used in the book is required.

Fluid refers to either a liquid or a gas or both. The word has no other special connotation.

The word *petroleum* refers to either a liquid or a gas in the sense of naturally occurring and predominately hydrocarbon.

The word *liquid* is used throughout the book to indicate a petroleum liquid. It means oil, crude oil, condensate, distillate, as the case may be. The various regulatory bodies which oversee the petroleum industry have attempted to attach special definitions to these terms. These definitions are often contradictory and add confusion to the study of reservoir fluid properties.

The word *oil* is used in phrases such as gas-oil ratio, stock-tank oil, oil formation volume factor, and oil compressibility because of common usage in the petroleum industry. In this context, *oil* simply refers to a petroleum liquid which may be the result of the production of reservoir liquid or condensation from the production of reservoir gas.

In some instances the phrases *surface liquid* or *stock-tank liquid* are used to mean a petroleum liquid which ends up in the stock tank.

The word *water* is used for liquid water. In the oil field, *water* is usually highly saline and perhaps should be called *brine*.

I use the word *gas* throughout the book to indicate petroleum gas. Limited use occurs of the term *natural gas,* meaning "a naturally occurring gas which is predominately hydrocarbon." Other terminologies (such as *gas-well gas, oil-well gas, casinghead gas*) have been given special meanings through regulatory or contractual definitions. Vacuous phrases such as these are not used in this book. Expressions like *air* or *water vapor* are used for gases other than petroleum gases.

Brenda G. Bridges edited this book. Her competence, professionalism, diligence, and positive attitude made working with her a pleasure. She made an important contribution to the completion of this huge task.

I am indebted to the Darnell Group, Fort Worth, for their excellent drafting and cheerful redrafting; to Charles Aufil of Texas A&M University for his careful editing of early drafts of the book; to the employees of Cawley, Gillespie & Associates, Inc. (especially Bob Ravnaas and Richard Alexander) for assistance and encouragement; to Betty Campbell (Now you've seen everything—a Campbell mentioned in a book written by a MacIain! They won't believe this in rocky Glencoe!) for her patience in translating my abominable handwriting; to Jim Murtha of Marietta College for taking the time to write a lengthy and very helpful critique of the first edition; to Phil Moses, retired from Core Labs, for many hours of friendly discussions and heated arguments from which I learned so much; and to John Lee of Texas A&M University, John Edd Parker of Mississippi State University, and Richard Strickland

and Aaron Cawley of Cawley, Gillespie & Associates Inc., for their continued friendship and encouragement.

All of us in the petroleum industry are indebted to the host of engineering scientists who have provided, through petroleum engineering literature, the data and correlations which are presented in this book.

Fort Worth, Texas   William D. McCain, Jr.
1990

The Properties of Petroleum Fluids

Second Edition

Components of Naturally Occurring Petroleum Fluids

The naturally occurring petroleum deposits which the petroleum engineer encounters are composed of organic chemicals. When the chemical mixture is composed of small molecules, it is a gas at normal temperatures and pressures. Table 1–1 gives the composition of typical naturally occurring hydrocarbon gases.

When the mixture contains larger molecules, it is a liquid at normal temperatures and pressures. A typical crude oil contains thousands of different chemical compounds, and trying to separate it into different chemicals is impractical. Therefore, the crude oil normally is separated into crude fractions according to the range of boiling points of the compounds included in each fraction. Table 1–2 gives a list of the typical fractions which are separated from crude oil.

Crude oils are classified chemically according to the structures of the larger molecules in the mixture. Classification methods use combinations of the words paraffinic, naphthenic, aromatic, and asphaltic. For instance, crude oil which contains a predominance of paraffinic molecules will yield very fine lubricating oils from the gas-oil fraction and paraffin wax from the residuum. On the other hand, if the larger molecules are aromatic and asphaltic, the heavier fractions of the crude oil are useful for pitch, roofing compounds, paving asphalts, and other such applications.

Liquids obtained from different petroleum reservoirs have widely different characteristics. Some are black, heavy, and thick, like tar, while others are brown or nearly clear with low viscosity and low specific gravity. However, nearly all naturally occurring petroleum liquids have elemental analyses within the limits given in Table 1–3.

TABLE 1-1

Components of typical petroleum gases

Natural gas

Hydrocarbon
Methane	70–98%
Ethane	1–10%
Propane	trace–5%
Butanes	trace–2%
Pentanes	trace–1%
Hexanes	trace–½%
Heptanes +	trace–½%

Nonhydrocarbon
Nitrogen	trace–15%
Carbon dioxide*	trace–5%
Hydrogen sulfide*	trace–3%
Helium	up to 5%, usually trace or none

*Occasionally natural gases are found which are predominately carbon dioxide or hydrogen sulfide.

Gas from a well which also is producing petroleum liquid

Hydrocarbon
Methane	45–92%
Ethane	4–21%
Propane	1–15%
Butanes	½–7%
Pentanes	trace–3%
Hexanes	trace–2%
Heptanes +	none–1½%

Nonhydrocarbon
Nitrogen	trace–up to 10%
Carbon dioxide	trace–4%
Hydrogen sulfide	none–trace–6%
Helium	none

TABLE 1-2

Typical crude oil fractions

Crude fraction	Boiling point, °F (melting point)	Approximate chemical composition	Uses
Hydrocarbon gas		C_1–C_2	Fuel gas
	to 100	C_3–C_6	Bottled fuel gas, solvent
Gasoline	100–350	C_5–C_{10}	Motor fuel, solvent
Kerosene	350–450	C_{11}–C_{12}	Jet fuel, cracking stock
Light gas oil	450–580	C_{13}–C_{17}	Diesel fuel, furnace fuel
Heavy gas oil	580–750	C_{18}–C_{25}	Lubricating oil, bunker fuel
Lubricants and waxes	750–950 (100)	C_{26}–C_{38}	Lubricating oil, paraffin wax, petroleum jelly
Residuum	950+ (200+)	C_{38}+	Tars, roofing compounds, paving asphalts, coke, wood preservatives

TABLE 1-3
Elemental analysis of typical crude oils

Element	Percentage by weight
Carbon	84 −87
Hydrogen	11 −14
Sulfur	0.06 − 2.0
Nitrogen	0.1 − 2.0
Oxygen	0.1 − 2.0

This consistency is not so remarkable: remember that the molecules of these organic chemicals consist of various structures built primarily of CH_2 groups.

Since the petroleum engineer spends his professional life working with mixtures of organic chemicals, he needs to understand the different types of organic compounds which make up the mixtures. He must know their nomenclatures, their relationships one to another, their degrees of volatility, and their degrees of reactivity. Therefore, our study of the properties of reservoir fluids begins with a review of the behaviors of the compounds which make up these naturally occurring petroleum mixtures.

Organic Chemistry

Organic chemistry is the chemistry of compounds of carbon. The misleading name *organic* is a relic of days when chemical compounds were divided into two classes, *inorganic* and *organic,* depending on their source. Inorganic compounds are obtained from minerals. Organic compounds are obtained from material produced by living organisms.

However, organic compounds now can be produced in the laboratory, so this definition has lost its significance. Nevertheless, the definition organic is still pertinent because the chemistry of carbon compounds is more important to everyday life than that of any other element.

The compounds derived from organic sources have one thing in common: all contain the element carbon. Today most compounds of carbon are synthesized rather than obtained from plant and animal sources. Organic compounds usually are synthesized from other organic compounds, although making organics from inorganic substances such as carbonates or cyanides is possible.

Two of the major sources of organic material from which organic compounds can be obtained are petroleum and coal. Both of these sources are organic in the old sense because both are products of the decay of plants and animals. Compounds from these sources are used as building blocks for the more complicated organic compounds so important to civilization today.

Curiously a whole branch of chemistry is centered on a single element. The main reason is the strength of the carbon-to-carbon bonds. Long chains of carbon atoms, one bonded to another, are possible.

There are other elements (such as boron, silicon, and phosphorus) which can form chains of atoms bonded to one another. Carbon is unique because it not only forms strong carbon-carbon bonds but also because these bonds remain strong when the carbon atoms are bonded with other elements. Carbon compounds are stable and relatively unreactive chemically. This is not true of the compounds of other chain-forming atoms.

Structural Theory

Basis of organic chemistry is *structural theory*. Structural theory is concerned with the way in which atoms are combined to form molecules. The hundreds of thousands of individual organic compounds have been arranged in a system based upon the structures of the molecules of the compounds.

Molecules which have similar structures exhibit similar physical and chemical properties. Thus organic chemicals can be grouped into families called *homologous series* in which molecular structures and, accordingly, physical and chemical properties are similar.

Chemical Bonding

Before considering the structures of molecules, we must begin with a discussion of *chemical bonds;* the forces that hold atoms together in molecules. There are two types of chemical bonds, the *ionic bond* and the *covalent bond*.

Remember that each atom consists of a positively charged nucleus surrounded by negatively charged electrons arranged in concentric shells. There is a maximum number of electrons that can be accommodated in each shell: two in the first shell, eight in the second shell, eight or eighteen in the third shell, etc.

The greatest stability is reached when the outer shell is full, as in the case of helium, which has two electrons in its single shell. Helium is unreactive.

Both ionic and covalent bonds arise from the tendency of atoms to seek this stable configuration of electrons.

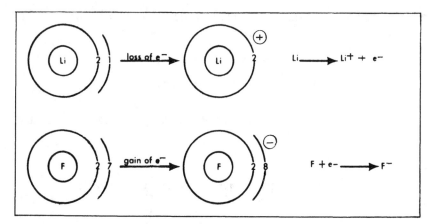

Fig. 1–1. Ionic bonding of lithium fluoride.

The Ionic Bond

The *ionic bond* results from a transfer of electrons from one atom to another. For example, consider the compound lithium fluoride. The lithium atom has two electrons in its inner shell and one electron in its outer shell. The loss of one electron from the outer shell would leave the lithium ion with only an inner shell with its maximum of two electrons.

The fluorine atom has two electrons in its inner shell and seven electrons in its outer shell. The gain of one electron would give fluorine a full eight electrons in its outer shell, Figure 1–1.

Lithium fluoride is formed by the transfer of one electron from lithium to fluorine. This results in each ion having a full outer shell and, thereby, gives each ion the stable configuration of electrons. The transfer of the electron leaves the lithium ion with a positive charge and gives the fluoride ion a negative charge. The electrostatic attraction between the oppositely charged ions holds them together. This connection between the ions is called an ionic bond.

The Covalent Bond

The *covalent bond* results when atoms share electrons, as in the formation of the hydrogen molecule. Each hydrogen atom has a single electron; thus, by sharing a pair of electrons, two hydrogen atoms can complete their shells of two. Likewise, two fluorine atoms, each with seven electrons in the outer shell, can complete their outer shells by sharing a pair of electrons. As with the ionic bond, the bonding force of the covalent bond is due to electrostatic attraction. However, in the

covalent bond, the attraction is between the electrons and the nuclei of
the atoms forming the compound.

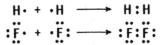

In a similar way we can visualize the covalent bonds of water and
methane.

$$\text{H}\cdot \; + \; \text{H}\cdot \; + \; \cdot \overset{\cdot}{\underset{\cdot\cdot}{\text{O}}}\text{:} \; \longrightarrow \; \text{H}\text{:}\overset{\text{H}}{\underset{\cdot\cdot}{\text{O}}}\text{:} \qquad \text{water}$$

$$\text{H}\cdot \; + \; \text{H}\cdot \; + \; \text{H}\cdot \; + \; \text{H}\cdot \; + \; \cdot \overset{\cdot}{\underset{\cdot}{\text{C}}}\cdot \; \longrightarrow \; \text{H}\text{:}\overset{\text{H}}{\underset{\cdot\cdot}{\text{C}}}\text{:}\text{H} \qquad \text{methane}$$

Bonding in Organic Compounds

Covalent bonds are the bonds with which we are concerned in our
study of organic chemistry. The building block of structural organic
chemistry is the tetravalent carbon atom. With few exceptions, carbon
compounds are formed with four covalent bonds to each carbon atom,
regardless of whether the combination is between two or more carbon
atoms or between carbon and some other element.

The bond which results from the sharing of two electrons, as
illustrated by carbon-to-hydrogen bonds and carbon-to-carbon bonds in
compounds such as ethane, is called a *single bond*. The carbon atom is
capable of forming four single bonds.

However, there exist compounds like ethylene in which two electrons
from each of the carbon atoms are mutually shared, thereby producing a
bond consisting of four electrons. This is called a *double bond*.

Also, in some compounds (such as acetylene) three electrons from
each carbon atom are mutually shared, thereby producing a six-electron
bond called a *triple bond*.

In the case of single, double, or triple bonds, each carbon atom ends
up with a full eight electrons in its outer shell.

Also, a carbon atom can share two or three electrons with atoms other
than carbon. Examples of this are the double bonds in carbon dioxide and
the triple bond in hydrogen cyanide.

$$\text{:}\overset{\cdot\cdot}{\text{O}}\text{::}\text{C}\text{::}\overset{\cdot\cdot}{\text{O}}\text{:} \qquad\qquad \text{H}\text{:}\text{C}\text{:::}\text{N}\text{:}$$

carbon dioxide   hydrogen cyanide

Normally a single straight line connecting the atomic symbols represents a single bond; two lines represent a double bond; and three lines represent a triple bond. Structural formulas of some common carbon compounds are given below.

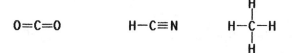

carbon dioxide   hydrogen cyanide   methane

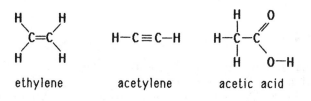

ethylene   acetylene   acetic acid

Sometimes condensed formulas are used in which the bonds are not shown.

CO_2 HCN   CH
carbon dioxide   hydrogen cyanide   methane

CH_2CH_2 CHCH CH_3COOH
ethylene   acetylene   acetic acid

For convenience, sometimes the carbon atoms in a long chain are grouped.

$CH_3CH_2CH_2CH_2CH_2CH_3$ $CH_3CH_2CH_2CH_2OH$
hexane   butanol
$CH_3(CH_2)_4CH_3$ $CH_3(CH_2)_2CH_2OH$

$$CH_3-\underset{\underset{CH_3}{|}}{CH}-CH_3$$

methyl propane
$(CH_3)_2CHCH_3$

Another system shows all bonds except the hydrogen bonds, which are understood to be single bonds.

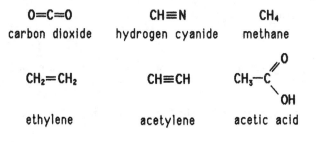

Bond Angles

Normally, the carbon atom forms its four bonds so that the four attached atoms lie at corners of a regular tetrahedron. The angles between the bonds are the same as the tetrahedral angle of 109.5°. Thus 109.5° is regarded as the normal valence angle of carbon, as shown in Figure 1–2. Carbon atoms strongly resist forces which alter their valence angle from the normal value.

Ball-and-stick models of organic substances provide a convenient means of studying the structures of various organic compounds. Balls with holes represent the atoms, and sticks represent the covalent bonds. Figure 1–3 shows two conformations of ethane in which the dark balls represent carbon and the light balls represent hydrogen. All bond angles are the normal 109.5°.

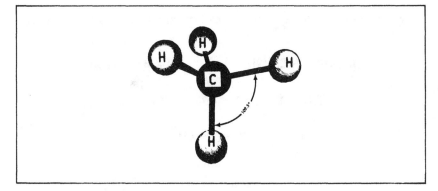

Fig. 1–2. Ball-and-stick model of methane.

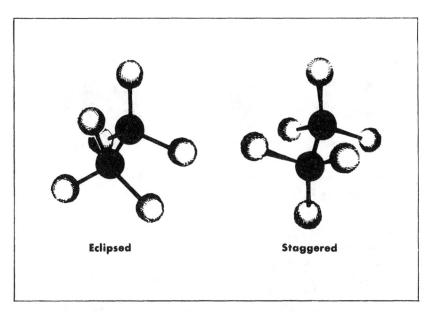

Fig. 1-3. Two rotational configurations of ethane.

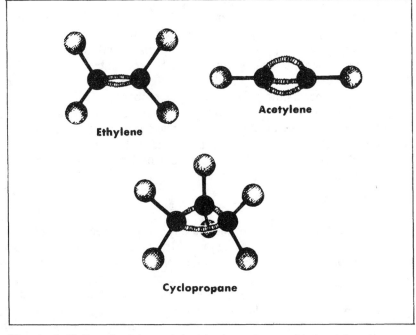

Fig. 1-4. Ball-and-stick model of hydrocarbons with bent bonds.

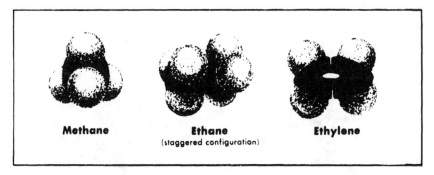

Methane   **Ethane**   **Ethylene**
                  (staggered configuration)

Fig. 1-5. Models of organic compounds.

Actual molecules appear to have essentially free rotation around the single bond joining the carbon atoms so that in ball-and-stick models one must allow the sticks to rotate freely in the holes.

The arrangements shown in Figure 1–3 are the two extreme conformations which one might expect from the ethane molecule. The staggered conformation is somewhat more stable than the eclipsed conformation because atoms in the staggered conformation are as far away from one another as possible and thus offer the least interaction.

Double and triple bonds are represented with bent bonds formed with flexible couplings. Substances that require models with bent bonds normally are found to be much less stable and, therefore, chemically more reactive than molecules which can be constructed with straight sticks. Figure 1–4 shows the double bond of ethylene, the triple bond of acetylene, and the distorted bonds of cyclopropane.

Actual molecules do not in any way look like the ball-and-stick models. The sharing of electrons in the covalent bonds requires that the atoms of the molecule overlap as represented in Figure 1–5. However, the ball-and-stick models will suffice for our understanding of structural organic chemistry.

Naming Organic Chemicals

One of the problems that we face in our study of organic chemistry is learning the names of various organic structures. The abundance and complexity of the varieties of organic compounds make this problem particularly significant. Ideally, every organic substance should have a completely descriptive and systematic name representative of its structure. A systematic nomenclature system is in use, and we will study it. Unfortunately, the system requires rather large groupings of words and symbols, making the names extremely unwieldy for conversational uses.

For instance, the systematic name of the compound 9-(2,6,6-tri-methyl-1-cyclohexenyl)-3,7-dimethyl-2,4,6,8-nonatetraen-1-ol would bog down a conversation hopelessly, although it is perfectly descriptive. This compound is usually called vitamin A.

We will base our study on the system developed by the International Union of Pure and Applied Chemistry. The system is called the IUPAC Rules. There are other semi-systematic nomenclature systems, and many individual compounds are known by nonsystematic or *trivial names*. These names will be given along with the IUPAC names when such trivial names are commonly used.

Hydrocarbons

We will first consider those organic compounds which contain only two elements, hydrogen and carbon. These compounds are known as *hydrocarbons*. Later we will consider organic compounds which contain oxygen, nitrogen, or sulfur atoms in addition to hydrogen and carbon.

On the basis of structure, hydrocarbons are divided into two main classes, *aliphatic* and *aromatic*. Aliphatic hydrocarbons are further divided into families: alkanes, alkenes, alkynes, and their cyclic analogs. Figure 1–6 shows the relationships between some of these classes and families of hydrocarbons.

Homologous Series

A family of organic chemicals is known as a *homologous series*. Members of a homologous series have similar molecular structures and

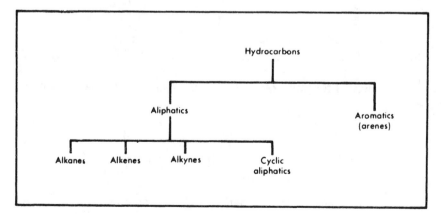

Fig. 1–6. Classes and homologous series of hydrocarbons.

have graded physical properties which differ from one another according to the number of carbon atoms in the structure. If the homologous series to which a particular compound belongs is known, the chemical and physical properties of the compound can be inferred from corresponding properties of the other compounds of the series. Therefore, we will study organic chemistry by studying the various families of organic chemicals.

Alkanes

The homologous series of hydrocarbons designated by the name *alkanes* has the general formula C_nH_{2n+2}. The alkanes are named through the combination of a prefix (which denotes the number of carbon atoms) and the suffix *-ane* (which classifies the compound as an alkane). Compounds of this family sometimes are called *saturated hydrocarbons* because the carbon atoms are attached to as many hydrogen atoms as possible, i.e., the carbons are saturated with hydrogen. These alkanes also are called *paraffin hydrocarbons*. Petroleum engineers normally call these *paraffins*. Table 1–4 gives examples.

TABLE 1–4
Alkanes (C_nH_{2n+2})

No. of carbon atoms, n	Name
1	Methane
2	Ethane
3	Propane
4	Butane
5	Pentane
6	Hexane
7	Heptane
8	Octane
9	Nonane
10	Decane
20	Eicosane
30	Triacontane

The structural formulas of methane and ethane and propane are given below.

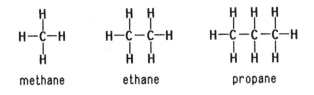

methane   ethane   propane

Notice that the carbon atoms form continuous chains. As the number of carbon atoms in the compound increases, the carbon atoms may be connected in continuous chains or may be connected as branches with more than two carbon atoms linked together. For instance, butane may take either of the two structures given below.

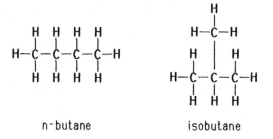

These different configurations are known as *structural isomers,* or simply as *isomers*.

The continuous chain hydrocarbons are known as normal hydrocarbons, and the prefix *n-* is usually attached to the name. The branched-chain hydrocarbons may have the prefix *iso-* attached to the name. Usually the prefix *iso-* is reserved for substances with two methyl groups attached to carbon atoms at the end of an otherwise straight chain. In a straight chain, each carbon atom is connected to no more than two other carbon atoms. The prefix *neo-* denotes three methyl groups on a carbon atom at the end of a chain. For example, the isomers of pentane are illustrated below.

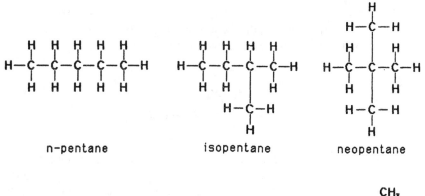

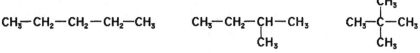

Obviously, the system of adding a prefix to the name of the compound to denote the structural configuration will become unwieldy as the number of different structural configurations increases. Consider decane, which has 75 isomers. A more systematic nomenclature is desirable.

The IUPAC system is based on *alkyl groups* substituted on *parent chains* of carbon atoms. An alkyl group is simply an alkane with one hydrogen atom missing. For instance, the methyl group $-CH_3$ looks like methane less one hydrogen atom. Alkyl groups are named by using the prefixes corresponding to the number of carbon atoms and the suffix *-yl*.

ethyl group

CH_3-CH_2-

propyl group

$CH_3-CH_2-CH_2-$

Nomenclature of Alkanes

The nomenclature system commonly in use is known as the IUPAC Rules. These rules are as follows:

1. The longest continuous chain of carbon atoms is taken as the framework on which the various alkyl groups are considered to be substituted. Thus, the following hydrocarbon is a pentane.

$$CH_3-\overset{\overset{\displaystyle CH_3}{|}}{CH}-\overset{\overset{\displaystyle CH_3}{|}}{CH}-CH_2-CH_3$$

2. The parent hydrocarbon chain then is numbered starting from the end, and the substituent groups are assigned numbers corresponding to their positions on the chain. The direction of numbering is chosen to give the lowest sum for the numbers of the side chain substituents. Thus, the above hydrocarbon is 2,3-dimethylpentane.

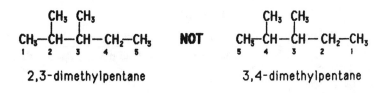

2,3-dimethylpentane   3,4-dimethylpentane

3. Where there are two identical substituents in one position, as in the compound below, numbers are supplied for both.

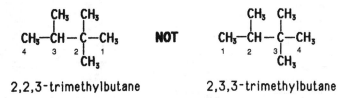

2,2,3-trimethylbutane   2,3,3-trimethylbutane

4. Branched-chain substituent groups are given appropriate names by a simple extension of the system used for branched-chain hydrocarbons. The longest chain of the substituent is numbered starting with the carbon attached directly to the parent hydrocarbon chain. Parentheses separate the numbering of the substituent from the main hydrocarbon chain.

$$CH_3-CH_2-CH_2-CH_2-CH_2-\underset{5}{CH}-CH_2-CH_2-CH_2-CH_3$$

with substituent:
$$\underset{1}{CH}\ \text{bearing}\ \underset{1}{CH_3}\ \text{and}\ \underset{2}{CH_2}-\underset{3}{CH_3}$$

5-(1-methylpropyl)-decane

5. When two or more different substituents are present, the common method is to list the substituents in alphabetical order. However, the substituents sometimes are listed in order of increasing complexity.

$$\underset{7}{CH_3}-\underset{6}{CH_2}-\underset{5}{CH_2}-\underset{4}{CH}-\underset{3}{CH}-\underset{2}{CH_2}-\underset{1}{CH_3}$$

with CH_3-CH_2 on carbon 4 and CH_3 on carbon 3

4-ethyl-3-methylheptane

TABLE 1–5

Physical properties of n–alkanes, $CH_3(CH_2)_{n-2}CH_3$

n	Name	Boiling point, °F	Melting point, °F	Specific gravity, 60°/60°
1	Methane	−258.7	−296.4	
2	Ethane	−127.5	−297.0	
3	Propane	−43.7	−305.7	0.507
4	Butane	31.1	−217.1	0.584
5	Pentane	96.9	−201.5	0.631
6	Hexane	155.7	−139.6	0.664
7	Heptane	209.2	−131.1	0.688
8	Octane	258.2	−70.2	0.707
9	Nonane	303.5	−64.3	0.722
10	Decane	345.5	−21.4	0.734
11	Undecane	384.6	−15	0.740
12	Dodecane	421.3	14	0.749
15	Pentadecane	519.1	50	0.769
20	Eicosane	648.9	99	
30	Triacontane	835.5	151	

Physical and Chemical Properties of Alkanes

The series of straight chain alkanes given in Table 1–5 shows a remarkably smooth gradation of physical properties. As molecular size increases, each additional CH_2 group contributes a fairly constant increment to boiling point and specific gravity. The additional CH_2 groups also cause increases in melting point, but the increments are not as constant as with the boiling points. Notice that the melting points of methane and ethane do not fit the pattern of the other melting points listed in Table 1–5. The smaller molecules of a homologous series often have physical properties which do not fit the pattern shown by the larger molecules.

The first four alkanes are gases at normal temperatures and pressures. As a result of the decrease in volatility with increasing number of carbon atoms, the next 13, pentane through heptadecane, are liquid. Alkanes containing 18 or more carbon atoms are solid at normal temperatures and pressures. Figure 1–7 shows the changes of boiling point, specific gravity, and melting point for the straight-chain alkanes.

The boiling points and melting points of the alkanes are fairly low, particularly in the case of the smaller members of the family. This is because the molecules are highly symmetrical, and therefore the attractions between molecules, known as *intermolecular forces,* are fairly small. Since the processes of boiling and melting involve overcoming these intermolecular forces, the boiling points and melting points of the alkanes are low. Boiling points and melting points are higher for the

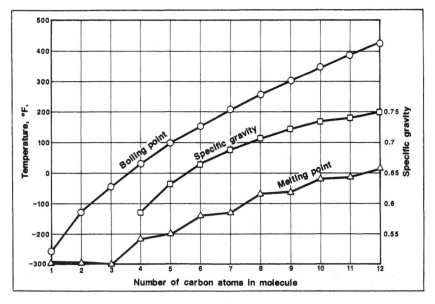

Fig. 1–7. Dependence of some physical properties on the number of carbon atoms for straight-chain alkanes. (Data from Table A-1)

larger molecules because the intermolecular forces are greater between larger molecules.

The increases in melting points are not as regular as the increases in boiling points because the intermolecular forces in a solid depend not only on the size of the molecules but on how well the molecules fit into the crystal lattice.

Branched-chain alkanes do not exhibit the same smooth gradation of physical properties as the straight-chain alkanes.

Differences in structure cause differences in intermolecular forces. Thus there are differences in boiling points and melting points between the isomers of any particular alkane. Table 1–6 shows the physical properties of the isomers of hexane. An increase in branching causes a decrease in intermolecular attraction, which results in a lower boiling point.

The increased branching changes the way the molecules fit into the crystal lattice of the solid, causing the melting points of the isomers to differ.

Chemically, alkanes are particularly unreactive. The name paraffin literally means *not enough affinity*. The strong single bonds between carbon and hydrogen and between carbon and carbon are attacked only by very strong reactants at ordinary temperatures.

TABLE 1–6

Some physical properties of hexane isomers

Isomer	Structure	Boiling point, °F	Melting point, °F	Specific gravity, 60°/60°
n-Hexane	$CH_3(CH_2)_4CH_3$	155.7	−139.6	0.664
3-Methylpentane	$\overset{\displaystyle CH_3}{\overset{\displaystyle \vert}{CH_3CH_2CHCH_2CH_3}}$	145.9	−180.4	0.669
2-Methylpentane (Isohexane)	$\overset{\displaystyle CH_3}{\overset{\displaystyle \vert}{CH_3CHCH_2CH_2CH_3}}$	140.5	−244.6	0.658
2,3-Dimethylbutane	$\overset{\displaystyle CH_3\ \ CH_3}{\overset{\displaystyle \vert\ \ \ \ \vert}{CH_3CH{-}CHCH_3}}$	136.4	−199.4	0.666
2,2-Dimethylbutane (Neohexane)	$\overset{\displaystyle CH_3}{\overset{\displaystyle \vert}{\underset{\displaystyle \underset{\displaystyle CH_3}{\vert}}{CH_3CCH_2CH_3}}}$	121.5	−147.7	0.654

All hydrocarbons are attacked by oxygen at elevated temperatures. If enough oxygen is present, complete combustion to carbon dioxide and water occurs. This is particularly important since alkanes are the major constituents of natural gas, and the principal use of natural gas is in combustion.

Alkenes

The homologous series known as *alkenes* is also called *unsaturated hydrocarbons* or *olefins*. The general formula for the alkene family is C_nH_{2n}. The distinguishing feature of the alkene structure is the carbon-carbon double bond which, as we have discussed previously, is a four-electron bond formed by the sharing of two electrons from each of two carbon atoms.

Nomenclature of Alkenes

The names of the alkenes are formed with the same prefixes used in naming the alkanes. These prefixes correspond to the number of carbon atoms in the compound. The suffix is *-ene,* which indicates that the compound belongs to the alkene family. Thus the simplest member of the alkene family, C_2H_4, should be called ethene. Ethene is commonly known as ethylene, Figure 1–8. The next larger member of the family, propene, commonly is called propylene.

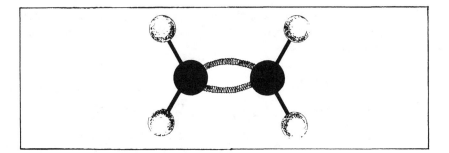

Fig. 1–8. Ball-and-stick model of ethylene.

The rules of the IUPAC system for naming the alkenes are as follows:
1. Select as the parent structure the longest continuous chain that contains the carbon-carbon double bond; then consider the compound to have been derived from this structure by replacement of hydrogen atoms by various alkyl groups.

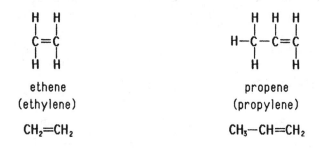

2. Indicate by a number the position of the double bond in the parent chain. Designate the position of the double bond by the number of the first doubly bonded carbon encountered when numbering from the end of the chain nearest the double bond. Thus, the following compound is 1-butene rather than 3-butene.

$$CH_3-CH_2-CH=CH_2$$
$$4321$$

1- butene

3. Indicate the positions of the alkyl groups attached to the parent chain by the numbers of the carbon atoms to which the alkyl group is attached.

$$CH_2-CH_3$$
$$CH_3-C=CH_2$$

$$CH_3$$
$$CH_3-C-CH=CH_2$$
$$CH_3$$

2-methyl-1-butene   3,3-dimethyl-1-butene

Consider the alkenes which contain four carbon atoms. There are several structural possibilities. The double bond can be located between the first and second carbons from the end of the chain or between the second and third carbons in the chain. Also, the four carbons can be in a straight chain or branched. The position of the double bond and the branching combine to give the three structures shown below.

$$CH_2=CH-CH_2-CH_3 \qquad CH_3-CH=CH-CH_3$$

$$CH_2=C-CH_3$$
$$CH_3$$

1-butene   2-butene   methyl propene
                                                (isobutylene)

However, there are actually four alkenes of the formula C_4H_8. Their physical properties are shown in Table 1–7.

We have developed an understanding of *structural isomerism,* i.e., straight chain and branched chain. We have seen structural isomerization connected with the location of the double bond. We must recognize another type of isomerism—*geometric isomerism,* also called *diastereomerism* or simply *cis-trans isomerism.*

Consider the two configurations of 2-butene in Figure 1–9 and remember that the carbon-carbon double bond does not permit the rotation of the carbon atoms. Notice that in one case the two methyl groups are located on the same side of the basic structure while in the other case the methyl groups are located on opposite sides of the structure. These two molecules exhibit geometric isomerization because of the way the atoms are oriented in space, even though they are alike

TABLE 1–7

Some physical properties of the butylenes

Name	Boiling point, °F	Melting point, °F	Specific gravity, 60°/60°	Refractive index (−12.7°)
Isobutylene	19.6	−220.6	0.600	1.3727
1-Butene	20.8	−301.6	0.601	1.3711
trans-2-Butene	33.6	−158.0	0.610	1.3778
cis-2-Butene	38.7	−218.1	0.627	1.3868

with respect to the connections of the atoms. These two compounds are known as *stereoisomers*. The configurations are differentiated in their names by the prefixes *cis-* and *trans-*.

The prefix *cis-* is Latin for on this side, and *trans* is Latin for across.

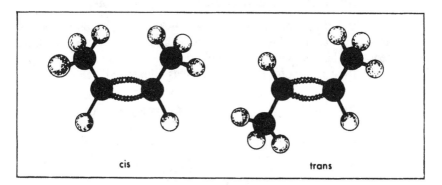

Fig. 1–9. Ball-and-stick models of cis- and trans-2-butene.

Physical and Chemical Properties of Alkenes

The physical properties of the alkenes are very much like the physical properties of the alkanes. Table 1–8 gives comparisons of the melting points, boiling points, and specific gravities of the simpler alkenes. As with the alkanes, these physical properties increase with increasing carbon content.

Figure 1–8 shows the ball–and–stick model of ethylene. Obviously, the carbon-to-carbon bonds do not satisfy the normal tetrahedron bonding angles of 109.5°. This indicates that the double bonds are less stable than the single bonds and are, therefore, more likely to be attacked by other chemicals.

Thus, alkenes, because of their double bonds, are more reactive than alkanes. Most alkene reactions involve the elimination of the double bond and the formation of two strong single bonds.

TABLE 1–8

Some physical properties of the alkenes

Name	Formula	Boiling point, °F	Melting point, °F	Specific gravity, 60°/60°
Ethylene	$CH_2{=}CH_2$	−154.7	−272.5	
Propylene	$CH_2{=}CHCH_3$	−53.8	−301.4	
1-Butene	$CH_2{=}CHCH_2CH_3$	20.8	−301.6	0.601
1-Pentene	$CH_2{=}CH(CH_2)_2CH_3$	85.9	−265.4	0.646
1-Hexene	$CH_2{=}CH(CH_2)_3CH_3$	146	−216	0.675
1-Heptene	$CH_2{=}CH(CH_2)_4CH_3$	199	−182	0.698
1-Octene	$CH_2{=}CH(CH_2)_5CH_3$	252	−155	0.716
1-Nonene	$CH_2{=}CH(CH_2)_6CH_3$	295		0.731
1-Decene	$CH_2{=}CH(CH_2)_7CH_3$	340		0.743

Alkadienes, Alkatrienes, and Alkatetraenes

Many hydrocarbon compounds contain two or more double bonds. These are known as *alkadienes, alkatrienes,* and *alkatetraenes,* with the suffix denoting the number of double bonds. The location of each double bond is specified by appropriate numbers as illustrated below.

$$CH_2{=}C{=}CH{-}CH_3 \qquad\qquad CH_2{=}CH{-}CH{=}CH_2$$

1, 2-butadiene   1, 3-butadiene

$$CH_2{=}C{=}C{=}CH{-}CH_3$$

1, 2, 3-pentatriene

These compounds, are known as *diolefins, triolefins,* etc., also. The alkadienes also are commonly called *dienes.* The dienes have the general formula C_nH_{2n-2}, which indicates a higher degree of unsaturation than the alkenes.

Chemically, these compounds are as reactive as alkenes. Physically, the properties of these compounds are similar to alkanes with the corresponding number of carbon atoms.

Alkynes

The distinguishing feature of the alkyne structure is the carbon-carbon triple bond. The general formula for the *alkynes* is C_nH_{2n-2}, which is the same as the general formula for the dienes. However, alkynes and dienes have different functional groups and hence different properties.

Nomenclature of Alkynes

The IUPAC Rules are exactly the same for naming alkynes as for naming of alkenes except that the suffix *-yne* replaces *-ene*.

The parent structure is the longest continuous chain that contains the triple bond, and the positions of both the substituents and the triple bond are indicated by numbers. According to this system the simplest alkyne should be named ethyne; however, this compound commonly is called acetylene. The other alkynes are usually named according to the IUPAC system.

$$CH\equiv CH \qquad\qquad CH\equiv C-\underset{\underset{\displaystyle CH_3}{|}}{\overset{\overset{\displaystyle CH_3}{|}}{C}}-CH_3$$

acetylene
(ethyne)   3,3-dimethyl-1-butyne

Physical and Chemical Properties of Alkynes

The alkynes have physical properties essentially the same as those of the alkanes and alkenes. Table 1–9 gives the physical properties of some of the alkynes. The trends in melting points and boiling points are again apparent.

Chemically the alkynes are very much like the alkenes. Both families are much more reactive than the alkanes. Since the triple bond appears to be strained somewhat more than the double bond, it would appear chemically more unstable. However, for reasons that are not fully understood, the carbon-carbon triple bond is less reactive than the carbon-carbon double bond toward some reactants and is more reactive toward other reactants.

As with the alkenes, most chemical reactions of alkynes involve the elimination of the triple bond in favor of a double and a single bond. Normally the reactions continue so that the double bond is eliminated in favor of two more single bonds.

TABLE 1–9

Some physical properties of the alkynes

Name	Formula	Boiling point, °F	Melting point, °F	Specific gravity, 60°/60°
Acetylene	HC≡CH	−120	−114	
Propyne	HC≡CCH₃	−9	−151	
1-Butyne	HC≡CCH₂CH₃	48	−188	
1-Pentyne	HC≡C(CH₂)₂CH₃	104	−144	0.695
1-Hexyne	HC≡C(CH₂)₃CH₃	162	−191	0.719
1-Heptyne	HC≡C(CH₂)₄CH₃	212	−112	0.733
1-Octyne	HC≡C(CH₂)₅CH₃	259	−94	0.747
1-Nonyne	HC≡C(CH₂)₆CH₃	304	−85	0.763
1-Decyne	HC≡C(CH₂)₇CH₃	360	−33	0.770
2-Butyne	CH₃C≡CCH₃	81	−11	0.694
2-Pentyne	CH₃C≡CCH₂CH₃	131	−150	0.714
3-Methyl-1-butyne	HC≡CCH(CH₃)₂	84		0.665
2-Hexyne	CH₃C≡C(CH₂)₂CH₃	183	−134	0.730
3-Hexyne	CH₃CH₂C≡CCH₂CH₃	178	−60	0.725
3,3-Dimethyl-1-butyne	HC≡CC(CH₃)₃	100	−114	0.669
4-Octyne	CH₃(CH₂)₂C≡C(CH₂)₂CH₃	268		0.748
5-Decyne	CH₃(CH₂)₃C≡C(CH₂)₃CH₃	347		0.759

Cyloaliphatic Hydrocarbons

The carbon atoms in the compounds we have studied in previous sections of this chapter are attached to one another to form chains. However, in many hydrocarbon compounds the carbon atoms are arranged in rings. These are called *cyclic compounds*. This section considers *cycloalkanes* and *cycloalkenes*.

Cycloalkanes

The cycloalkanes also are known as *naphthenes, cycloparaffins,* or *alicyclic hydrocarbons.* In the petroleum industry, this class of hydrocarbons is known as *naphthenes.* Naphthenes have saturated rings. The general formula for the ring without substituents is C_nH_{2n}. This is the same as the general formula for the alkene series; however, the structural configurations differ completely and, thus, the physical and chemical properties are not at all similar.

The most common cycloalkanes are cyclohexane and cyclopentane.

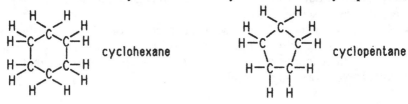

Nomenclature of Cycloalkanes

Cycloalkanes are named by prefixing *cyclo-* to the name of the alkane having the same number of carbon atoms.

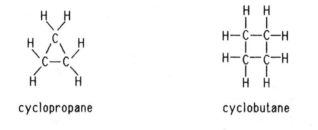

cyclopropane   cyclobutane

Substituents on the rings are named and their positions indicated by numbers. The lowest possible combination of numbers is used.

          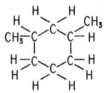

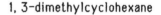

1, 4-dimethylcyclohexane   1, 3-dimethylcyclohexane

For convenience, paraffinic rings often are represented by simple geometric figures as shown below. Each corner of a figure represents a carbon atom, and the carbon valence of four is satisfied with hydrogen atoms which are not shown.

cyclohexane   cyclopentane   1, 2-dimethylcyclopentane

Physical and Chemical Properties of Cycloalkanes

Table 1–10 gives the physical properties of the cycloalkanes. Again we see a homologous series with reasonably regular changes in boiling points, melting points, and specific gravities.

TABLE 1-10

Some physical properties of cyclic aliphatic hydrocarbons

Name	Boiling point, °F	Melting point, °F	Specific gravity, 60°/60°
Cyclopropane	−27	−197	
Cyclobutane	55	−112	
Cyclopentane	121	−137	0.750
Cyclohexane	177	44	0.783
Cycloheptane	244	10	0.810
Cyclooctane	300	57	0.830
Methylcyclopentane	161	−224	0.753
cis-1,2-Dimethylcyclopentane	210	−80	0.772
trans-1,2-Dimethylcyclopentane	198	−184	0.750
Methylcyclohexane	214	−196	0.774
Cyclopentene	115	−135	0.774
1,3-Cyclopentadiene	108	−121	0.798
Cyclohexene	181	−155	0.810
1,3-Cyclohexadiene	177	−144	0.840
1,4-Cyclohexadiene	189	−56	0.847

Unlike most homologous series, the different members of the cycloalkane family exhibit different chemical reactivities. We already have seen that chemical reactivity is related to the strain in the carbon-carbon bonds and that ideally the carbon bonds should have bond angles of 109.5°.

Since there is a difference in reactivity between the different cycloparaffins, we will begin with the least reactive—cyclohexane. If cyclohexane existed as a flat hexagon, the carbon-carbon bond angle would be 120°. However, cyclohexane takes on a puckered structure as shown in the two conformations in Figure 1–10. The bonds in these conformations of cyclohexane have bond angles of 109.5°. Thus, the stability of the bonds is the same as in the straight-chain alkanes.

The five carbon atoms of cyclopentane form a flat regular pentagon with internal angles of 108°, which are very close to normal carbon bond angles. Actually, cyclopentane molecules are not exactly flat; however, the bond angles are close enough to the normal bond angle that these bonds are substantially as stable as those of cyclohexane.

Cyclobutane and cyclopropane have bond angles considerably smaller than the normal 109.5°, and the angle strain is great. See Figure 1–4. These compounds are much more reactive than cyclopentane and cyclohexane; usually the reaction involves cleavage of a carbon-carbon bond to give an open-chain compound having normal bond angles. Cyclobutane and cyclopropane are not found in naturally occurring petroleum mixtures.

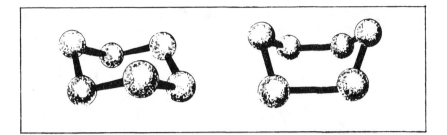

Fig. 1–10. two configurations of cyclohexane with 109.5° bond angles (hydrogen omitted).

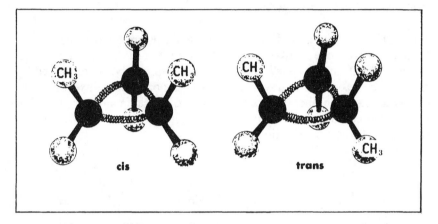

Fig. 1–11. Ball-and-stick models of cis and trans isomers of 1,2-dimethyl cyclopropane.

Rings of seven carbons and higher can achieve conformations which give nearly normal bond angles. This indicates that the large rings are reasonably stable chemically. However, rings with seven or more carbons are not particularly common, probably because there is little possibility that a long straight chain with the proper reactive groups at both ends would be folded into a position which would permit a ring-forming reaction.

Cis-trans isomerism occurs in disubstituted cycloalkanes as shown in Figure 1–11.

Condensed Rings

Cycloparaffins can have more than one ring, with rings sharing carbon atoms. These are called *condensed rings*. One common compound of this type is bicyclodecane, also called decalin. Other simple condensed ring structures commonly found in petroleum are shown.

decalin   bicyclooctane   dodecahydrofluorene
(bicyclodecane)

Cycloalkenes and Cycloalkadienes

Organic compounds in which the carbon atoms are bonded together to form a ring with one or two double bonds present in the ring are known as *cycloalkenes* and *cycloalkadienes*. These hydrocarbons are named by prefixing *cyclo-* to the names of corresponding open-chain hydrocarbons having the same number of carbon atoms and the same number of double bonds. The carbon atoms in the ring are numbered so a double bond is considered to occupy positions 1 and 2.

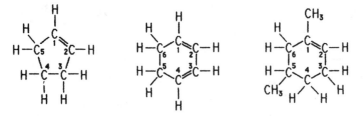

cyclopentene   1, 3-cyclohexadiene   1, 5-dimethylcyclohexene

Cycloalkenes and cycloalkadienes are about as reactive chemically as their open-chain analogs. Cycloalkenes can undergo addition reactions in which the double bond is eliminated and also can undergo cleavage reactions which cause the ring structure to be opened into a chain.

Aromatics

Aromatic compounds (also called *arenes*) include benzene and compounds that resemble benzene in chemical behavior. Those properties of benzene that distinguish it from aliphatic hydrocarbons are called aromatic properties. Usually the structures of aromatic compounds are formed with benzene as the basic building block, although a few compounds that possess aromatic properties have structures that differ from the structure of benzene.

Obviously, any study of the chemistry of aromatic compounds must begin with a study of benzene. Benzene is unique in that the bonds between the carbon atoms in benzene do not seem to follow the theory of covalent bonds we have developed so far.

Benzene

Benzene is a flat molecule with six carbon atoms arranged in a hexagonal ring. Six hydrogen atoms, one associated with each carbon, radiate out from the ring. The molecule is flat and symmetrical so all bond angles are 120°. At first chemists thought benzene must have a structure such that it appeared to be cyclohexatriene. However, as we have seen from our study of the cycloalkenes and cycloalkadienes, if benzene has this structure, its chemical properties should be similar to the corresponding alkatriene: 1,3,5-hexatriene.

If benzene had a structure as postulated, it would be very reactive and undergo reactions which would cause the ring to become saturated or break at the positions of the double bonds.

This, in fact, does not happen. The benzene ring is a very stable structure that can be involved in many reactions in which the reaction products retain the benzene ring. Thus the bonds in benzene rings must be much more stable than double bonds.

Benzene Bonds

Many bond structures have been proposed for benzene, and no single one may be accepted as fully satisfactory. Probably the best explanation is that the electrons are delocalized over all the six carbon nuclei. Thus benzene does not contain three carbon-carbon single bonds and three carbon-carbon double bonds. Rather, the benzene molecule contains six identical bonds, each one intermediate between a single and a double bond. This type of bond has been called a *hybrid bond,* a *one and one-half bond*, or simply a *benzene bond*. We normally draw the benzene ring as below. Each of the six corners represents a carbon atom, and each carbon atom is bonded to one hydrogen atom.

Further, it is understood that the three double bonds and three single bonds indicated in the drawing do not represent the actual bonding structure, but that the bonds are hybrid bonds.

Other symbols as shown below are representations of the benzene ring.

However, the normal convention is to use the double-single bond representation.

Nomenclature of Aromatic Hydrocarbons

Unfortunately, a completely systematic method of naming aromatic compounds is not in use. The system used is a combination of trivial names and the IUPAC system. Sometimes compounds that contain benzene rings are considered to be substituted benzenes, in which case the word benzene appears in the name of the compound along with the name of the substituent.

ethylbenzene   bromobenzene   nitrobenzene

Sometimes the benzene ring is considered to be the substituent; in these cases it is called a *phenyl group*.

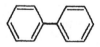

biphenyl
(phenylbenzene)

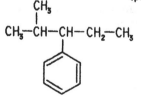

2-methyl-3-phenylpentane

3-phenylpropene
(allylbenzene)

Many of the simple aromatic compounds have trivial names.

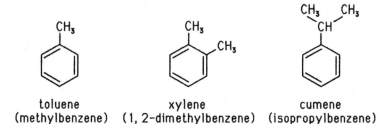

toluene
(methylbenzene)

xylene
(1, 2-dimethylbenzene)

cumene
(isopropylbenzene)

When there are two or more substituents on a benzene ring, structural isomerization can result. For instance, there are three possible isomers of xylene (dimethylbenzene). The isomers are commonly designated as ortho, meta, and para. A letter (o, m, or p) designates the isomeric position.

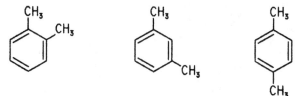

ortho-xylene
(1, 2-dimethylbenzene)

meta-xylene
(1, 3-dimethylbenzene)

para-xylene
(1, 4-dimethylbenzene)

A wide range of aromatic compounds have benzene rings with common ortho positions. These are called *condensed rings*. Note the numbering of the carbon atoms for the purpose of locating substituent groups. Condensed aromatics have the general formula $C_{4r+2}H_{2r+4}$ for rings without substituents, where r = number of rings.

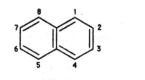

naphthalene

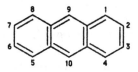

anthracene

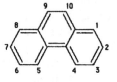

phenanthrene

TABLE 1-11

Some physical properties of aliphatic–aromatic hydrocarbons

Name	Formula	Boiling point, °F	Melting point, °F	Specific gravity, 60°/60°
Benzene	C_6H_6	176	42	0.884
Toluene	$C_6H_5CH_3$	231	-139	0.872
o-Xylene	$1,2\text{-}C_6H_4(CH_3)_2$	292	-14	0.885
m-Xylene	$1,3\text{-}C_6H_4(CH_3)_2$	282	-54	0.869
p-Xylene	$1,4\text{-}C_6H_4(CH_3)_2$	281	56	0.866
Hemimellitene	$1,2,3\text{-}C_6H_3(CH_3)_3$	349	-13	0.895
Pseudocumene	$1,2,4\text{-}C_6H_3(CH_3)_3$	336	-47	0.876
Mesitylene	$1,3,5\text{-}C_6H_3(CH_3)_3$	331	-49	0.864
Prehnitene	$1,2,3,4\text{-}C_6H_2(CH_3)_4$	401	20	0.902
Isodurene	$1,2,3,5\text{-}C_6H_2(CH_3)_4$	387	-11	
Durene	$1,2,4,5\text{-}C_6H_2(CH_3)_4$	383	176	
Pentamethylbenzene	$C_6H(CH_3)_5$	448	127	
Hexamethylbenzene	$C_6(CH_3)_6$	507	329	
Ethylbenzene	$C_6H_5C_2H_5$	277	-139	0.872
n-Propylbenzene	$C_6H_5CH_2CH_2CH_3$	318	-146	0.862
Cumene	$C_6H_5CH(CH_3)_2$	306	-141	0.866
n-Butylbenzene	$C_6H_5(CH_2)_3CH_2$	361	-114	0.860
Isobutylbenzene	$C_6H_5CH_2CH(CH_3)_2$	340		0.867
sec-Butylbenzene	$C_6H_5CH(CH_3)C_2H_5$	344	-177	0.864
tert-Butylbenzene	$C_6H_5C(CH_3)_3$	336	-72	0.867
p-Cymene	$1,4\text{-}CH_3C_6H_4CH(CH_3)_2$	351	-94	0.857
Biphenyl	$C_6H_5C_6H_5$	491	158	
Diphenylmethane	$C_6H_5CH_2C_6H_5$	505	79	
Triphenylmethane	$(C_6H_5)_3CH$	680	199	
1,2-Diphenylethane	$C_6H_5CH_2CH_2C_6H_5$	543	126	
Styrene	$C_6H_5CH{=}CH_2$	293	-23	0.911
trans-Stilbene	$trans\text{-}C_6H_5CH{=}CHC_6H_5$	585	255	
cis-Stilbene	$cis\text{-}C_6H_5CH{=}CHC_6H_5$		43	
unsym-Diphenylethylene	$(C_6H_5)_2C{=}CH_2$	531	48	1.02
Triphenylethylene	$(C_6H_5)_2C{=}CHC_6H_5$		163	
Tetraphenylethylene	$(C_6H_5)_2C{=}C(C_6H_5)_2$	797	441	0.930
Phenylacetylene	$C_6H_5C{\equiv}CH$	288	-49	
Diphenylacetylene	$C_6H_5C{\equiv}CC_6H_5$	572	144	

Physical and Chemical Properties of the Aromatic Hydrocarbons

Many of the compounds containing benzene rings have very pleasant odors and for this reason are called aromatic hydrocarbons. These compounds, however, are generally quite toxic; some are carcinogenic. Inhalation of aromatic hydrocarbon vapors should be avoided. A list of the most common aromatic compounds and their physical properties is given in Table 1-11.

The usual relationships between physical properties and molecular size are evident. Also, the effect of molecular symmetry can be seen, particularly in the case of xylenes. The melting point of p-xylene is considerably greater than the melting points of o-xylene and m-xylene.

As already noted the hybrid bonds of benzene rings have the same stability as the carbon-carbon single bonds found in the alkanes. The aromatic compounds can enter into many reactions which do not affect the ring structure. The volatile aromatics are highly flammable and burn with a luminous, sooty flame in contrast to alkanes and alkenes, which burn with a bluish flame, leaving little carbon residue.

Cycloalkanoaromatics

Many of the large molecules found in petroleum are condensed rings consisting of cycloparaffins and aromatics. These compounds are called *cycloalkanoaromatics* or *naphtheno-aromatics*. The two smallest compounds of this class are lindane and tetrahydronaphthalene (tetralin).

lindane

tetralin
(tetrahydronaphthalene)

Compounds of this type often have a paraffinic component in which one or more alkyl groups, usually methyl, are attached to the ring carbons.

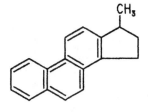

methylfluorene

methylcyclopentanophenanthrene

Other Organic Compounds

The organic chemicals studied thus far in this text are known as hydrocarbons, since the molecules contain only carbon and hydrogen atoms. There are many other families of organic compounds which are composed of the basic structures that we have studied; however, their molecules contain atoms other than carbon and hydrogen. Table 1–12 lists these families along with the functional groups which characterize the families. These families will not be considered in detail since they are present only to a limited extent in naturally occurring petroleum deposits.

TABLE 1–12

Classification of organic compounds according to functional groups

Class of compound	Functional group	Typical example Formula	Name
Alkene	$>C{=}C<$	$CH_3CH_2CH{=}CH_2$	1-Butene
Alkyne	$-C{\equiv}C-$	$CH_3C{\equiv}CH$	Methylacetylene
Alcohol	$-OH$	⬠—OH	Cyclobutanol
Ether	$-O-$	$CH_3OCH_2CH_3$	Methyl ethyl ether
Halide	F, Cl, Br, or I	⬠—Cl	Cyclopentyl chloride
Aldehyde	$-C{\scriptstyle\overset{\displaystyle O}{\diagdown H}}$	CH_3CH_2CHO	Propionaldehyde
Ketone	$>C{=}O$	$CH_3\overset{O}{\overset{\|}{C}}CH_2CH_3$	2-Butanone
Carboxylic acid	$-C{\overset{\displaystyle O}{\diagdown OH}}$	$CH_3CH_2CH_2CH_2COOH$	n-Pentanoic acid
Amine	$-NH_2$	CH_3NH_2	Methylamine
Nitro compound	$-N\overset{O}{\underset{O}{}} \leftrightarrow -N\overset{O}{\underset{O}{}}$	CH_3CHCH_3 NO_2	2-Nitropropane
Nitrile	$-C{\equiv}N$	CH_3CN	Acetonitrile
Organo metallic	$-\overset{\|}{\underset{\|}{C}}-Metal$	$CH_3CH_2CH_2CH_2Li$	n-Butyllithium

The classes listed in Table 1–12 are families which exhibit the same regularity of boiling points, melting points, densities, and other properties seen in the hydrocarbon families we have already studied. Some of the families are named with characteristic suffixes while others have prefixes, or even separate words in the names. For instance, alcohols are named with the suffix *-ol*. Ketones are named with the suffix *-one*. Amine and nitriles are named with the full suffix according to the family name. Ethers and halides usually have the full family name as a separate word, and *nitro-* and organometallic compounds have the prefix nitro- or the prefix corresponding to the hydrocarbon part of the organometallic molecule.

Further information on these organic compounds can be obtained in any basic organic chemistry textbook.

Many compounds have two or more of the functional groups given in Table 1–12. Several examples of these *polyfunctional* compounds are given below.

acetamide

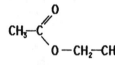

ethyl acetate

$CH_2{=}CH{-}Cl$

vinyl chloride

$HO-CH_2-CH_2-O-CH_2-CH_2-OH$

diethylene glycol

$HO-CH_2-CH_2-NH_2$

ethanolamine

$HO-CH_2-CH_2-O-CH_2-CH_2-O-CH_2-CH_2-OH$

triethylene glycol

Nonhydrocarbon Components of Petroleum

Nitrogen, carbon dioxide, and hydrogen sulfide are common nonhydrocarbon constituents of petroleum. All three are light molecules and mainly are part of the gas at the surface. Hydrogen and helium are found in some natural gases. Table 1–1 shows the quantities of these nonhydrocarbons typically found in naturally occurring petroleum gases.

Petroleums also contain compounds in which sulfur, oxygen, and/or nitrogen atoms are combined with carbon and hydrogen. These elements usually are combined with the complex ring structures that make up the larger molecules of petroleums. These larger nonhydrocarbon compounds form a class of chemicals generally called *resins and asphaltanes*. The quantity of these compounds in petroleum is often very small; however, as much as 50% of the total molecules in some heavy crude oils are resins and asphaltines.

Sulfur Compounds

Sulfur compounds form the largest group of nonhydrocarbons in petroleum. Crude oils vary considerably in their sulfur content. Some have extremely low sulfur contents with less than 0.1 weight percent of sulfur. However, high sulfur crudes can contain as much as five to seven weight percent sulfur. Since the sulfur atom is only a small part of a large molecule, a crude oil with a sulfur content of five weight percent may actually have sulfur atoms as a part of more than half of the total molecules.

In general, the quantity of sulfur increases as the density of the crude increases. Sulfur compounds poison the metallic catalysts used in the refining process. There are legal and contractual limits on the amount of sulfur contained in most products. Thus, the refiner must remove or destroy the sulfur compounds in the crude to reach an acceptable level.

Hydrogen Sulfide

Hydrogen sulfide, H_2S, is a colorless gas which has a boiling point of $-76.5°F$ and has an extremely bad odor. Natural gases which contain very small quantities of hydrogen sulfide have very disagreeable odors. Hydrogen sulfide is poisonous; breathing gases with moderate quantities of hydrogen sulfide can be fatal.

Hydrogen sulfide is usually removed from natural gas by absorption with ethanolamines. Some natural gases which contain high concentrations of hydrogen sulfide are used in the production of sulfur.

Mercaptans

The *mercaptans* have a general formula, RSH, in which the R represents any organic group. The mercaptans are known also as *thiols*.

In general, the mercaptans have a more disagreeable odor than hydrogen sulfide. For example, butanethiol is a component of skunk secretion, and propanethiol is a component of the vapor from freshly chopped onions.

$$H-S-CH_3$$

methanethiol
(methyl mercaptan)

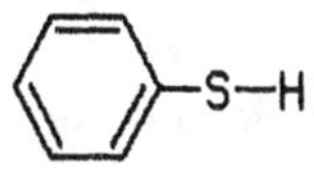

thiophenyl

$$H-S-CH_2-CH_2-CH_2-CH_3$$

butanethiol
(butyl mercaptan)

$$H-S-CH_2-CH_2-CH_3$$

propanethiol
(n-propyl mercaptan)

Most natural organic materials which have strong odors, either agreeable or disagreeable, contain either mercaptans or alkyl sulfide. Various mercaptans have been found in crude oil, and the odor contributed by small quantities of these materials led to the name *sour crude*.

Alkyl Sulfides

The *alkyl sulfides* have the general formula RSR. Aklyl sulfides also are called *thio ethers* or *monosulfides*. Alkyl sulfides form the chief ingredient in many of the spices used in cooking. For instance, allylsulfide is a chief constituent of garlic.

$$CH_2=CH-CH_2-S-CH_2-CH=CH_2$$

allylsulfide

Alkyl sulfides are found in crude oil and cause problems for the refiner, as do mercaptans and hydrogen sulfide.

Sulfides can take the form RSSR. These are called *disulfides*.

$$CH_3-S-S-CH_2-CH_3$$

ethylmethyldisulfide

Other Sulfur Compounds

Sulfur can also be present in complex ring structures found in crude oils. The sulfur atoms can either be part of the ring or be attached to the ring. These molecules are usually very large, containing 30 or more carbon atoms. However, some simple examples will illustrate the types of sulfur compounds found in crude oils.

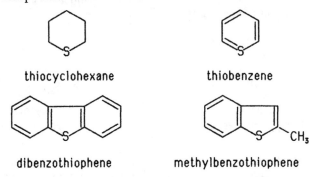

thiocyclohexane   thiobenzene

dibenzothiophene   methylbenzothiophene

Oxygen Compounds

Less is known about the oxygen compounds found in crude oils. However, large compounds with several rings, either aromatic or naphthenic, containing oxygen, have been isolated from many crudes. The oxygen can be a part of the ring or can take the form of aldehydes, alcohols, ketones, or acids. Oxygen normally is associated with very large molecules. However, some simple examples to show the types of structures are given below.

  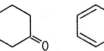

phenol   cyclopentane   cyclohexanone   benzaldehyde
                carboxylic acid

The carboxylic acid group usually is attached to a naphthenic ring rather than an aromatic ring. These organic acids generally are known by the rather loose term *naphthenic acids*. These acids may be neutralized with common bases. For instance, the *acid number* of a crude oil is the number of milligrams of potassium hydroxide required to neutralize the carboxylic acid in one gram of crude oil. Acids in crude oil perform an important part of the enhanced oil recovery scheme known as caustic flooding.

Often oxygen occurs in crude oil as a part of the ring structure. One simple example is coumarone.

coumarone

Nitrogen Compounds

Very little is known about nitrogen compounds found in crude oils. Apparently, the nitrogen atoms are included in the complex ring structures. Simple examples are given below.

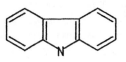   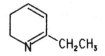

carbazole   ethylpyridine

Organometallic Compounds

Vanadium and nickel are present in parts-per-million quantities in most crude oils, usually in large, oil-soluble organometallic compounds termed porphyrins. The chemical structure of porphyrins is closely akin to the coloring matter in blood and to chlorophyll in plants.

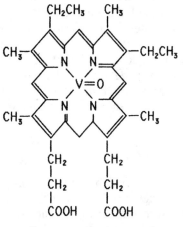

vanadium porphyrin complex

Lessor quantities of other metals, including copper and iron, have been observed in petroleum.

Resins and Asphaltenes

The chemicals in petroleum are classified as paraffins, naphthenes, aromatics, and resins-asphaltenes. We have discussed the first three classes, which are hydrocarbons. Now we turn to *resins and asphaltenes*.

Resins and asphaltenes are large molecules, primarily hydrogen and carbon, with one to three sulfur, oxygen, or nitrogen atoms per molecule. The basic structure is composed of rings, primarily aromatic, with from three to ten or more rings in each molecule. The nonhydrocarbon atoms can be a part of the ring structure or can be located in links connecting the rings.

The basic structures of resins and asphaltenes are similar. Both can be formed by oxidation of polycyclic aromatic hydrocarbons. On the other hand, both can be reduced to hydrocarbons by hydrogenation, which yields moderate to large hydrocarbon molecules, hydrogen sulfide, and water. Further, resins can be converted to asphaltenes by oxidation.

There are, however, important differences between resins and asphaltenes. Asphaltenes do not dissolve in petroleum but are dispersed as *colloids*. Resins readily dissolve in petroleum. Pure asphaltenes are solid, dry, black powders and are nonvolatile. Pure resins are heavy liquids or sticky solids and are as volatile as hydrocarbons of the same size. The resins of high molecular weight are red; the lighter resins are less colored.

Thus, when petroleum is separated into fractions by distillation, the asphaltenes remain in the heaviest fraction, the residuum, while the resins are distributed through the various fractions according to volatility. The color of these distillates almost entirely depends on the presence of resins. The color of the heaviest fraction, residuum, is determined to a great extent by the presence of asphaltenes.

The color of petroleum is determined largely by the quantity of resins and asphaltenes present, although the greenish cast of some crude oils is probably due to the presence of molecules containing six or more rings.

Classification of Crude Oils

Crude oils may be classified by physical properties or by chemical structure of the constituent molecules. Chemical structures are much more difficult to measure than are physical properties.

Physical Classification

Commercial value of a petroleum liquid can be estimated quickly through measurement of the following physical characteristics:
    specific gravity,
    gasoline and kerosene content,
    sulfur content,
    asphalt content,
    pour point,
    and cloud point.
The first four of these properties have been discussed. Pour point is the lowest temperature, expressed as a multiple of 5°F, at which the liquid is observed to flow when cooled under prescribed conditions. Cloud point is the temperature at which paraffin wax begins to solidify and is identified by the onset of turbidity as the temperature is lowered. Both tests qualitatively measure the paraffin content of the liquid.

Chemical Classification

Chemical classifications of petroleums relate to the molecular structures of the molecules in the oil. Of course the smaller molecules, six carbon atoms and less, are predominately paraffins. So chemical classification is usually based on analysis of the petroleum after most of the light molecules are removed.

Terms such as paraffinic, naphthenic, naphthenic-aromatic, and aromatic-asphaltic are used in the several classification methods which have been proposed. These terms obviously relate to the molecular structure of the chemical species most prominent in the crude oil mixture. However, such classification is made difficult because the large molecules usually consist of condensed aromatic and naphthenic rings with paraffinic side chains. The characteristic properties of the molecules depend on the proportions of these structures.

One classification method treats a large molecule as aromatic if it has a single benzene ring regardless of the other content. Another method considers the fraction of each molecule that is aromatic, naphthenic, or paraffinic. Obviously, in either case the analysis procedure is tedious. A third classification method simply measures the specific gravities of several fractions separated by distillation and attempts to relate chemical structure to specific gravity.

The petroleum industry has not agreed on a standard classification system. Further, the classification of crude oil is of little importance in its production, with the exception that paraffinic crudes can precipitate wax and plug the production string. Consequently, we will not examine the classification systems in detail. Kinghorn gives a good review of chemical classification systems.

Exercises

1-1. A partial listing of some of the compounds that have been isolated from a crude oil taken from the South Ponca Field in Oklahoma are listed below. All told, a total of 141 compounds, accounting for about 14% of the crude oil volume, have been isolated and identified in this particular oil sample.

PARAFFINS	NAPHTHENES	AROMATICS
All normal paraffins to $C_{10}H_{22}$	Cyclopentane	Benzene
Isobutane	Cyclohexane	Toluene
2-Methylbutane	Methylcyclopentane	Ethylbenzene
2,3-Dimethylbutane	1,1-Dimethylcyclopentane	Xylene
2-Methylpentane	Methylcyclohexane	1,2,4-Trimethyl benzene
3-Methylpentane	1,3,-Dimethylcyclohexane	
2-Methylhexane	1,2,4-Trimethylcyclohexane	
3-Methylhexane		
2-Methylheptane		
2,6-Dimethylheptane		
2-Methyloctane		

Give the structural formulas of each of these compounds.

1-2. The table below gives the hydrocarbons which have been identified in the gasoline fraction of a crude oil.
Identify the paraffins, naphthenes, and aromatics and draw the molecular structure of each.

Hydrocarbon
Cyclopentane
2,2-Dimethylbutane
2,3-Dimethylbutane
2-Methylpentane
3-Methylpentane
n-Hexane
Methylcyclopentane

2,2- and 2,4-Dimethylpentane
Cyclohexane
1,1-Dimethylcyclopentane
2,3-Dimethylpentane and 2-methylhexane
trans-1,3-Dimethylcyclopentane
trans-1,2-Dimethylcyclopentane
3-Methylhexane
n-Heptane
Methylcyclohexane
Benzene
Toluene
Ethylbenzene
p-Xylene
m-Xylene
o-Xylene
Isopropylbenzene
n-Propylbenzene

1–3. Table 10–2 gives the composition of a heptanes plus fraction of a crude oil. Separate the components into paraffins, naphthenes, and aromatics. Identify the number of carbon atoms in the paraffins and draw the structure of the naphthenes and aromatics.

1–4. Explain each type of chemical bonding and give an example of each.

1–5. Draw structural formulas for the isomers of heptane. Name the isomers by the IUPAC system.

1–6. There are 18 isomers of octane. Draw the molecular structure and name each.

1–7. Draw structural formulas for:
a. Pristane (2,6,10,14-tetramethylpentadecane)
b. Phytane (2,6,10,14-tetramethylhexadecane)

1–8. What is the IUPAC system name for neopentane?

1–9. Define and give an example of each:
a. Homologous series
b. Olefin

c. Diolefin
d. Cyclic hydrocarbon
e. Unsaturated hydrocarbon

1–10. Define and give two examples of each:
a. Stereoisomerism
b. Structural isomerism

1–11. What are the general formulas for the following hydrocarbon series?
a. Paraffin series
b. Olefin series
c. Diolefin series
d. Acetylene series
e. Naphthene series

1–12. Predict which one of each pair below will have the higher boiling point temperature. Justify your prediction in each case.
a. 3-Methylpentane and 3-methylhexane
b. 3-Methylpentane and n-hexane
c. n-Pentane and 1-pentene
d. 1-Pentene and 2-methyl–1-butene

1–13. Name the following compounds:

CH_3CH_2SH

$CH_3CH_2NH_2$

$CH_3CH_2CH_2CH_2COOH$

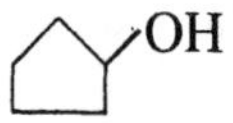

CH_3CCH_3
$\overset{\|}{O}$

1–14. Draw structural formulas for the following compounds:

Biphenyl
Chlorobenzene
Naphthalene
Cumene
Dimethyl disulfide

General References

Roberts, J.D. and Caserio, M.C.: *Modern Organic Chemistry*, Benjamin, New York (1967).

"Naming Hydrocarbons," *Chem. and Eng. News* (February 25, 1957) 72.

Sachanen, A.N.: "Hydrocarbons in Petroleum," *The Science of Petroleum*, v.V, part I, *Crude Oils Chemical and Physical Properties*, B.T. Brooks and A.E. Dunstan (eds.), Oxford University Press, London (1950).

Smith, H.K.: "Crude Oil: Qualitative and Quantitative Aspects. The Petroleum World," *Information Circular 8286* U.S. Dept. of the Interior, Bureau of Mines (1966).

Kinghorn, R.R.F.: *An Introduction to the Physics and Chemistry of Petroleum*, John Wiley, New York (1983).

Barker, C.: "Origin, Composition and Properties of Petroleum," *Enhanced Oil Recovery I*, Fundamentals and Analysis, E.C. Donaldson, G.V. Chilingarian, and T.F. Yen (eds.), Elsevier, New York (1985).

Methods for Analysis and Testing, The Institute of Petroleum, London, John Wiley and Sons, New York (1984).

2

Phase Behavior

Before studying the properties of gases and liquids, we need to understand the relationship between the two phases. The starting point will be a study of vapor pressure and the development of the definition of the critical point. Then we will look in detail at the effects of pressure and temperature on one of the intensive properties of particular interest to petroleum engineers: specific volume.

The term *phase* defines any homogeneous and physically distinct part of a system which is separated from other parts of the system by definite bounding surfaces. For example, ice, liquid water, and water vapor are three phases. Each is physically distinct and homogeneous, and there are definite boundaries between ice and water, between ice and water vapor, and between liquid water and water vapor. Thus, we say that we have a three-phase system: solid, liquid, and gas. One particular phase need not be continuous. For instance, the ice may exist as several lumps in the water.

The definition of phase can be carried further. Depending on temperature and pressure, solids can take on different crystalline forms. These constitute separate phases. The petroleum engineer normally is not concerned with the crystalline forms of solids; therefore, we will consider only three phases: gas, liquid, and solid. For the purpose of our discussion we will use the words *vapor* and *gas* interchangeably. Some authors try to distinguish between these two words, but the difference is unimportant to us.

Physical properties are termed either intensive or extensive. *Intensive properties* are independent of the quantity of material present. Density, specific volume, and compressibility factor are examples. Properties such as volume and mass are termed *extensive;* their values are determined by the total quantity of matter present.

The manner in which hydrocarbons behave when pressure and temperature are changed is explained best by a consideration of the behavior of the individual molecules. Three factors are important to the physical behavior of molecules:

Pressure—a reflection of the number of molecules present and their motion, *temperature*—a reflection of the kinetic energy of the molecules, and *molecular attraction and repulsion.* Pressure and molecular attraction tend to confine the molecules and pull them together. Temperature and molecular repulsion tend to separate the molecules.

When a material appears to be at rest, that is, not changing volume or changing phases, the forces confining the molecules are balanced with the forces tending to throw them apart. When the material is not at rest, in the case of boiling water for instance, the forces are unbalanced and the material is changing so these forces can regain their balance.

Temperature is simply a physical measure of the average kinetic energy of the molecules of the material. As heat is added to the material, the kinetic energy of the molecules is increased, and, as a result, the temperature is increased. The increase in kinetic energy causes an increase in molecular motion, which results in a tendency for the molecules to move apart.

Pressure is a reflection of the number of times the molecules of a gas strike the walls of its container. If the molecules are forced closer together the pressure increases.

Intermolecular forces are the forces of attraction and repulsion between molecules. These forces change as the distance between molecules changes. The attractive force increases as the distance between the molecules decreases until the molecules get so close together that their electronic fields overlap. Any further decrease of the distance between the molecules will cause a repulsive force between them. This repulsive force will increase as the molecules are forced closer together.

Gases, in which the molecules are relatively far apart, have an attractive force between the molecules as the distance between molecules decreases. However, in a liquid, in which the molecules are fairly close together, there is a repelling force between molecules which causes the liquid to resist further compression.

Pure Substances

We will first consider systems which consist of a single, pure substance. These systems behave differently from systems made up of two or more components. In particular, we will be interested in *phase behavior,* that is, the conditions of temperature and pressure for which different phases can exist.

We will be concerned with three variables: pressure, temperature, and volume. Pressure and temperature are imposed on the system and determine the phase or phases which exist. The phases which exist are identified by their specific volumes or densities.

After a careful examination of the phase behavior of a pure substance, we will discuss the behavior of systems which contain two or more components and point out the differences between multicomponent behavior and pure substance behavior.

Phase Diagram for a Pure Substance

A *phase diagram* is a graph of pressure plotted against temperature showing the conditions under which the various phases of a substance will be present. Figure 2–1 shows a phase diagram for a single-component system. Phase diagrams are often called *pressure-temperature diagrams*.

The Vapor-Pressure Line

Line $\overline{TC}$ on Figure 2–1 is called the *vapor-pressure line*. This line separates the pressure-temperature conditions for which the substance is a liquid from the conditions for which the substance is a gas. Pressure-temperature points which lie above this line indicate conditions for which the substance is a liquid. Similarly, points below the vapor-pressure line represent conditions for which the substance is a gas. Pressure-temperature points which fall exactly on the line indicate conditions for which gas and liquid coexist.

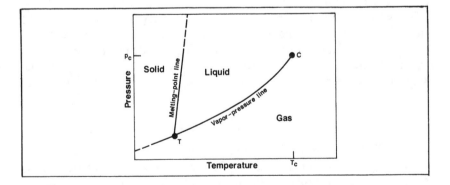

Fig. 2–1. Typical phase diagram of a pure substance.

The Critical Point

The upper limit of the vapor-pressure line is the *critical point,* indicated by point C. The temperature and pressure represented by this point are called the *critical temperature,* T_c, and the *critical pressure,* p_c.

For a pure substance, the critical temperature may be defined as the temperature above which the gas cannot be liquefied, regardless of the pressure applied. Similarly, the critical pressure of a pure substance is defined as the pressure above which liquid and gas cannot coexist, regardless of the temperature. These definitions of critical properties are invalid for systems with more than one component.

The Triple Point

Point T on the vapor-pressure line is called the *triple point.* This point represents the pressure and temperature at which solid, liquid, and gas coexist under equilibrium conditions.

The Sublimation-Pressure Line

At temperatures below the triple-point temperature, the vapor-pressure line divides the conditions for which the substance is solid from the conditions for which the substance is gas. This line also is called the *sublimation-pressure line*. Theoretically, this line extends to a temperature of absolute zero and a pressure of absolute zero.

The Melting Point Line

The *melting point line* is the nearly vertical line above the triple point. This line separates solid conditions from liquid conditions. Again, pressure-temperature points which fall exactly on this line indicate a two-phase system—in this case coexistence of solid and liquid. Phase diagrams for some pure materials show other lines within the solid region, indicating changes of phase of the solid brought about by changes in crystalline structure. The upper limit of the melting point line has not been obtained experimentally.

Use of Phase Diagrams

In order to gain a better understanding of the usefulness of phase diagrams, consider a cylinder in which temperature can be controlled and volume varied by injection or removal of mercury as shown in Figure 2–2. Figure 2–2A shows that a pure substance has been trapped in the

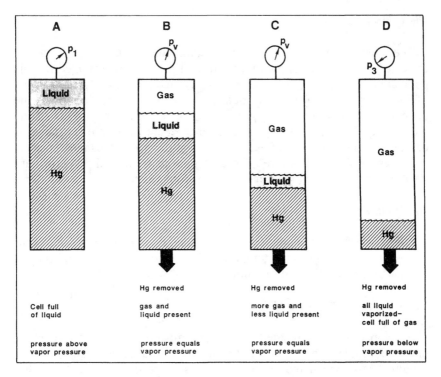

Fig. 2–2. Vaporization of a pure substance at constant temperature.

cylinder at pressure p_1 and at some temperature below the critical temperature of the substance. We will hold temperature constant and increase the volume by removing mercury, thereby causing pressure to decrease.

The process will follow the path of line $\overline{123}$ on Figure 2–3. As mercury is removed, the pressure decreases rapidly until it reaches a value of p_v, the vapor pressure of the substance. At this point gas begins to form as molecules leave the liquid. Pressure, which has forced the molecules together, has been reduced so that those molecules with the highest kinetic energy can escape the liquid and form gas.

As mercury removal continues, the volume of gas increases and the volume of liquid decreases; however, pressure remains constant at a value of p_v. See Figures 2–2B, 2–2C, and point 2 on Figure 2–3.

Once the liquid disappears, further mercury removal causes a decrease in pressure as the gas expands. Eventually, the pressure reaches point p_3. The above description *only* applies for a pure substance. Later we will see how this process works for a mixture.

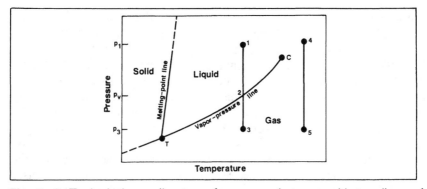

Fig. 2-3. Typical phase diagram of a pure substance with two lines of isothermal expansion: 123 below critical temperature, 45 above critical temperature.

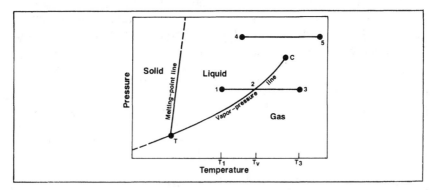

Fig. 2-4. Typical phase diagram of a pure substance with two lines of isobaric temperature change: 123 below critical pressure, 45 above critical pressure.

If the same process is followed for a temperature above the critical temperature, for instance, line 45 on Figure 2-3, the removal of mercury will cause pressure to decrease. However, there will not be a sudden change in the density of the substance. The vapor-pressure line will not be crossed. There is no abrupt phase change.

Note that the terms gas and liquid simply designate the less dense and denser phases, respectively. These phases merge and lose their identity at the critical point.

We will use the cylinder described previously to study the process of increasing the temperature at constant pressure. See line 123 on Figure 2-4. Temperature will be increased by adding heat. This addition of energy also will cause pressure to increase so that increases in volume (removal of mercury) will be required to maintain constant pressure.

Figure 2–5A shows the cell full of liquid at temperature T_1, which is less than the vapor-pressure temperature of the substance. In Figure 2–5B, the substance has been heated at constant pressure to the vapor-pressure temperature. The injection of heat has caused the kinetic energy of the molecules to increase so that the molecules with the highest kinetic energy can escape the attractive forces to form a gas. Gas and liquid coexist.

After the vapor-pressure temperature is reached, heat put into the cylinder does not cause an increase in temperature; rather, it causes vaporization of the liquid. The temperature will remain constant as long as gas and liquid coexist. Figure 2–5D indicates that enough heat has been put into the cylinder to evaporate all the liquid and that additional heat has caused an increase in temperature to T_3.

The same process at pressures above the critical pressure, for instance, line $\overline{45}$ on Figure 2–4, does not show the abrupt change in phase that the process below the critical pressure shows.

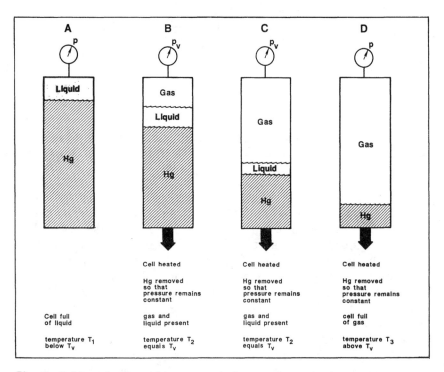

Fig. 2–5. Vaporization of a pure substance at constant pressure.

Vapor Pressure of a Pure Substance

Since the petroleum engineer is rarely concerned with solid hydrocarbons, in the following discussion we will consider only the vapor-pressure line and the liquid and gas portions of the phase diagram.

The Clausius-Clapeyron Equation

The Clausius-Clapeyron Equation expresses the relationship between vapor pressure and temperature. It is the equation for the vapor-pressure line. We will develop this equation with the Clapeyron Equation, which was developed using thermodynamic theory.

$$\frac{dp_v}{dT} = \frac{L_v}{T(V_{Mg} - V_{ML})} \qquad (2-1)$$

The relationship dp_v/dT is the rate of change of vapor pressure with temperature. Thus it represents the slope of the vapor-pressure line. L_v is the heat of vaporization of one mole of liquid, T is the absolute temperature, and $V_{Mg} - V_{ML}$ represents the change in volume of one mole as it goes from liquid to gas.

Normally, the molar volume of a liquid is somewhat smaller than the molar volume of the gas, so we will neglect the molar volume of the liquid and rewrite Equation 2-1.

$$\frac{dp_v}{dT} = \frac{L_v}{TV_{Mg}} \qquad (2-2)$$

We will see later that the pressure, volume, and temperature of an ideal gas may be related by Equation 3-13.

$$p_v V_{Mg} = RT \qquad (3-13)$$

Combination of Equation 2-2 and Equation 3-13 results in

$$\frac{dp_v}{dT} = \frac{p_v L_v}{RT^2} . \qquad (2-3)$$

This is known as the Clausius-Clapeyron equation.

If we assume that L_v is a constant, we can rearrange and integrate Equation 2-3,

$$\int \frac{dp_v}{p_v} = \frac{L_v}{R} \int \frac{dT}{T^2} , \qquad (2-4)$$

resulting in

$$\ln p_v = -\frac{L_v}{R}\left(\frac{1}{T}\right) + C, \qquad (2-5)$$

where C is the constant of integration, or

$$\boxed{\ln \frac{p_{v2}}{p_{v1}} = \frac{L_v}{R}\left(\frac{1}{T_1} - \frac{1}{T_2}\right),} \qquad (2-6)$$

where the subscripts 1 and 2 indicate different conditions of temperature and pressure.

Equation 2–5 indicates that a straight line results when the logarithm of vapor pressure is plotted against the reciprocal of absolute temperature. The slope of this line is $-L_v/R$, and the intercept is C.

EXAMPLE 2–1: *Plot the vapor pressure of n-hexane in a way that will result in a straight line.*

Temperature, °F	Vapor pressure of n-hexane, psia
155.7	14.7
199.4	29.4
269.1	73.5
331.9	147.0
408.9	293.9
454.6	435.0

Solution

See Figure 2–6.

This is an application of theory that you will encounter throughout your professional career. Many times we start from basic principles and derive equations through the use of mathematical manipulations. Often, the resulting equations are not used to make calculations but are simply used to indicate a method of handling data. In this case the equation provides a method of plotting vapor pressure data so that a straight line results.

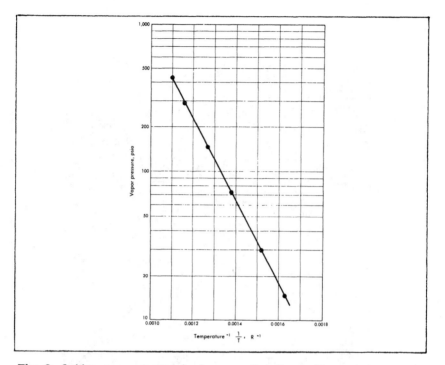

Fig. 2–6. Vapor pressures of n-hexane. Solution to Example 2–1. (Data from Stull, *Ind. and Eng. Chem., 39,* 517, 1947.)

Cox Charts

The assumption that the heat of vaporization is constant is not necessarily valid. Also, at temperatures near the critical temperature, the assumption about the molar volume of the liquid is invalid. Thus, a vapor-pressure graph of the type given in Figure 2–6 usually results in a line with some curvature over a long temperature range.

This problem was solved by plotting the logarithm of vapor pressure against an arbitrary temperature scale which is related to the reciprocal of the absolute temperature. The temperature scale was constructed by drawing a straight line on the chart and adjusting the temperature scale so that the vapor pressure of water corresponded to the straight line. This modified graph of vapor pressure data is known as a Cox Chart.

The vapor pressures of most hydrocarbons will plot as straight lines on this chart. The Cox Charts for normal paraffin and isometric paraffin hydrocarbons are given in Figures 2–7 and 2–8.[1] The dot on each vapor-pressure line indicates the critical point. The vapor-pressure line should

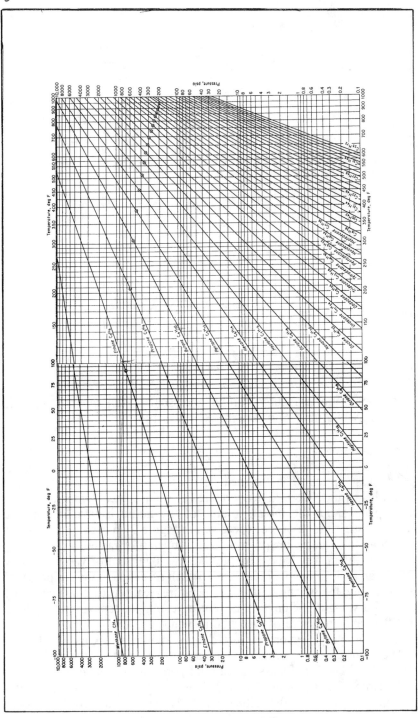

Fig. 2–7. Vapor pressures of normal paraffins. (From *Handbook of Natural Gas Engineering* by Katz et al. Copyright 1959 by McGraw-Hill Book Co. Used with permission of McGraw-Hill Book Co.)

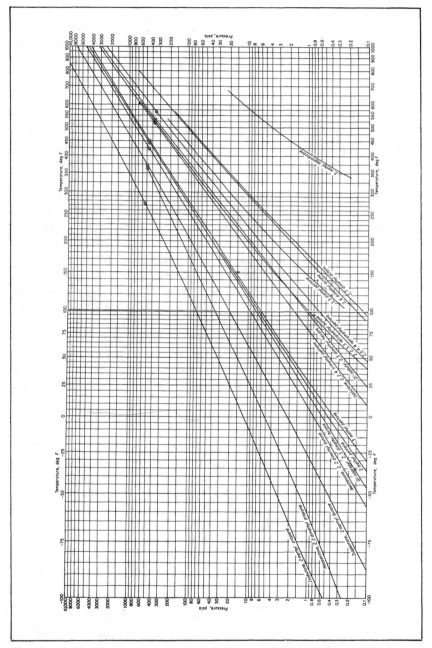

Fig. 2–8. Vapor pressures of isomeric paraffins. (From *Handbook of Natural Gas Engineering* by Katz et al. Copyright 1959 by McGraw-Hill Book Co. Used with permission of McGraw-Hill Book Co.)

not extend to temperatures higher than the critical temperature. However, these lines have been extrapolated past the critical temperatures because certain approximate calculations require the use of artificially extrapolated vapor pressures.

Pressure-Volume Diagram for a Pure Substance

The results of the process described in Figure 2–2 may be presented in the form of a pressure-volume diagram. Figure 2–9 shows two isotherms of a typical pressure-volume diagram for a pure substance. Processes 1–3 and 4–5 correspond to the processes indicated in Figure 2–3.

Consider a process starting at point 1 with a substance in the liquid phase. Temperature is held constant, and volume is increased by removal of the mercury. This causes a reduction in pressure from p_1 to p_v. A relatively large change in pressure results from a small change in volume. This is because liquids are relatively incompressible.

When the pressure is reduced to the vapor pressure, p_v, gas begins to form, and further increases in volume cause vaporization of the liquid. This continues at constant pressure until all the liquid is vaporized. Point 2 on Figure 2–3 is a straight horizontal line on Figure 2–9. This shows that pressure remains constant as long as liquid and gas coexist at constant temperature. After all the liquid is vaporized, continued increase in volume causes expansion of the gas and reduction in pressure to p_3. Since the gas is highly compressible, the slope of the isotherm is much less steep in the gas region than in the liquid region.

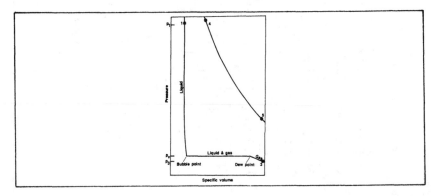

Fig. 2–9. Typical pressure-volume diagram of a pure substance showing two isotherms: 13 below critical temperature, 45 above critical temperature.

Line $\overline{45}$ on Figure 2–9 illustrates the same process at a temperature above the critical temperature of the substance. The line shows that there is simply an expansion of the substance and that no abrupt phase change occurs.

Bubble Point and Dew Point

Again, consider the constant temperature expansion illustrated by line $\overline{13}$ on Figure 2–9. The point at which the first few molecules leave the liquid and form a small bubble of gas is called the *bubble point*. The point at which only a small drop of liquid remains is known as the *dew point*. The bubble point and dew point are indicated by the sharp changes in slope along the isotherm. For a pure substance, the pressure at the bubble point and dew point is equal to the vapor pressure of the substance at the temperature of interest.

Saturation Envelope

Figure 2–10 shows a more nearly complete pressure-volume diagram.[2] The dashed line shows the locus of all bubble points and dew points. The area within the dashed line indicates conditions for which liquid and gas coexist. Often this area is called the *saturation envelope*. The bubble-point line and dew-point line coincide at the critical point. Notice that the isotherm at the critical temperature shows a point of horizontal inflection as it passes through the critical pressure.

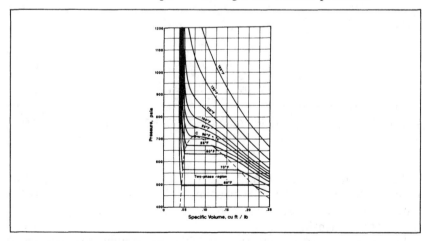

Fig. 2–10. Pressure-volume diagram of ethane. (Data from Brown et al., *Natural Gasoline and the Volatile Hydrocarbons*, NGAA, Tulsa, 1947.)

EXAMPLE 2-2: *Use Figure 2-10 to estimate the critical temperature and critical pressure of ethane. Also estimate the specific volumes of ethane liquid and gas at 70°F.*

Solution

T_c = 90°F and p_c = 710 psia, point C, Figure 2-10.
v of liquid at 70°F = 0.047 cu ft/lb, and
v of gas at 70°F = 0.178 cu ft/lb, 70°F isotherm, Figure 2-10.

Density-Temperature Diagram for a Pure Substance

The shape of a typical density-temperature diagram is given in Figure 2-11. The line shows the densities of the liquid and gas that coexist in the two-phase region. Often these are called the saturated densities. Notice that the densities of the liquid and gas are identical at the critical point.

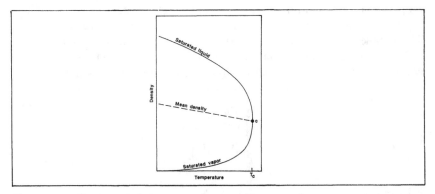

Fig. 2-11. Typical density-temperature diagram of a pure substance.

The average densities of the liquid and gas will plot as a straight line which passes through the critical point. This property is known as the Law of Rectilinear Diameters. The dashed line on Figure 2-11 shows these average densities.

Figure 2-12 gives the saturated densities for a number of substances of interest to the petroleum engineer.[2]

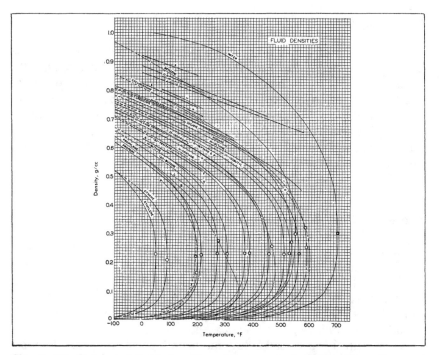

Fig. 2–12. Saturated fluid densities. (Brown, et al. *Natural Gasoline and the Volatile Hydrocarbons,* NGAA, 1947, 48, with permission.)

Two-Component Mixtures

Next we will consider the phase behavior of mixtures of two components. The petroleum engineer does not normally work with two-component systems; usually mixtures consisting of many components are encountered. However, it is instructive to observe the differences in phase behavior between two-component mixtures and pure substances. These differences are amplified in multicomponent mixtures.

We will first consider phase diagrams. Then we will define the critical point for a two-component mixture. This will be the correct definition for multicomponent mixtures. Also, we will look at an important concept called retrograde condensation. Then the pressure-volume diagram will be discussed, and differences between pure substances and two-component mixtures in the two-phase region will be illustrated. Finally, the effects of temperature and pressure on the compositions of the coexisting liquid and gas will be illustrated.

Phase Diagrams of Two-Component Mixtures

The behavior of a mixture of two components is not as simple as the behavior of a pure substance. Instead of a single line representing the vapor-pressure curve, there is a broad region in which two phases coexist. This region is called the *saturation envelope, phase envelope,* or *two-phase region.* Figure 2–13 shows the typical shape of the phase diagram for a mixture of two components. The two-phase region of the phase diagram is bounded on one side by a bubble-point line and the other side by a dew-point line. The two lines join at the critical point.

Bubble Point and Dew Point

Consider the constant temperature expansion illustrated on Figure 2–13 by line $\overline{12}$. At pressure p_1 the mixture is liquid. As pressure is decreased, the liquid expands until the pressure reaches a point at which a few molecules are able to leave the liquid and form a small bubble of gas. This is the *bubble point.* The pressure at which the first gas is formed is the *bubble-point pressure,* p_b.

As pressure is decreased below the bubble-point pressure, additional gas appears. Finally, only a minute amount of liquid remains. This is the *dew point.* The pressure at this point is known as the *dew-point pressure,* p_d. Further reduction of pressure to point 2 simply causes an expansion of the gas.

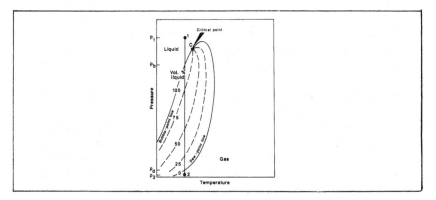

Fig. 2–13. Typical phase diagram of a two-component mixture with line of isothermal expansion, 12.

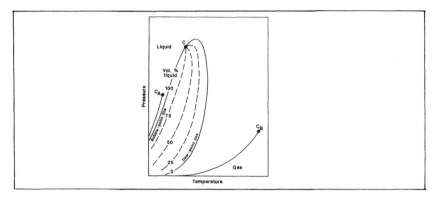

Fig. 2–14. Typical phase diagram of a two-component mixture with vapor-pressure lines of the two pure components.

The Critical Point

The definition of the critical point as applied to a pure substance *does not* apply to a two-component mixture. In a two-component mixture, liquid and gas can coexist at temperatures and pressures above the critical point. Notice that the saturation envelope exists at temperatures higher than the critical temperature and at pressures higher than the critical pressure. We see now that the definition of the critical point is simply the point at which the bubble-point line and the dew-point line join. A more rigorous definition of the *critical point* is that it is the point at which all properties of the liquid and the gas become identical.

Figure 2–14 shows the vapor-pressure lines of the two components of a mixture superimposed on the phase diagram of the mixture. The saturation envelope for the mixture lies between the vapor pressure lines of the two components. The critical temperature of the mixture lies between the critical temperatures of the two pure components. However, the critical pressure of the mixture is above the critical pressures of both of the components. The critical pressure of a two-component mixture usually will be higher than the critical pressure of either of the components.

Figure 2–15 shows phase data for eight mixtures of methane and ethane, along with the vapor-pressure lines for pure methane and pure ethane.[3] Again, observe that the saturation envelope of each of the mixtures lies between the vapor pressure lines of the two pure substances and that the critical pressures of the mixtures lie well above the critical pressures of the pure components. The dashed line is the *locus of critical points* of mixtures of methane and ethane.

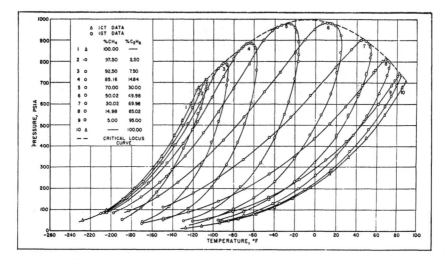

Fig. 2–15. Phase diagrams of mixtures of methane and ethane. (Bloomer, et al., *Institute of Gas Technology, Research Bulletin 22*, 1953. Reproduced courtesy of Institute of Gas Technology, Chicago.)

EXAMPLE 2–3: *Determine the critical temperature and critical pressure of a mixture of 50.02 mole percent methane and 49.98 mole percent ethane. Also determine the bubble-point pressure and dew-point pressure of this mixture at −20°F.*

Solution

T_c = 15°F and p_c = 975 psia, line 6, Figure 2–15
p_b = 842 psia, p_d = 339 psia at −20°F, line 6, Figure 2–15.

The *critical loci* of binary systems composed of normal paraffin hydrocarbons are shown in Figure 2–16.[2] Obviously, the critical pressures of mixtures are considerably higher than the critical pressures of the components of the mixtures. In fact, a larger difference in molecular size of the components causes the mixtures to have very large critical pressures.

Cricondentherm and Cricondenbar

The *highest temperature* on the saturation envelope is called the *cricondentherm*. The *highest pressure* on the saturation envelope is called the *cricondenbar*. These conditions are illustrated on Figure 2–17.

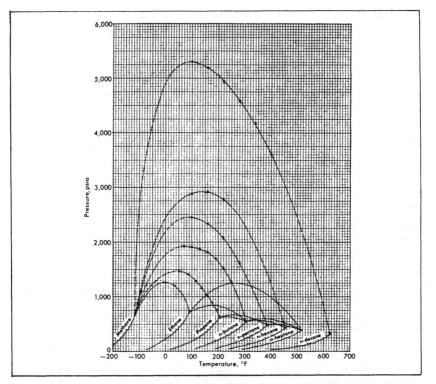

Fig. 2–16. Critical loci of binary n-paraffin mixtures. (Brown et al., *Natural Gas and the Volatile Hydrocarbons,* NGAA, 1947, 4, with permission.)

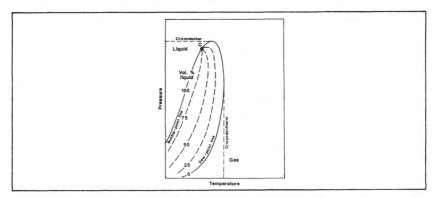

Fig. 2–17. Typical phase diagram of a two-component mixture with definitions of cricondenbar and cricondentherm.

EXAMPLE 2–4: *Determine the cricondentherm and cricondenbar of a mixture of 50.02 mole percent methane and 49.98 mole percent ethane.*

Solution

cricondentherm = 24°F, cricondenbar is very near critical pressure of 975 psia, line 6, Figure 2–15.

Retrograde Condensation

Another feature of the two-component system is illustrated on Figure 2–18. Remember that for a pure substance a decrease in pressure causes a change of phase from liquid to gas at the vapor-pressure line. Likewise, in the case of a two-component system a decrease in pressure causes a change of phase from liquid to gas at temperatures below the critical temperature. An example is process 1–2 on Figure 2–13.

However, consider the isothermal decrease in pressure illustrated by line 123 on Figure 2–18. As pressure is decreased from point 1, the dew-point line is crossed and liquid begins to form. At the position indicated by point 2, the system is 25 percent liquid by volume and 75 percent gas. A decrease in pressure has caused a change from gas to liquid. This is exactly the reverse of the behavior one would expect, hence the name *retrograde condensation*. As pressure is decreased from point 2 toward point 3, the amount of liquid decreases, the dew-point line is crossed a second time, and the system again becomes gas.

Note that an *upper dew point* and a *lower dew point* exist. The upper dew point sometimes is called the *retrograde dew point*.

The region of retrograde condensation occurs at temperatures between the critical temperature and the cricondentherm. A similar retrograde

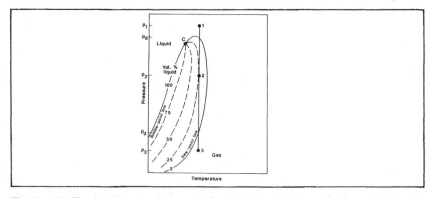

Fig. 2–18. Typical phase diagram of a two-component mixture with line of isothermal expansion, 123, in the retrograde region.

situation occurs when temperature is changed at constant pressure between the critical pressure and the cricondenbar.

EXAMPLE 2–5: *Determine the temperature range in which retrograde condensation will occur for a mixture of 50.02 mole percent methane and 49.98 mole percent ethane.*

Solution

Retrograde condensation will occur at temperatures between the critical temperature, 15°F, and the cricondentherm, 24°F.

Pressure-Volume Diagram for a Two-Component Mixture

Figure 2–19 shows a single isotherm on a pressure-volume diagram of a two-component mixture with a constant overall composition. The feature that distinguishes this diagram from a pressure-volume diagram of a pure substance, Figure 2–9, is that pressure decreases as the process passes from the bubble point to the dew point. The line from bubble point to dew point is not horizontal and is not necessarily straight. The decrease in pressure is caused by the changes in the compositions of the liquid and the gas as the process passes through the two-phase region.

At the bubble point, the composition of the liquid is essentially equal to the overall composition of the mixture, but the infinitesimal amount of gas is richer in the more volatile component. Likewise, at the dew point the composition of the vapor is essentially equal to the overall composition of the system and the infinitesimal amount of liquid is richer in the less volatile component. The changes in slope of the line at the bubble point and dew point are not as sharp as for a pure substance.

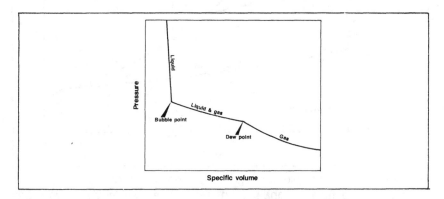

Fig. 2–19. Typical pressure-volume diagram of a two-component mixture showing one isotherm below the critical temperature.

Figure 2–20 gives the pressure-volume diagram for a mixture of n-pentane and n-heptane, showing several isotherms and the saturation envelope.[4] Notice that at lower temperatures the changes in slope of the isotherms at the dew points are almost nonexistent. Also notice that the critical point is not at the top of the saturation envelope as it was for pure substances.

EXAMPLE 2–6: *Consider a mixture of 47.6 weight percent n-pentane and 52.4 weight percent n-heptane. Estimate the specific volume of the liquid at its bubble point at 400°F. Also estimate the specific volume of the gas at its dew point at 400°F.*

Solution

v of bubble-point liquid at 400°F and 332 psia = 0.036 cu ft/lb,
v of dew-point gas at 400°F and 275 psia = 0.249 cu ft/lb,
from 400°F isotherm of Figure 2–20.

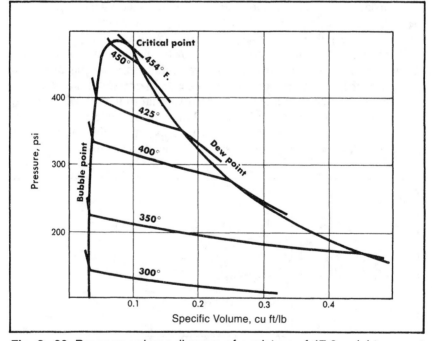

Fig. 2–20. Pressure-volume diagram of a mixture of 47.6 weight percent n-pentane and 52.4 weight percent n-heptane. (From *Volumetric and Phase Behavior of Hydrocarbons,* Bruce H. Sage and William N. Lacey. Copyright 1949, Gulf Publishing Co., Houston. Used with permission.)

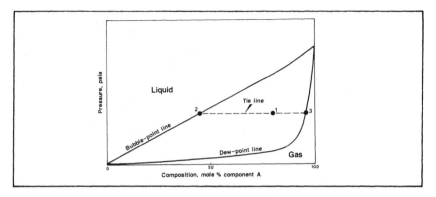

Fig. 2–21. Typical pressure-composition diagram of a two-component mixture with one tie line, $\bar{1}23$.

Composition Diagrams

A series of laboratory experiments with a pure substance (shown in Figure 2–2) will result in data for pressure, temperature, and volume. A similar series of experiments with a two-component system will result in data for additional variables. The composition of the overall mixture, the composition of the equilibrium liquid, and the composition of the equilibrium gas are all important. Therefore, in addition to plotting combinations of temperature, pressure, and volume, additional graphs with these variables plotted against composition are possible.

Pressure-Composition Diagrams for Two-Component Mixtures

Figure 2–21 gives a typical *pressure-composition diagram* for a two-component mixture at a single temperature. Combinations of composition and pressure which plot above the envelope indicate conditions at which the mixture is completely liquid. Combinations of composition and pressure which plot below the envelope indicate conditions at which the mixture is gas. Any combinations of pressure and composition which plot within the envelope indicate that the mixture exists in two phases, gas and liquid. Bubble point and dew point have the same definitions as previously discussed.

The bubble-point line is also the locus of compositions of the liquid when two phases are present. The dew-point line is the locus of compositions of the gas when gas and liquid are in equilibrium. The line which ties the composition of the liquid with the composition of gas in equilibrium is known as an *equilibrium tie-line*. *Tie-lines* are always horizontal for two-component mixtures.

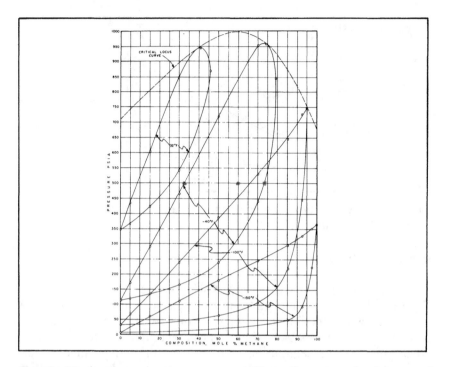

Fig. 2–22. Isothermal pressure-composition diagrams of mixtures of methane and ethane. (Bloomer, et al., *Institute of Gas Technology, Research Bulletin 22,* 1953. Reproduced courtesy of Institute of Gas Technology, Chicago.)

Consider that a mixture of composition represented by point 1 is brought to equilibrium at the indicated pressure and the temperature of the diagram. The composition of the equilibrium liquid is indicated by point 2, and the composition of the equilibrium gas is given by point 3.

The tie-line can also be used to determine the quantities of gas and liquid present at point 1. The length of line $\overline{12}$ divided by the length of the tie-line, $\overline{23}$, is the ratio of moles of gas to total moles of mixture. The length of line $\overline{13}$ divided by $\overline{23}$ is the ratio of moles of liquid to total moles of mixture.

Figure 2–22 gives pressure-composition diagrams for mixtures of methane and ethane.[3] There are four saturation envelopes corresponding to four different temperatures.

The edge of the diagram labeled 100 mole percent methane represents vapor pressures of methane. The edge of the diagram labeled zero mole percent methane gives vapor pressures of ethane.

When the temperature exceeds the critical temperature of one component, the saturation envelope does not go all the way across the diagram; rather, the dew-point and bubble-point lines join at a critical point. For instance, when the critical temperature of a mixture of methane and ethane is minus 100°F, the critical pressure is 750 psia, and the composition of the critical mixture is 95 mole percent methane and 5 mole percent ethane.

Notice that the locus of the critical points connects the critical pressure of ethane, 708 psia, to the critical pressure of methane, 668 psia. When the temperature exceeds the critical temperature of both components, it is not possible for any mixture of the two components to have two phases.

EXAMPLE 2–7: *Determine the compositions and quantities of gas and liquid formed when 3 lb moles of mixture of 70 mole percent methane and 30 mole percent ethane are brought to equilibrium at −100°F and 400 psia.*

Solution

First, plot 70 mole percent methane and 400 psia on the −100°F saturation envelope on Figure 2–22. Second, draw the tie-line (see Fig. 2–23) and read the composition of the equilibrium liquid on the

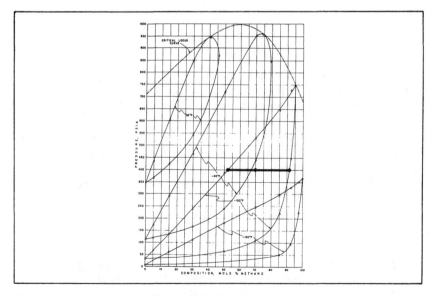

Fig. 2–23. Solution to Example 2–7.

bubble-point line and the composition of equilibrium gas on the dew-point line.

Component	Composition of liquid, mole percent	Composition of gas, mole percent
Methane	52.2	91.8
Ethane	47.8	8.2
	100.0	100.0

Third, calculate fractions of gas and liquid from length of the tie-line.

$$\text{fraction gas} = \frac{70.0 - 52.2}{91.8 - 52.2} = 0.45 \text{ lb mole of gas/lb mole total}$$

$$\text{fraction liquid} = \frac{91.8 - 70.0}{91.8 - 52.2} = 0.55 \text{ lb mole of liquid/lb mole total}$$

$$\text{quantity of gas} = (0.45)(3 \text{ lb mole}) = 1.35 \text{ lb mole gas}$$
$$\text{quantity of liquid} = (0.55)(3 \text{ lb mole}) = 1.65 \text{ lb mole liquid}$$

Temperature-Composition Diagrams for Two-Component Mixtures

Temperature-composition diagrams for mixtures of methane and ethane are given in Figure 2–24.[3] Six saturation envelopes corresponding to six different pressures are shown.

The lower line of a saturation envelope is the bubble-point line, and the upper line is the dew-point line. Composition-temperature conditions which plot below the saturation envelope indicate that the mixture is entirely liquid.

When pressure is less than the critical pressures of both components, the bubble-point and dew-point lines join at the vapor pressures of the pure components at either side of the diagram. When the pressure exceeds the critical pressure of one of the components, the bubble-point line and the dew-point line join at a critical point. For instance, a mixture of 98 mole percent methane and 2 mole percent ethane has a critical temperature of minus 110°F at a critical pressure of 700 psia.

When the pressure of interest exceeds the critical pressures of both components, the phase envelope exhibits two critical points. For instance, mixtures of methane and ethane exhibit critical points at 900 psia and minus 62°F and at 900 psia and 46°F.

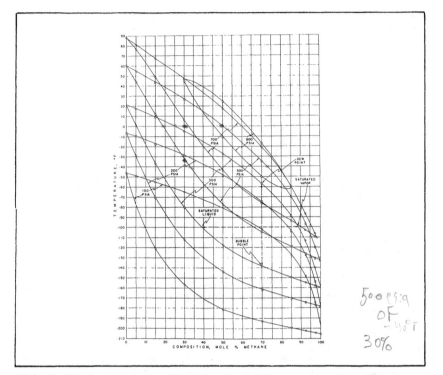

Fig. 2–24. Isobaric temperature-composition diagrams of mixtures of methane and ethane. (Bloomer, et al., *Institute of Gas Technology, Research Bulletin 22,* 1953. Reproduced courtesy of Institute of Gas Technology, Chicago.)

Tie-lines giving the compositions of the liquid and gas in equilibrium are again horizontal. The bubble-point line gives the composition of the equilibrium liquid, and the dew-point line gives the composition of the equilibrium gas. The lengths of the tie-lines represent the quantities of gas and liquid at equilibrium in the same manner as for the pressure-composition diagram.

Three-Component Mixtures

Compositional phase diagrams for three-component mixtures must be plotted in such a way that the compositions of all three components can be displayed. Diagrams formed from equilateral triangles are convenient for this purpose. These are called *ternary diagrams*.

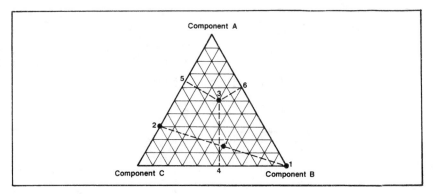

Fig. 2–25. Ternary diagram.

Ternary Diagrams

Figure 2–25 shows a ternary diagram. Each apex of the triangle corresponds to 100 percent of a single component. The usual convention is to plot the lightest component at the top and the heaviest component at the lower left. Each side of the triangle represents two-component mixtures. For instance, the left side of the triangle represents all possible mixtures of the light and the heavy components. Points within the triangle represent three-component mixtures. Composition is usually plotted in terms of mole fraction or mole percent. For a single diagram, both pressure and temperature are constant; only composition changes.

Point 1 on Figure 2–25 represents pure component B. Point 2 represents a mixture of 30 mole percent component A and 70 mole percent component C. Point 3 represents a mixture which consists of 50 mole percent A, 30 mole percent B, and 20 mole percent C. The composition of the mixture represented by point 3 is best determined by imagining three lines from point 3 perpendicular to the sides of the triangular diagram. The length of line $\overline{43}$ represents the composition of component A in the mixture. The length of line $\overline{53}$ represents the composition of component B, and the length of line $\overline{63}$ represents the composition of component C.

Line $\overline{21}$ represents a process of interest to the petroleum engineer. As we have seen, point 2 represents the composition of a mixture of component A and component C with no component B present. Line $\overline{12}$ represents the compositions of all mixtures formed by the addition of component B to the original mixture of components A and C. For instance, point 7 represents a mixture of equal parts of the original mixture of A and C with component B. The composition is 50 percent

component B, 15 percent component A, and 35 percent component C.
The ratio of components A to C, 15:35, is the same as the ratio of A to C
in the original mixture, 30:70.

Three-Component Phase Diagrams

Figure 2–26 shows a typical three-component phase diagram.[5]
Methane is the lightest component and is plotted at the top. Thus, the
dew-point line lies along the top of the saturation envelope, and the
bubble-point line lies along the bottom. This diagram is for a single
temperature and a single pressure. *Equilibrium tie-lines* are straight but
not horizontal, as was the case for the two-component composition
diagrams. In the case of the three-component mixtures, the tie-lines must
be determined experimentally and given on the diagram.

Point 1 represents a mixture of methane, propane, and n-pentane
which exhibits equilibrium gas and liquid at the temperature and pressure
indicated by the diagram. Point 2 represents the composition of the
equilibrium gas, and point 3 represents the composition of the equilibrium
liquid. The quantity of gas, in fraction of total moles of overall
mixture, is represented by the length of line $\overline{13}$ divided by the length of
line $\overline{23}$. The quantity of liquid in terms of fraction of total moles of
overall mixture is represented by the length of line $\overline{12}$ divided by the
length of line $\overline{23}$.

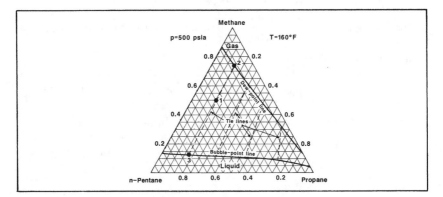

Fig. 2–26. Ternary phase diagram of mixtures of methane, propane, and
n-pentane at 500 psia and 160°F with equilibrium tie-line, $\overline{123}$. (Data from
Dourson et al., *Trans.* AIME, *151*, 206.)

EXAMPLE 2–8: *Determine the compositions and quantities of equilibrium gas and liquid when 6 lb moles of a mixture of 50 mole percent methane, 15 mole percent propane, and 35 mole percent n-pentane are brought to equilibrium at 160°F and 500 psia.*

Solution

First, plot composition of the mixture on the ternary diagram for the given temperature and pressure, (see Figure 2–26, point 1).
Second, read composition of equilibrium gas at point where the tie-line through point 1 connects with dew-point line. See point 2.

composition of gas: 74 mole percent methane
14 mole percent propane
12 mole percent n–pentane
100 mole percent

Third, read composition of equilibrium liquid at point where tie-line through point 1 connects with bubble-point line. See point 3.

composition of liquid: 13 mole percent methane
17 mole percent propane
70 mole percent n-pentane
100 mole percent

Fourth, calculate fraction of mix which is gas

$$\text{fraction gas} = \frac{0.65 \text{ inches}}{1.07 \text{ inches}} = 0.607 \text{ lb mole of gas/lb mole total}$$

and quantity of gas

quantity of gas = (0.607) (6 lb mole) = 3.6 lb moles

Fifth, calculate fraction of mix which is liquid

$$\text{fraction liquid} = \frac{0.42 \text{ inches}}{1.07 \text{ inches}} = 0.393 \text{ lb mole liquid/lb mole total}$$

and quantity of liquid

quantity of liquid = (0.393) (6 lb mole) = 2.4 lb moles

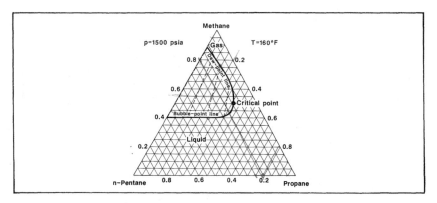

Fig. 2–27. Ternary phase diagram of mixtures of methane, propane, and n-pentane at 1500 psia and 160°F. (Data from Dourson et al., *Trans.* AIME, *151*, 206.)

Figure 2–27 gives the saturation envelope for mixtures of methane, propane, and n-pentane at the same temperature as Figure 2–26 but at a higher pressure.[5] The bubble-point and dew-point lines join at a critical point. The critical point gives the composition of the mixture, which has a critical pressure of 1500 psia and a critical temperature of 160°F.

Figure 2–28 shows the various positions the saturation envelopes of mixtures of methane, propane, and n-pentane can take at 160°F as pressure is increased from atmospheric to 2350 psia. Reference to the binary mixtures shown in Figure 2–29 will assist in understanding the reasons for the changes in the shapes of the saturation envelopes as pressure is increased.[2] The numbered dots on Figure 2–29 correspond to the numbered diagrams of Figure 2–28.

At atmospheric pressure, all mixtures of these components will be gas. See Figure 2–28(1). The temperature is well above the critical temperature of methane, and atmospheric pressure is well below the vapor pressures of propane and n-pentane at 160°F.

Consider a pressure above the vapor pressure of n-pentane and below vapor pressure of propane, for instance, 200 psia. See dot 2 on Figure 2–29 and Figure 2–28(2). All mixtures of methane and propane are gas. Both the methane-n-pentane binary and the propane-n-pentane binary are in their two-phase regions. Their bubble-point and dew-point compositions appear along the sides of the ternary diagram as the ends of the bubble-point and dew-point lines of the ternary mixtures.

At the vapor pressure of propane, 380 psia at this temperature, the bubble-point and dew-point lines of the saturation envelope converge at

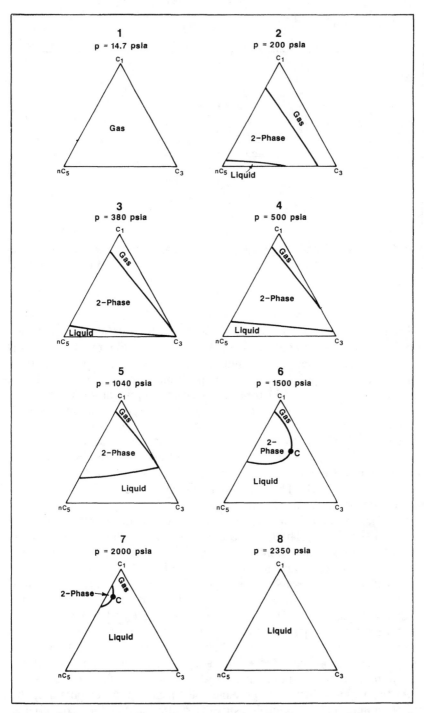

Fig. 2–28. Ternary phase-diagrams of mixtures of methane, propane, and n-pentane at 160°F and various pressures, showing typical changes in shapes of the diagrams as pressure is changed. (Diagram numbers are keyed to the numbered points on Figure 2–29.)

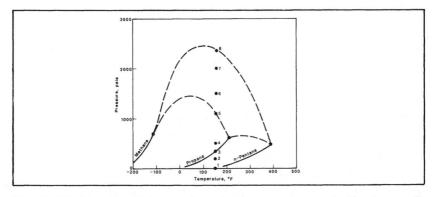

Fig. 2–29. Vapor pressures of methane, propane, and n-pentane with critical loci of binary mixtures. (The numbered points correspond to the diagram numbers on Figure 2–28.)

100 percent propane. See dot 3 on Figure 2–29 and Figure 2–28 (3). The only binary mixtures which exhibit two-phase behavior are those of methane and n-pentane.

At pressures above the vapor pressure of propane and less than the critical locus of mixtures of methane and n-pentane, for instance 500 psia, dot 4, the methane-propane and methane-n-pentane binaries exhibit two-phase behavior, and propane-n-pentane mixtures are all liquid. Thus the saturation envelope appears as in Figure 2–28 (4).

The critical point of a specific mixture of methane and propane occurs at 1040 psia at this temperature, dot 5. The dew-point and bubble-point lines of the ternary intersect the methane-propane side of the diagram at the composition of this critical point.

Above this pressure, dot 6, all mixtures of methane and propane are single phase. Thus only the methane-n-pentane binaries have two-phase behavior, and only the methane-n-pentane side of the ternary diagram can show a bubble point and a dew point. The bubble-point and dew-point lines of the saturation envelope do not intercept another side of the diagram, rather the two lines join at a critical point, i.e., the composition of the three-component mixture that has a critical pressure of 1500 psia at 160°F.

As pressure is increased, the size of the two-phase region decreases, Figure 2–28(7), until the critical pressure of a methane-n-pentane mixture is reached, 2350 psia at this temperature, dot 8. At this pressure and at all higher pressures, all mixtures of methane, propane, and n-pentane are single-phase.

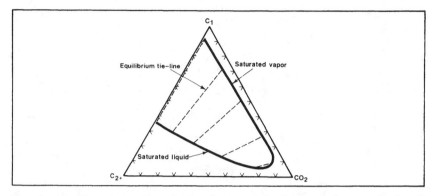

Fig. 2–30. Pseudoternary phase diagram of mixtures of a synthetic oil with carbon dioxide. The oil is represented as a mixture of methane and ethane plus. (Data from Leach and Yellig, *Trans.*, AIME, *271*, 89.)

Uses of Ternary Diagrams

A common use of three-component phase diagrams is in analysis of miscible displacement. For instance, Figure 2–30 gives the phase envelope of an oil mixed with carbon dioxide.[6] The oil is plotted as an artificial two-component mixture, with methane as one component and all other constituents added together as the other component.

Another technique separates the oil into two pseudo-components.[7] One contains all compounds of 12 carbon atoms or less. The other contains all compounds of 13 carbon atoms or more. Carbon dioxide is the third component of the diagram. This is illustrated in Figure 2–31.

EXAMPLE 2–9: *The crude oil of Figure 2–30 contains 35 mole percent methane. Will all mixtures of this oil and carbon dioxide be single phase (miscible)?*

Solution

No, a straight line from the point 35 mole percent methane, 65 mole percent ethane plus to the apex at pure carbon dioxide passes through the two-phase envelope.

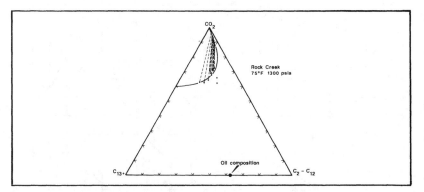

Fig. 2–31. Pseudoternary phase diagram of mixtures of a separator oil with carbon dioxide. (Data from Silva and Orr, SPE *Res. Eng.*, Nov. 1987, 468.)

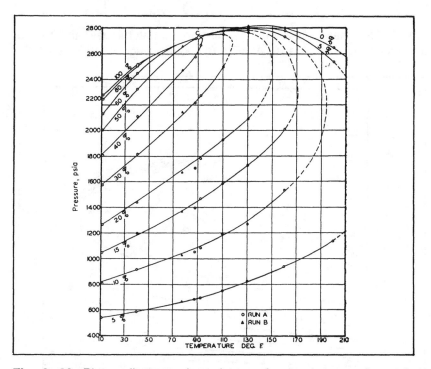

Fig. 2–32. Phase diagram of a mixture of natural gas and gasoline. (Reprinted with permission from *Industrial and Engineering Chemistry 32*, 817. Copyright 1940 American Chemical Society.)

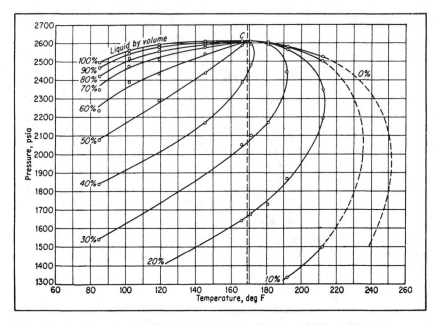

Fig. 2–33. Phase diagram of a mixture of natural gas and natural gasoline. (Katz et al., *Trans.,* AIME *136*, 106. Copyright 1940 SPE-AIME.)

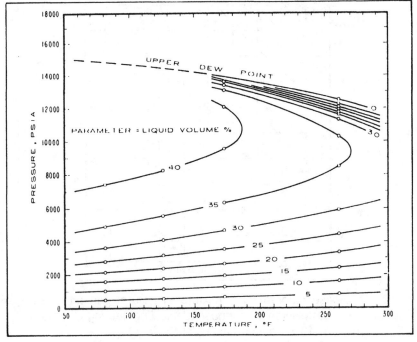

Fig. 2–34. Phase diagram of a reservoir fluid. (Kilgren, *Trans.,* AIME, *237*, 1001. Copyright 1966 SPE-AIME.)

Multicomponent Mixtures

As the number and complexity of the molecules in a mixture increase, the separation between the bubble-point and dew-point lines on the phase diagram becomes greater. Phase diagrams of several petroleum mixtures are shown in Figures 2–32 through 2–36.[8,9,10,11,12]

Note the wide variety of critical pressures and critical temperatures and the different positions that the critical points take on the saturation envelopes. Also note the very large separation between the critical temperature and the cricondentherm in all instances and the separation between cricondenbar and critical pressure for the lighter hydrocarbon mixtures in Figures 2–35 and 2–36.

In Chapter 5 we will attempt to make sense of the wide variety of shapes and sizes of phase diagrams for various petroleum mixtures.

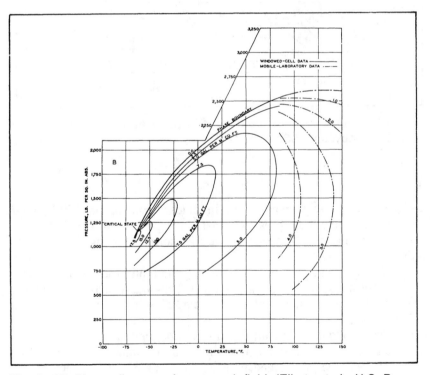

Fig. 2–35. Phase diagram of a reservoir fluid. (Eilerts et al., *U.S. Bureau of Mines, Monograph 10*, 1957, 303. Courtesy Bureau of Mines, U.S. Department of the Interior.)

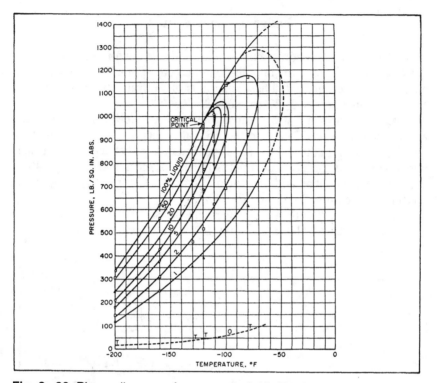

Fig. 2–36. Phase diagram of a reservoir fluid. (Davis et al., *Trans.,* AIME *201,* 245. Copyright 1954 SPE-AIME.)

Exercises

2–1. Pure 2-methylpentane is held in a closed container at 150°F. Both gas and liquid are present. What is the pressure in the container?

2–2. Determine the temperature at the normal boiling point of n-heptadecane.

2–3. Under certain conditions natural gas and liquid water will combine to form a solid called a gas hydrate. How many phases are present during the formation of the hydrate? How many components are present?

2–4. Pure n-butane is contained in a laboratory cell in which the volume can be varied. The cell is maintained at a temperature of 200°F. The volume is changed until a few drops of liquid n-butane are present. What is the pressure in the cell? The volume is reduced until the cell is essentially full of liquid n-butane with only a few bubbles of n-butane gas remaining. What is the pressure?

2–5. Draw the vapor-pressure line for ethane on Cartesian coordinates for a temperature range of 50°F to the critical point. Consider that the ethane is contained in a laboratory cell such as is described in Figure 2–2. Plot the following process on your graph. Start at 80°F and 560 psia; increase the pressure to 800 psia at constant temperature; increase the temperature to 100°F at constant pressure; decrease the pressure to 560 psia at constant temperature; and, finally, reduce the temperature to 80°F at constant pressure. Describe the changes between gas and liquid that occur during each step of the process. Now plot the same process on Figure 2–10. How does this help you understand the change from liquid to gas between steps two and three?

2–6. Vapor pressure data of carbon dioxide are given below. Estimate the vapor pressure of carbon dioxide at 40°F.

Vapor Pressure of Carbon Dioxide	
Temperature, °F	Vapor pressure, psia
−39.1	147.0
−2.0	293.9
22.5	440.9

2–7. The densities of methane liquid and gas in equilibrium along the vapor-pressure line are given below. Estimate the density of methane at its critical point of −116.7°F.

Temperature, °F	Density of saturated methane	
	liquid, lb/cu ft	gas, lb/cu ft
−253	26.17	0.1443
−235	25.25	0.2747
−217	24.28	0.4766
−199	23.18	0.7744
−181	21.89	1.202
−163	20.48	1.810

2–8. Thirty pounds of methane are held in a sealed container at 100°F. The volume of the container is two cu ft. Calculate the volume of gas in the container.

2–9. A sealed container with volume of one cu ft holds 40 lbs of isopentane at 150°F. Calculate the volume of liquid in the container.

2–10. A sealed container with volume of three cu ft holds seven lbs of n-butane at 300°F. What is the volume of liquid in the container?

2–11. Twenty-five pounds of propane are held in a sealed container at 350 psia. The volume of the container is three cu ft. Calculate the weights and volumes of the two phases in the container.

2–12. Estimate the critical pressure of a mixture of methane and n-hexane if the critical temperature of the mixture is known to be 200°F. Compare your answer with the critical pressures of pure methane and pure n-hexane.

2–13. Figure 2–37 gives the phase diagrams of eight mixtures of ethane and n-heptane along with the vapor pressure lines of the two hydrocarbons.[13] Use the figure to determine the following temperatures and pressures for the 29.91 weight percent ethane mixture: curves

Bubble-point pressure at 100°F
Critical temperature
Dew-point temperature at 400 psia
Critical pressure
Bubble-point temperature at 600 psia
Cricondentherm
Dew-point pressure at 250°F
Cricondenbar
Dew-point pressure at 400°F

2–14. Estimate the critical temperature and critical pressure for a mixture of 80 weight percent ethane and 20 weight percent n-heptane.

2–15. Estimate the temperature range and the corresponding pressures in which retrograde condensation will occur for a mixture of 50.25 weight percent ethane and 49.75 weight percent n-heptane.

2–16. Use the data given in Figure 2–20 to prepare a pressure-temperature diagram of a 52.4 weight percent n-heptane in n-pentane mixture.

2–17. Determine the compositions and quantities of gas and liquid when 1 lb mole of a mixture of 60 mole percent methane and 40 mole percent ethane is brought to equilibrium at 500 psia and −40°F.

2–18. Determine the compositions and quantities of gas and liquid when 1 lb mole of a mixture of 40 mole percent methane and 60 mole percent ethane is brought to equilibrium at 500 psia and −40°F. Compare your answer with the answer to problem 2–17. What was the effect of changing the composition of the mixture?

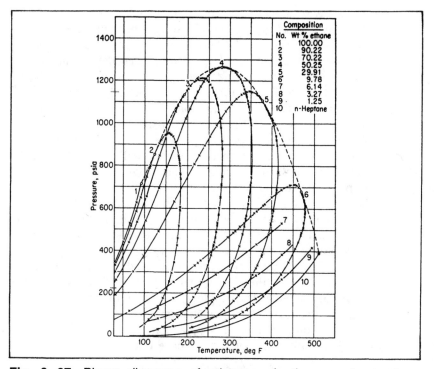

Fig. 2–37. Phase diagrams of mixtures of ethane and n-heptane. (Reprinted with permission from *Industrial and Engineering Chemistry 30, 461*. Copyright 1938 American Chemical Society.)

2–19. Determine the compositions and quantities of gas and liquid when 1 lb mole of a mixture of 30 mole percent methane and 70 mole percent ethane is brought to equilibrium at 0°F and 500 psia.

2–20. Determine the compositions and quantities of gas and liquid when 1 lb mole of a mixture of 30 mole percent methane and 70 mole percent ethane is brought to equilibrium at −20°F and 200 psia.

2–21. Replot three isobars of the data given in Figure 2–37 as temperature against composition in weight percent. Use 300 psia, 600 psia, and 900 psia. This is called a temperature composition diagram. Label the bubble-point lines and dew-point lines.

2–22. Replot two isotherms of the data given in Figure 2–37 as pressure against composition in weight percent. Use temperatures of 75°F and 300°F. This is called a pressure-composition diagram. Label the bubble-point lines and dew-point lines.

2–23. Determine the compositions and quantities of gas and liquid when 10 lb moles of a mixture of 55 mole percent methane, 20 mole percent propane and 25 mole percent n-pentane is brought to equilibrium at 160°F and 1500 psia.

2–24. A liquid of 80 mole percent propane and 20 mole percent n-pentane is to be diluted with methane. Will all mixtures of the liquid and methane be single phase at 160°F and 1500 psia? Explain the reason for your answer.

2–25. A liquid of 80 mole percent propane and 20 mole percent n-pentane is to be diluted with methane. Will all mixtures of the liquid and methane be single phase at 160°F and 500 psia? Explain the reason for your answer.

2–26. A liquid of 20 mole percent propane and 80 mole percent n-pentane is to be diluted with methane. Will all mixtures of the liquid and methane be single phase at 160°F and 1500 psia? Explain the reason for your answer.

2–27. Mixtures of methane, ethane, and n-butane at 500 psia and 160°F exhibit behavior like (choose one)

a. Figure 2–28 (2)
b. Figure 2–28 (4)
c. Figure 2–28 (6)

2–28. A petroleum reservoir is discovered at 3000 psia and 130°F. If the reservoir fluid is represented by the phase diagram of Figure 2–32, will the reservoir contain liquid or gas? What about the fluid of Figure 2–33? Figure 2–35? Figure 2–36? Which of the four fluids will exhibit retrograde behavior at reservoir conditions?

References

1. Katz, D.L., *et al.: Handbook of Natural Gas Engineering,* McGraw-Hill Book Co., New York City (1959).
2. Brown, G.G., Katz, D.L., Oberfell, G.G., and Alden, R.C.: *Natural Gasoline and the Volatile Hydrocarbons,* Natural Gasoline Association of America, Tulsa (1947).
3. Bloomer, O.T., Gami, D.C., and Parent, J.D.: *Physical-Chemical Properties of Methane-Ethane Mixtures,* Research Bulletin, Institute of Gas Technology, Chicago (1953) *22.*
4. Sage, B.H. and Lacey, W.N.: *Volumetric and Phase Behavior of Hydrocarbons,* Gulf Publishing Co., Houston (1949).
5. Dourson, R.H., Sage, B.H., and Lacey, W.N.: "Phase Behavior in the Methane-propane-n-pentane System," *Trans.,* AIME (1942) *151,* 206–215.
6. Leach, M.P. and Yellig, W.F.: "Compositional Model Studies— CO_2 Oil-Displacement Mechanisms," *Soc. Pet. Eng. J.* (Feb. 1981) *21,* 89–97; *Trans.,* AIME, *271.*
7. Silva, M.K. and Orr, F.M. Jr.: "Effect of Oil Composition on Minimum Miscibility Pressure—Part I: Solubility of Hydrocarbons in Dense CO_2," *SPE Res. Eng.* (1987) *2,* 468–478.
8. Katz, D.L. and Kurata, F.: "Retrograde Condensation," *Ind. Eng. Chem.* (1940) *32,* 817–823.
9. Katz, D.L., Vink, D.J., and David, R.A.: "Phase Diagram of a Mixture of Natural Gas and Natural Gasoline Near the Critical Conditions," *Trans.,* AIME (1940) *136,* 106–118.
10. Kilgren, K.H.: "Phase Behavior of a High-Pressure Condensate Reservoir Fluid," *Trans.,* AIME (1966) *237,* 1001–1005.
11. Eilerts, C.K., Barr, V.L., Mullins, M.B., and Hanna, B.: "Phase Relations of a Gas-Condensate Fluid at Low Temperatures, Including the Critical State," *Petrol. Engr.* (1948) *19,* 154–180.
12. Davis, P.C., Bertuzzi, A.F., Gore, T.L., and Kurata, F.: "The Phase and Volumetric Behavior of Natural Gases at Low Temperatures and High Pressures," *Trans.,* AIME (1954) *201,* 245–251.
13. Kay, W.B.: "Liquid-Vapor Phase Equilibrium Relations in the Ethane-n-Heptane System," *Ind. Eng. Chem.* (1938) *30,* 459–465.

Equations of State

3

A gas may be defined as a homogeneous fluid of low density and low viscosity, which has neither independent shape nor volume but expands to fill completely the vessel in which it is contained. The properties of gases differ considerably from the properties of liquids, mainly because the molecules in gases are much farther apart than molecules in liquids. For instance, a change in pressure has a much greater effect on the density of a gas than of a liquid.

In this chapter and the next, we will consider several equations used to describe the relationship between the volume of a gas and its pressure and temperature. We will use the term *equation of state* to mean an equation which relates volume to pressure and temperature.

The Ideal Gas

As a starting point in the study of equations of state of real gases, we will consider a hypothetical gas known as an ideal gas. We will develop the equation of state of an ideal gas in two ways, first from experimental evidence and then from kinetic theory. The *form* of the equation for ideal gases will be used as the basis of equations for real gases.

An ideal gas has these properties.
1. The volume occupied by the molecules is insignificant with respect to the volume occupied by the gas.
2. There are no attractive or repulsive forces between the molecules or between the molecules and the walls of the container.
3. All collisions of molecules are perfectly elastic, that is, there is no loss of internal energy upon collision.

We will see later why the molecules of an ideal gas must have these properties.

Boyle's Equation

Boyle experimentally observed that the volume of an ideal gas is inversely proportional to pressure for a given mass of gas when temperature is maintained constant. This may be expressed as

$$V \sim \frac{1}{p} \text{ or } pV = \text{constant.} \tag{3-1}$$

Charles' Equation

Unpublished experimental work attributed to Charles led to the discovery that the volume of an ideal gas is directly proportional to temperature for a given mass of gas when pressure is maintained constant. Symbolically,

$$V \sim T \text{ or } \frac{V}{T} = \text{constant.} \tag{3-2}$$

Avogadro's Law

Avogadro's law states that, under the same conditions of temperature and pressure, equal volumes of all ideal gases contain the same number of molecules. This is equivalent to the statement that at a given temperature and pressure one molecular weight of any ideal gas occupies the same volume as one molecular weight of any other ideal gas.

There are 2.73×10^{26} molecules per pound mole of ideal gas.

The Equation of State for an Ideal Gas

The equations of Boyle, Charles, and Avogadro can be combined to give an equation of state for an ideal gas.

We must imagine a two-step process in order to combine Charles' equation and Boyle's equation to describe the behavior of an ideal gas when both temperature and pressure are changed. We start with a given mass of gas with a volume of V_1 at pressure p_1 and temperature T_1 and end with a volume of V_2 at pressure p_2 and temperature T_2.

$$(V_1 \text{ at } p_1, T_1) \xrightarrow[T_1 = \text{constant}]{\text{Step 1}} (V \text{ at } p_2, T_1)$$

$$\xrightarrow[p_2 = \text{constant}]{\text{Step 2}} (V_2 \text{ at } p_2, T_2). \tag{3-3}$$

The first step is a change in pressure from a value of p_1 to a value of p_2 while the temperature is held constant. This causes volume to change from V_1 to V. In the second step, the pressure is maintained constant at a value of p_2. Temperature is changed to a value of T_2, causing volume to change to V_2. The change in volume of the gas during the first step may be described through the use of Boyle's equation since the mass of gas and the temperature are held constant. Thus,

$$p_1 V_1 = p_2 V \text{ or } V = \frac{p_1 V_1}{p_2}, \qquad (3\text{-}4)$$

where V represents the volume at pressure p_2 and temperature T_1.

Charles' equation applies to the change in volume of the gas during the second step since pressure and the mass of gas are maintained constant. Therefore,

$$\frac{V}{T_1} = \frac{V_2}{T_2} \text{ or } V = \frac{V_2 T_1}{T_2}. \qquad (3\text{-}5)$$

Elimination of volume V between Equations 3–4 and 3–5 gives

$$\frac{p_1 V_1}{p_2} = \frac{V_2 T_1}{T_2} \qquad (3\text{-}6)$$

or

$$\frac{p_1 V_1}{T_1} = \frac{p_2 V_2}{T_2}. \qquad (3\text{-}7)$$

Thus for a given mass of gas, pV/T is a constant. We will designate the constant with the symbol R when the quantity of gas is equal to one molecular weight.

$$\frac{p V_M}{T} = R \qquad (3\text{-}8)$$

At this point we do not know if R is the same for all ideal gases. So we need to consider Avogadro's law, which states that one molecular weight of an ideal gas occupies the same volume as one molecular weight of another ideal gas at the same pressure and temperature. That is,

$$V_{MA} = V_{MB}, \qquad (3-9)$$

where V_{MA} represents the volume of one molecular weight of gas A and V_{MB} represents the volume of one molecular weight of gas B, both at pressure p and temperature T.

Therefore, from Equation 3–8

$$\frac{pV_{MA}}{T} = R_A \quad \text{or} \quad V_{MA} = \frac{R_A T}{p}, \qquad (3-10)$$

and

$$\frac{pV_{MB}}{T} = R_B \quad \text{or} \quad V_{MB} = \frac{R_B T}{p}, \qquad (3-11)$$

where R_A represents the gas constant for gas A and R_B represents the gas constant for gas B. The combination of Equation 3–9 and Equations 3–10 and 3–11 reveals that

$$\frac{R_A T}{p} = \frac{R_B T}{p} \quad \text{or} \quad R_A = R_B. \qquad (3-12)$$

Thus, the constant R is the same for all ideal gases and is called the *universal gas constant*.

Therefore, we have the equation of state for one molecular weight of any ideal gas,

$$pV_M = RT, \qquad (3-13)$$

where V_M is the volume of one molecular weight of the gas, the *molar volume*. For n moles of ideal gas this equation becomes

$$pV = nRT, \qquad (3-14)$$

where V is the volume of n moles of gas at temperature T and pressure p. Since n is the mass of gas divided by the molecular weight, Equation 3–14 can be written as

$$pV = \frac{m}{M} RT \text{ or as } pv = \frac{RT}{M}, \qquad (3-15)$$

where m is mass and v is the volume of one unit of mass, the *specific volume*.

This expression is known by various names such as the *Ideal Gas Law*, the *General Gas Law*, or the *Perfect Gas Law*. We will call it the *equation of state of an ideal gas* or the *ideal gas equation*.

This equation has limited practical value since no known gas behaves as an ideal gas; however, the equation does describe the behavior of most real gases at low pressures. Also, it gives us a starting point for developing equations of state which describe more adequately the behavior of real gases at elevated pressures.

The numerical value of the constant R depends on the units used to express the variables. Table 3–1 gives numerical values of R for various systems of units.

EXAMPLE 3–1: *Calculate the mass of methane gas contained at 1000 psia and 68°F in a cylinder with volume of 3.20 cu ft. Assume that methane is an ideal gas.*

Solution

$$m = \frac{pMV}{RT} \qquad (3-15)$$

$$m = \frac{(1000 \text{ psia}) \left(16.04 \ \frac{lb}{lb \ mole} \right) (3.20 \text{ cu ft})}{\left(10.732 \ \frac{psia \ cu \ ft}{lb \ mole \ °R} \right) (528°R)}$$

$$m = 9.1 \text{ lb}$$

Density of an Ideal Gas

Since density is defined as the mass of gas per unit volume, an equation of state can be used to calculate the densities of a gas at various

TABLE 3-1

Values of the universal gas constant, R (from *Engineering Data Book*, GPSA, 1987, with permission)

Basis of units listed below is 22.4140 liters at 0°C and 1 atm for the volume of 1g mole. All other values were calculated from conversion factors.

n	Temperature	Pressure	Volume	R	n	Temperature	Energy	R
g mole	K	atm	liter	0.082057477	g mole	K	calorie	1.9859
g mole	K	atm	cc	82.057	g mole	K	joule	8.3145
g mole	K	mm Hg	liter	62.364				
g mole	K	bar	liter	0.083145	lb mole	°R	BTU	1.9859
g mole	K	kg/sq cm	liter	0.084784	lb mole	°R	hp-hr	0.00078048
g mole	K	k Pa	cu m	0.0083145	lb mole	°R	kw-hr	0.00058200
lb mole	°R	atm	cu ft	0.73024	lb mole	°R	ft-lb	1545.3
lb mole	°R	in Hg	cu ft	21.850				
lb mole	°R	mm Hg	cu ft	554.98	k mole	K	joule	8314.5
lb mole	°R	lb/sq in	cu ft	10.732				
lb mole	°R	lb/sq ft	cu ft	1545.3				
lb mole	K	atm	cu ft	1.3144				
lb mole	K	mm Hg	cu ft	998.97				
k mole	K	k Pa	cu m	8.3145				
k mole	K	bar	cu m	0.083145				

temperatures and pressures. The equation for the density of an ideal gas follows from Equation 3–15,

$$\rho_g = \frac{m}{V} = \frac{pM}{RT} . \qquad (3–16)$$

EXAMPLE 3–2: *Calculate the density of methane at the conditions given in Example 3–1. Assume that methane is an ideal gas.*

Solution

$$\rho_g = \frac{pM}{RT} \qquad (3–16)$$

$$\rho_g = \frac{(1000 \text{ psia}) \left(16.04 \ \dfrac{\text{lb}}{\text{lb mole}} \right)}{\left(10.732 \ \dfrac{\text{psia cu ft}}{\text{lb mole } °R} \right) (528°R)}$$

$$\rho_g = 2.83 \frac{\text{lb}}{\text{cu ft}}$$

Kinetic Theory of Gases

The previous equations which describe the behavior of an ideal gas now will be verified using the *kinetic theory of gases*. This will illustrate the reasons for the previously given three conditions imposed on the molecules of an ideal gas. Also, you will gain an understanding of the meanings of pressure and temperature.

According to kinetic theory, the molecules of any substance are in a constant state of motion at all temperatures above absolute zero. The molecules of a solid are restricted in their movement by attractive forces which hold them near a fixed position so that the motion corresponds to a molecular vibration rather than to actual movement of the molecules. This is true to a lesser extent in liquids in which the molecules both vibrate and move around.

The molecules of an ideal gas, however, are completely separated from each other and move with an average velocity, v. Molecular velocity increases as the temperature of the gas increases. Temperature, therefore, is simply a measure of the velocity or kinetic energy of the molecules.

The molecules of the gas undergo a tremendous number of collisions with other molecules and with the walls of the container. Consequently, the directions and velocities of the molecules are constantly changing in a random manner. Pressure is simply the combined effect of the collisions of the molecules with the walls of the containing vessel.

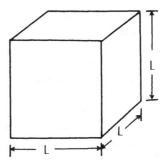

Consider a hollow cube with sides of length L. The cube contains n' molecules, each having the mass m'. We will assume that these molecules behave in a manner which fits the conditions required of the molecules of an ideal gas. Although temperature is held constant, the molecules are moving with many different velocities. However, we will take an average velocity as characteristic of all the molecules at the specified temperature.

As a result of collisions, the molecules move in random directions. However, if these directions are resolved along three different axes at right angles to each other, it may be supposed that one-third of the molecules are always moving in the direction perpendicular to one of the walls of the cube. This is strictly true only if the molecules move in straight lines, i.e., attraction between molecules does not affect the path.

We can now determine the impact pressure on the walls of the vessel. On the average, a single wall of the cube will be struck by a particular molecule each time it has traveled a distance equal to the round trip across the inside of the cube, 2L.

The number of collisions per unit time that this molecule strikes a single wall of the cube can be expressed as v/2L, where v is the average velocity of the molecule. In order for this to be strictly correct, the volume occupied by the molecules must be insignificant with respect to the volume of the container and the molecules must travel in straight lines between collisions.

Remember that these are the first and second requirements of the molecules of an ideal gas. The molecules striking the wall with velocity v will rebound with velocity –v, having suffered no loss in kinetic energy since the collisions are perfectly elastic. This is the third requirement of the molecules of an ideal gas.

Momentum is defined as the product of mass and velocity. The momentum of a molecule before collision with a wall is $m'v$, and after collision is $-m'v$. Thus the change in momentum per molecule collision is $2m'v$.

The momentum change per molecule per unit time is the product of the change in momentum per collision and the number of collisions per unit time. Thus $(2m'v)(v/2L)$ or $m'v^2/L$ is the change in momentum per molecule per unit time.

Since one-third of the molecules are traveling in a direction which will permit them to collide with a single wall, the total change in momentum per unit time at one wall is $(n'/3)(m'v^2/L)$.

Force is defined as the change in momentum per unit time, so this represents the average force on each wall of the cube.

$$\text{Force on each wall of the cube } = \left(\frac{n'}{3}\right)\left(\frac{m'v^2}{L}\right) \qquad (3-17)$$

Pressure is simply force per unit area so that the pressure on each wall of the container is

$$p = \frac{\text{Force}}{\text{Area}} = \frac{\left(\dfrac{n'}{3}\right)\left(\dfrac{m'v^2}{L}\right)}{L^2} = \frac{n'm'v^2}{3L^3}, \qquad (3-18)$$

and, since the volume of the container is L^3,

$$pV = \frac{n'm'v^2}{3}. \qquad (3-19)$$

Remember that temperature is measure of the average kinetic energy of the molecules. The kinetic energy of each molecule is equal to $m'v^2/2$, so that

$$T \sim \frac{m'v^2}{2} \text{ or } \frac{m'v^2}{2} = CT, \qquad (3-20)$$

where C is a constant of proportionality independent of the particular gas.

Combination of Equations 3–19 and 3–20 yields

$$pV = \frac{2}{3} Cn'T. \qquad (3-21)$$

If the temperature T and the number of molecules n' remain constant,

$$pV = \text{constant}, \qquad (3-1)$$

which is Boyle's equation.

If the pressure p and the number of molecules n' remain constant,

$$\frac{V}{T} = \text{constant}, \tag{3-2}$$

which is Charles' equation.

Finally, an expression equal to Avogadro's law can also be deduced from Equation 3–19. Consider two ideal gases A and B with molecules of different weights. At the same pressure and volume

$$pV = \frac{n'_A m'_A v^2_A}{3} \text{ and } pV = \frac{n'_B m'_B v^2_B}{3}. \tag{3-22}$$

Therefore,

$$\frac{n'_A m'_A v^2_A}{3} = \frac{n'_B m'_B v^2_B}{3}. \tag{3-23}$$

Since kinetic energy is only affected by temperature, the kinetic energies of the two gases are considered to be equal at the same temperature. Thus, for gases A and B,

$$\frac{m'_A v^2_A}{2} = \frac{m'_B v^2_B}{2}. \tag{3-24}$$

A combination of Equations 3–23 and 3–24 yields

$$n'_A = n'_B. \tag{3-25}$$

This, of course, is the statement of Avogadro's law.

Avogadro's number of molecules per molecular weight of gas may be denoted as A and then incorporated into Equation 3–21 so that

$$pV = \frac{n'}{A} \left(\frac{2}{3} CA \right) T \tag{3-26}$$

or

$$pV = nRT, \text{ where } n = \frac{n'}{A} \text{ and } R = \frac{2}{3} CA. \tag{3-14}$$

Since C and A are both constants which are not properties of a particular ideal gas, R is a constant independent of the particular ideal gas. This is the equation of state for an ideal gas which was previously derived from the equations of Boyle, Charles, and Avogadro.

Thus, the equation of state of an ideal gas can be derived from kinetic theory. All three properties of an ideal gas were used in the development of the equation.

We have arrived at the equation

$$pV = nRT \qquad (3-14)$$

both empirically and theoretically.

Mixtures of Ideal Gases

Since the petroleum engineer primarily is concerned with gas mixtures, the laws governing the behavior of mixtures of ideal gases will now be introduced. This will later lead to an understanding of the behavior of mixtures of real gases.

Dalton's Law of Partial Pressures

The total pressure exerted by a mixture of gases is equal to the sum of the pressures exerted by its components. The pressure exerted by each of the component gases is known as its *partial pressure*.

Dalton postulated that the partial pressure of each component in a mixture of gases is equal to the pressure the component would exert if it alone were present in the volume occupied by the gas mixture. This is valid only when the mixture and each component of the mixture act as ideal gases. Dalton's law sometimes is called the *law of additive pressures*.

The partial pressure exerted by each component of a gas mixture can be calculated using the ideal gas equation. Consider a mixture containing n_A moles of component A, n_B moles of component B, n_C moles of component C, and so on. The partial pressure exerted by each component of the gas mixture may be determined as

$$p_A = n_A \frac{RT}{V}, \; p_B = n_B \frac{RT}{V}, \; p_C = n_C \frac{RT}{V}, \ldots \quad (3-27)$$

According to Dalton, the total pressure is the sum of the partial pressures

$$p = p_A + p_B + p_C + \ldots \qquad (3-28)$$

Thus,

$$p = n_A \frac{RT}{V} + n_B \frac{RT}{V} + n_C \frac{RT}{V} + \ldots \qquad (3-29)$$

and

$$p = \frac{RT}{V} \sum_j n_j = \frac{RT}{V} n. \qquad (3-14)$$

The ratio of the partial pressure of component j, p_j, to the total pressure of the mixture p is

$$\frac{p_j}{p} = \frac{n_j}{\sum_j n_j} = \frac{n_j}{n} = y_j, \qquad (3\text{--}30)$$

where y_j is defined as the mole fraction of the jth component in the gas mixture. Therefore, the partial pressure of a component of a mixture of ideal gases is the product of its mole fraction times the total pressure.

$$p_j = y_j p. \qquad (3\text{--}30)$$

Remember that this is valid only for ideal mixtures of ideal gases.

EXAMPLE 3–3: *Calculate the partial pressure exerted by methane in the following gas when the gas is at a pressure of 750 psia. Assume that the gas is a mixture of ideal gases.*

Component	Composition, mole fraction
Methane	0.85
Ethane	0.10
Propane	0.05
	1.00

Solution

$$p_{C1} = p y_{C1} \qquad (3\text{--}30)$$
$$p_{C1} = (750 \text{ psia})(0.85)$$
$$p_{C1} = 638 \text{ psia}$$

Amagat's Law of Partial Volumes

Amagat postulated that the total volume occupied by a gas mixture is equal to the sum of the volumes that the pure components would occupy at the same pressure and temperature. This is sometimes called the *law of additive volumes*. Amagat's law of partial volumes is analogous to Dalton's law of partial pressures.

The volumes occupied by the individual components are known as *partial volumes*. This equation is correct only if the mixture and each of the components obey the ideal gas equation.

Again consider a gas mixture consisting of n_A moles of component A, n_B moles of component B, and so on. The partial volume occupied by each component can be calculated using the ideal gas equation.

$$V_A = n_A \frac{RT}{p}, \; V_B = n_B \frac{RT}{p}, \; V_C = n_C \frac{RT}{p}, \; \ldots \quad (3\text{--}31)$$

According to Amagat the total volume is

$$V = V_A + V_B + V_C + \ldots \quad (3\text{--}32)$$

Thus,

$$V = n_A \frac{RT}{p} + n_B \frac{RT}{p} + n_C \frac{RT}{p} + \ldots \quad (3\text{--}33)$$

and

$$V = \frac{RT}{p} \sum_j n_j = \frac{RT}{p} n. \quad (3\text{--}14)$$

The ratio of the partial volume of component j to the total volume of the mixture is

$$\frac{V_j}{V} = \frac{n_j \dfrac{RT}{p}}{n \dfrac{RT}{p}} = \frac{n_j}{n} = y_j. \quad (3\text{--}34)$$

Equation 3–34 states that for an ideal gas the volume fraction of a component is equal to the mole fraction of that component.

Apparent Molecular Weight of a Gas Mixture

Since a gas mixture is composed of molecules of various sizes, saying that a gas mixture has a molecular weight is not strictly correct. However, a gas mixture behaves as if it has a definite molecular weight. This molecular weight is known as the *apparent molecular weight* and is defined as

$$M_a = \sum_j y_j M_j. \quad (3\text{--}35)$$

EXAMPLE 3–4: *Dry air is a gas mixture consisting essentially of nitrogen, oxygen, and small amounts of other gases. Compute the apparent molecular weight of air given its approximate composition.*

Component	Composition, mole fraction
Nitrogen	0.78
Oxygen	0.21
Argon	0.01
	1.00

Solution

$$M_a = y_{N2}M_{N2} + y_{O2}M_{O2} + y_A M_A \qquad (3-35)$$

$$M_a = (0.78)(28.01) + (0.21)(32.00) + (0.01)(39.94) =$$

$$28.97 \text{ lb/lb mole}$$

A value of 29.0 lb/lb mole usually is considered sufficiently accurate for engineering calculations.

Specific Gravity of a Gas

The *specific gravity of a gas* is defined as the ratio of the density of the gas to the density of dry air with both measured at the same temperature and pressure. Symbolically

$$\gamma_g = \frac{\rho_g}{\rho_{air}}, \qquad (3-36)$$

where γ_g is the specific gravity of the gas.

On the assumption that the behavior of both the gas and air may be represented by the ideal gas equation, specific gravity may be given as

$$\gamma_g = \frac{\dfrac{pM_g}{RT}}{\dfrac{pM_{air}}{RT}} = \frac{M_g}{M_{air}} = \frac{M_g}{29}, \qquad (3-37)$$

where M_{air} is the apparent molecular weight of air and M_g is the molecular weight of the gas. If the gas is a mixture, this equation becomes

$$\gamma_g = \frac{M_a}{M_{air}} = \frac{M_a}{29} , \qquad (3-38)$$

where M_a is the apparent molecular weight of the gas mixture. Note that this equation is strictly true only if both the gas and air act like ideal gases.

Often, specific gravity is called *gravity* or *gas gravity*; however, *specific gravity* is the correct term.

EXAMPLE 3-5: *Calculate the specific gravity of a gas of the following composition.*

Component	Composition, mole fraction
Methane	0.850
Ethane	0.090
Propane	0.040
n-Butane	0.020
	1.000

Solution

First, calculate apparent molecular weight, Equation 3-35.

Component	Mole fraction, y_i	Molecular weight, M_i	y_iM_i
C_1	0.850	16.04	13.63
C_2	0.090	30.07	2.71
C_3	0.040	44.10	1.76
n-C_4	0.020	58.12	1.16
	1.000		19.26 = M_a

Second, calculate specific gravity.

$$\gamma_g = \frac{M_a}{29} = \frac{19.26}{29} = 0.664 \qquad (3-38)$$

Behavior of Real Gases

Researchers have proposed hundreds of equations of state for real gases. We will consider first the *compressibility equation of state*. This equation of state is the one used most commonly in the petroleum industry. This equation does have some limitations; therefore, we will examine later several other equations of state which are used to a lesser extent by petroleum engineers.

The Compressibility Equation of State

We have shown from kinetic theory that the ideal gas equation has the correct form. The behavior of most real gases does not deviate drastically from the behavior predicted by this equation. So one way of writing an equation of state for a real gas is to insert a correction factor into the ideal gas equation.[1] This results in

$$pV = znRT, \quad pV_M = zRT, \quad pv = \frac{zRT}{M}, \quad \text{and } \rho_g = \frac{pM}{zRT}, \quad (3\text{–}39)$$

where the correction factor, z, is known as the *compressibility factor* and the equation is known as the *compressibility equation of state*. The equation has various names, such as the *compressibility equation* or the *real gas equation*.

Compressibility factor is also known as *gas deviation factor*, *supercompressibility*, or *z-factor*. Sometimes the reciprocal of compressibility factor is called *supercompressibility*.

The z-factor is the ratio of the volume actually occupied by a gas at given pressure and temperature to the volume the gas would occupy at the same pressure and temperature if it behaved like an ideal gas.

$$z = \frac{V_{actual}}{V_{ideal}} \quad (3\text{–}40)$$

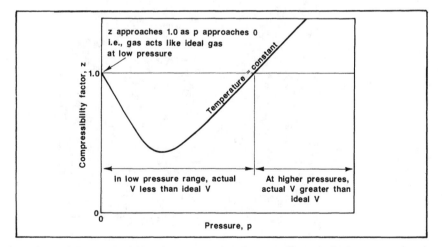

Fig. 3–1. Typical shape of z-factor as a function of pressure at constant temperature.

The z-factor is not a constant. It varies with changes in gas composition, temperature, and pressure. It must be determined experimentally. The results of experimental determinations of z-factors usually take the form shown in Figure 3–1.

The shape of the curve is consistent with our knowledge of the behavior of gases. At very low pressure the molecules are relatively far apart, and the conditions of ideal gas behavior are more likely to be met. Experiments show that at very low pressures the z-factor approaches a value of 1.0. This indicates that ideal gas behavior does in fact occur at very low pressures.

At moderate pressures, the molecules are close enough to exert some attraction between molecules. This attraction causes the actual volume to be somewhat less than the volume predicted by the ideal gas equation, that is, the z-factor will be less than 1.0.

At higher pressures the molecules are forced close together, repulsive forces come into play, the actual volume is greater than ideal volume, and z-factor is greater than 1.0.

z-Factors for several hydrocarbon gases are given in Figures 3–2, 3–3, and 3–4.[2]

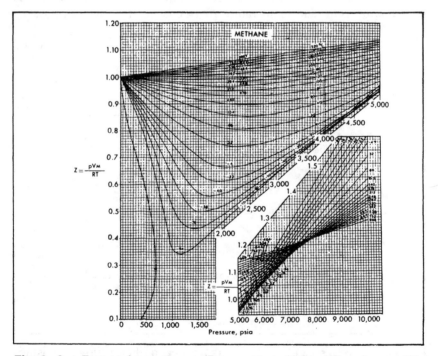

Fig. 3–2. z-Factors for methane. (Brown, et al., *Natural Gasoline and the Volatile Hydrocarbons,* NGAA, 1947, 24, with permission.)

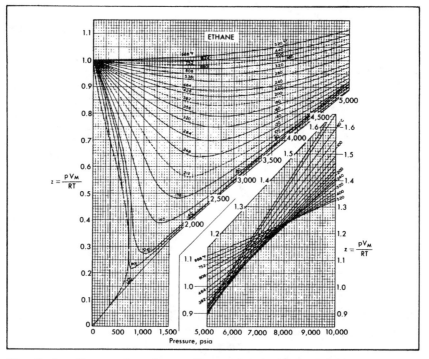

Fig. 3–3. z-Factors for ethane. (Brown, et al., *Natural Gasoline and the Volatile Hydrocarbons,* NGAA, 1947, 26, with permission.)

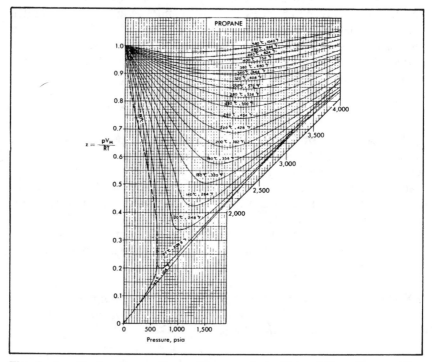

Fig. 3–4. z-Factors for propane. (Brown, et al., *Natural Gasoline and the Volatile Hydrocarbons,* NGAA, 1947, 27, with permission.)

EXAMPLE 3–6: *Calculate the mass of methane gas contained at 1,000 psia and 68°F in a cylinder with volume of 3.20 cu ft. Do not assume that methane is an ideal gas.*

Solution

$$m = \frac{pMV}{zRT} \qquad (3-39)$$

$$z = 0.890, \text{ Figure } 3-2$$

$$m = \frac{(1,000 \text{ psia}) \left(16.04 \ \dfrac{lb}{lb \text{ mole}}\right) (3.20 \text{ cu ft})}{(0.89) \left(10.732 \ \dfrac{psia \text{ cu ft}}{lb \text{ mole}°R}\right) (528°R)}$$

$$m = 10.2 \text{ lb}$$

Note that the assumption that methane acts as an ideal gas in Example 3–1 resulted in an error in mass of about 11%.

The Law of Corresponding States[3]

Notice that the shapes of the isotherms of compressibility factors for the three gases given in Figures 3–2, 3–3, and 3–4 are very similar. The realization that this is true for nearly all real gases led to the development of the *Law of Corresponding States* and the definition of the terms *reduced temperature* and *reduced pressure*. Reduced temperature and reduced pressure are defined as

$$T_r = \frac{T}{T_c} \text{ and } p_r = \frac{p}{p_c}. \qquad (3-41)$$

The Law of Corresponding States says that all pure gases have the same z-factor at the same values of reduced pressure and reduced temperature. Figure 3–5 gives a test of this theory for compressibility data of methane, propane, n-pentane, and n-hexane.[4] Some of the deviation between lines at constant reduced temperatures may be due to experimental error and some due to inexactness of the theory.

Data for pure hydrocarbon gases such as those presented in Figures 3–2, 3–3, and 3–4 have been put on a reduced basis and are given as Figure 3–6.[4]

EXAMPLE 3–7: *Determine the specific volume of ethane at 918 psia and 117°F. Use Figure 3–6 to determine the compressibility factor.*

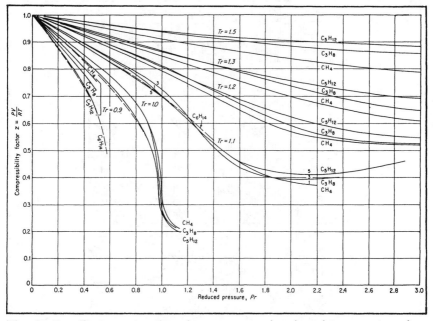

Fig. 3–5. z-Factors at reduced pressure and reduced temperature for methane, propane, n-pentane, and n-hexane. (From the *Handbook of Natural Gas Engineering* by Katz et al. Copyright 1959 by McGraw-Hill Book Co. Used with permission of McGraw-Hill Book Co.)

Solution

First, calculate reduced temperature and reduced pressure and determine z.

$$T_r = \frac{T}{T_c} = \frac{577°R}{549.9°R} = 1.05 \tag{3-41}$$

$$p_r = \frac{p}{p_c} = \frac{918 \text{ psia}}{706.5 \text{ psia}} = 1.30 \tag{3-41}$$

$$z = 0.390, \text{ Figure } 3-6$$

Second, calculate specific volume.

$$v = \frac{zRT}{pM} \tag{3-39}$$

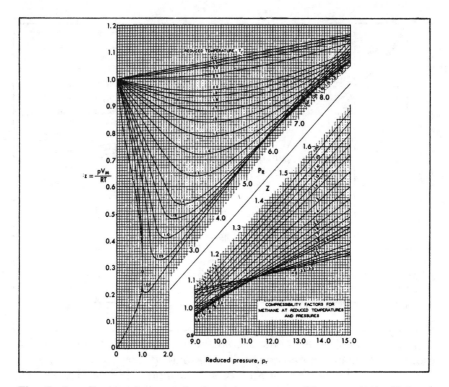

Fig. 3–6. z-Factors for pure hydrocarbon gases. (Brown, et al., *Natural Gasoline and the Volatile Hydrocarbons,* NGAA, 1947, 32, with permission.)

$$v = \frac{(0.390)\left(10.732 \; \frac{\text{psia cu ft}}{\text{lb mole}^{\circ}\text{R}}\right)(577^{\circ}\text{R})}{(918 \; \text{psia})\left(30.07 \; \frac{\text{lb}}{\text{lb mole}}\right)}$$

$$v = 0.0875 \; \frac{\text{cu ft}}{\text{lb}}$$

Note the error that would occur if ethane was assumed to be an ideal gas at this pressure and temperature.

The Law of Corresponding States is more accurate if the gases have similar molecular characteristics. Fortunately most of the gases the petroleum engineer deals with are composed primarily of molecules of the same class of organic compounds known as paraffin hydrocarbons.

The Compressibility Equation of State for Gas Mixtures[5]

The Law of Corresponding States has been extended to cover mixtures of gases which are closely related. As was brought out in Chapter 2, obtaining the critical point for multicomponent mixtures is somewhat difficult; therefore, *pseudocritical temperature* and *pseudocritical pressure* have been invented.

These quantities are defined as

$$T_{pc} = \sum_j y_j T_{cj} \text{ and } p_{pc} = \sum_j y_j p_{cj} . \qquad (3-42)$$

These pseudocritical properties were devised simply for use in correlating physical properties. Pseudocritical properties *are not* equal to the actual critical properties of a gas mixture. Equations 3–42 are often called Kay's mixture rules.[5] A somewhat more accurate method of calculating pseudocritical properties is given in Appendix B.

EXAMPLE 3–8: *Calculate the pseudocritical temperature and pseudocritical pressure of the gas given in Example 3–5. Use the critical constants given in Appendix A.*

  Solution

$$T_{pc} = \sum_j y_j T_{cj} \text{ and } p_{pc} = \sum_j y_j p_{cj} \qquad (3-42)$$

Component	Mole fraction, y_j	Critical temperature, °R T_{cj}	$y_j T_{cj}$	Critical pressure, psia p_{cj}	$y_j p_{cj}$
C_1	0.850	343.3	291.8	666.4	566.4
C_2	0.090	549.9	49.5	706.5	63.6
C_3	0.040	666.1	26.6	616.0	24.6
$n\text{-}C_4$	0.020	765.6	15.3	550.6	11.0
	1.000	T_{pc} =	383.2°R	p_{pc} =	665.6 psia

$$T_{pc} = 383°R, \ p_{pc} = 666 \text{ psia}$$

Physical properties of *gas mixtures* are correlated with *pseudoreduced temperature* and *pseudoreduced pressure* in the same manner that properties of pure gases are correlated with reduced temperature and reduced pressure.

$$T_{pr} = \frac{T}{T_{pc}} \text{ and } p_{pr} = \frac{p}{p_{pc}} \qquad (3-43)$$

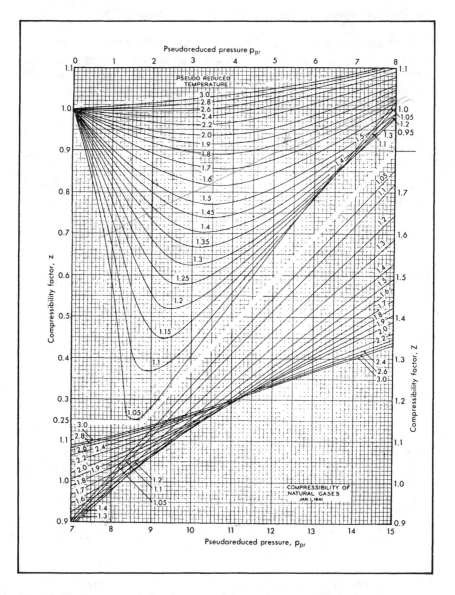

Fig. 3-7. Compressibility factors of natural gases. (Standing and Katz, *Trans.,* AIME, *146*, 140. Copyright 1942 SPE-AIME).

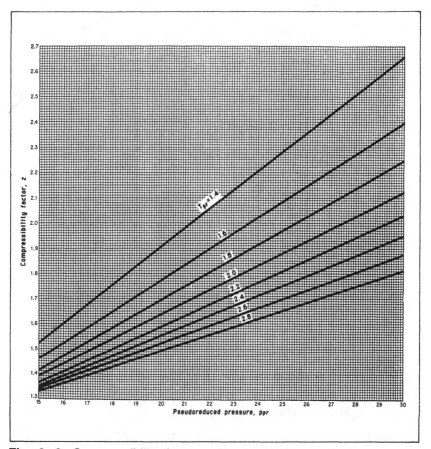

Fig. 3–8. Compressibility factors of natural gases at high pressures. (From the *Handbook of Natural Gas Engineering* by Katz et al. Copyright 1959 by McGraw-Hill Book Co. Used with permission of McGraw-Hill Book Co.)

The compressibility factors for natural gases have been correlated using pseudoreduced properties.[2,6] The correlations are presented in Figures 3–7, 3–8, and 3–9.

Remember that z-factors are a function of the type of gas as well as of temperature and pressure. Fortunately, most of the components of natural gases are hydrocarbons of the same family. Therefore, a correlation of this type is possible. The correlation is also successful because the components of most natural gases appear in approximately the same ratio to one another.

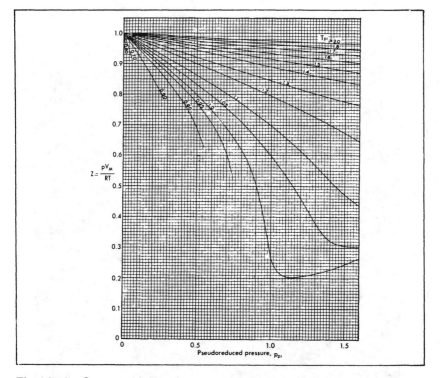

Fig. 3–9. Compressibility factors of natural gases at low pressure. (Brown, et al., *Natural Gasoline and the Volatile Hydrocarbons,* NGAA, 1947, 33, with permission.)

EXAMPLE 3–9: *Calculate the mass in lb moles of the gas of Example 3-8 which is contained in 43,560 cu ft at 9,300 psia and 290°F.*

Solution

First, calculate the pseudoreduced temperature, pseudoreduced pressure, and determine z.

$$T_{pr} = \frac{T}{T_{pc}} = \frac{750°R}{383°R} = 1.96 \qquad (3-43)$$

$$p_{pr} = \frac{p}{p_{pc}} = \frac{9300 \text{ psia}}{666 \text{ psia}} = 14.0 \qquad (3-43)$$

$$z = 1.346, \text{ Figure } 3-7$$

Second, calculate mass in lb moles.

$$n = \frac{pV}{zRT} \tag{3-39}$$

$$n = \frac{(9300 \text{ psia})(43,560 \text{ cu ft})}{(1.346)\left(10.732 \ \dfrac{\text{psia cu ft}}{\text{lb mole °R}}\right)(750°R)}$$

$$n = 37,400 \text{ lb moles}$$

Pseudocritical Properties of Heptanes Plus

We will see in Chapter 10 that the composition of a petroleum fluid is often given with all components heavier than hexane lumped together as *heptanes plus*. Pseudocritical pressures and pseudocritical temperatures for heptanes plus for use in Equations 3–42 can be obtained from Figure 3–10.[7]

EXAMPLE 3–10: *Determine a value of z-factor for the dry gas given below for use at 3810 psia and 194°F.*

Component	Composition, mole percent
Methane	97.12
Ethane	2.42
Propane	0.31
i- Butane	0.05
n-Butane	0.02
i- Pentane	trace
n-Pentane	trace
Hexanes	0.02
Heptanes plus	0.06
	100.00

Properties of heptanes plus	
Specific gravity	0.758
Molecular weight	128 lb/lb mole

Solution

First, calculate pseudocritical properties, Equations 3–42.

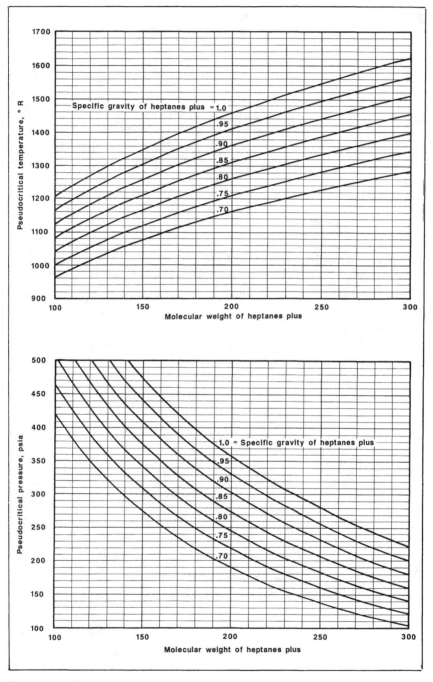

Fig. 3–10. Pseudocritical properties of heptanes plus.

Component	Mole fraction, y_i	Critical temperature, °R T_{ci}	$y_i T_{ci}$	Critical pressure, psia p_{ci}	$y_i p_{ci}$
C_1	0.9712	343.33	333.44	666.4	647.21
C_2	0.0242	549.92	13.31	706.5	17.10
C_3	0.0031	666.06	2.06	616.0	1.91
i- C_4	0.0005	734.46	0.37	527.9	0.26
n-C_4	0.0002	765.62	0.15	550.6	0.11
C_6	0.0002	913.60	0.18	436.9	0.09
C_{7+}	0.0006	*1082.	0.65	*372.	0.22
	1.0000		T_{pc} = 350.17°R		p_{pc} = 666.90 psia

*Pseudocritical properties for C_{7+} from Figure 3–10.

Second, calculate pseudoreduced properties.

$$T_{pr} = \frac{T}{T_{pc}} = \frac{654°R}{350.2°R} = 1.87 \qquad (3-43)$$

$$p_{pr} = \frac{p}{p_{pc}} = \frac{3810 \text{ psia}}{666.9 \text{ psia}} = 5.71 \qquad (3-43)$$

Third, determine z.

$$z = 0.951, \text{ Figure } 3-7$$

The z-factor of this gas was measured in the laboratory as 0.962, a difference of just over 1%.

Pseudocritical Properties of Gas When Composition is Unknown

Often the composition of a natural gas is unknown. However, specific gravity, being more easily measured, is usually available for a gas of interest. In this situation the correlation presented in Figure 3–11 can be used to determine the pseudocritical properties of the gas.[8]

EXAMPLE 3–11: *Determine a z-factor of a natural gas with specific gravity of 1.26 for use at 256°F and 6025 psia.*

Solution

First, determine pseudocritical properties.

$$T_{pc} = 492°R \text{ , Figure } 3-11$$

$$p_{pc} = 587 \text{ psia, Figure } 3-11$$

Second, calculate pseudoreduced properties.

$$T_{pr} = \frac{T}{T_{pc}} = \frac{716}{492}°R = 1.46 \qquad (3-43)$$

$$p_{pr} = \frac{p}{p_{pc}} = \frac{6025 \text{ psia}}{587 \text{ psia}} = 10.26 \qquad (3-43)$$

Third, determine z.

$$z = 1.154, \text{ Figure } 3-7$$

The z-factor of this gas was measured in the laboratory as 1.140, a difference of just over 1%.

The accuracy of the compressibility equation of state is not any better than the accuracy of the values of the z-factors used in the calculations. The accuracy of Figures 3–7 and 3–8 was tested with data from 634 natural gas samples of known composition.[8] Experimentally determined z-factors of these gases were compared with z-factors obtained from the charts using Kay's rules for calculating the pseudocritical properties and Figure 3–10 for properties of heptanes plus.

Gases with specific gravities of 1.0 or less showed average absolute errors in calculated z-factor of 1.5% or less. Gases with specific gravities greater than 1.0 had average absolute errors of calculated z-factors of up to 8%.

The alternate method of calculating the pseudocriticals given in Appendix B should be used for gases with specific gravities greater than 1.0. With it and Figures 3–7 and 3–8, the calculated z-factors had average absolute errors of less than 1.5%.

Measured z-factors for the same natural gases were compared with compressibility factors calculated with Figures 3–11, 3–7, and 3–8. The average absolute error was less than 2% throughout the range of specific gravities from 0.57 to 1.68.

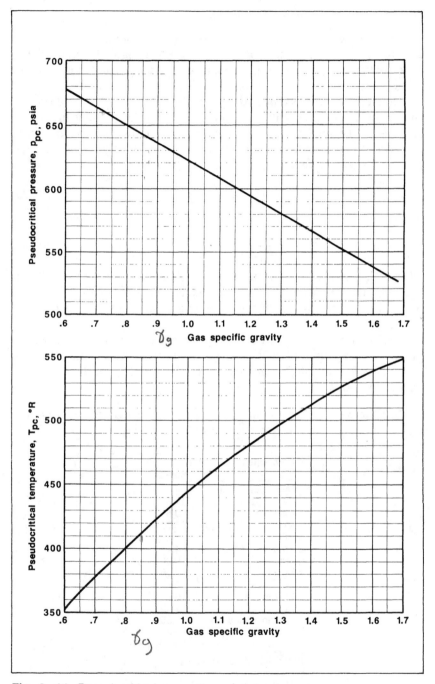

Fig. 3–11. Pseudocritical properties of natural gases.

Effect of Nonhydrocarbon Components

Natural gases commonly contain hydrogen sulfide, carbon dioxide, and nitrogen. The presence of nitrogen does not greatly affect the z-factor obtained by methods previously described; z-factor increases by about 1% for each 5% of nitrogen in the gas.[9]

However, the presence of hydrogen sulfide and carbon dioxide causes large errors in compressibility factors obtained by the methods previously discussed. The remedy to this problem is to adjust the pseudocritical properties to account for the unusual behavior of these acid gases.[10] The equations used for this adjustment are

$$T'_{pc} = T_{pc} - \epsilon \qquad (3-44)$$

and

$$p'_{pc} = \frac{p_{pc}T'_{pc}}{T_{pc} + y_{H2S}(1 - y_{H2S})\epsilon} , \qquad (3-45)$$

where T'_{pc} and p'_{pc} are used to calculate T_{pr} and p_{pr} for acid gases. The pseudocritical temperature adjustment factor, ϵ, is obtained from Figure 3–12. The symbol y_{H2S} represents the mole fraction of hydrogen sulfide in the gas.

Pseudocritical properties obtained by either of the methods described previously are adjusted in this manner. These adjustments bring the accuracy of calculated compressibility factors to within the limits expressed above.

EXAMPLE 3–12:   *Determine values of pseudocritical temperature and pseudocritical pressure for the gas given below.*

Component	Composition, mole percent
Hydrogen sulfide	4.91
Carbon dioxide	11.01
Nitrogen	0.51
Methane	57.70
Ethane	7.22
Propane	4.45
i- Butane	0.96
n-Butane	1.95
i- Pentane	0.78
n-Pentane	0.71
Hexanes	1.45
Heptanes plus	8.35
	100.00

Properties of heptanes plus
Specific gravity   0.807
Molecular weight   142 lb/lb mole

Solution

First, determine value of pseudocritical properties in the usual manner.

Component	Mole fraction, y_i	Critical temperature, °R T_{ci}	y_iT_{ci}	Critical pressure, psia p_{ci}	y_ip_{ci}
H_2S	0.0491	672.4	33.0	1300.0	63.8
CO_2	0.1101	547.9	60.3	1071.0	117.9
N_2	0.0051	227.5	1.2	493.1	2.5
C_1	0.5770	343.3	198.1	666.4	384.5
C_2	0.0722	549.9	39.7	706.5	51.0
C_3	0.0445	666.1	29.6	616.0	27.4
i- C_4	0.0096	734.5	7.1	527.9	5.1
n-C_4	0.0195	765.6	14.9	550.6	10.7
i- C_5	0.0078	829.1	6.5	490.4	3.8
n-C_5	0.0071	845.8	6.0	488.6	3.5
C_6	0.0145	913.6	13.2	436.9	6.3
C_{7+}	0.0835	*1157	96.6	*367	30.7
	1.0000		T_{pc} = 506.2°R		p_{pc} = 707.3 psia

* Pseudocritical properties for C_{7+} from Figure 3–10.

Second, determine the value of ϵ and adjust the pseudocritical properties.

$$\epsilon = 19.9°R, \text{ Figure } 3-12$$

$$T'_{pc} = T_{pc} - \epsilon \qquad (3-44)$$

$$T'_{pc} = 506.2 - 19.9 = 486.3°R$$

$$p'_{pc} = \frac{p_{pc}T'_{pc}}{T_{pc} + y_{H2S}(1 - y_{H2S})\epsilon} \qquad (3-45)$$

$$p'_{pc} = \frac{(707.3)(486.3)}{506.2 + (0.0491)(0.9509)(19.9)}$$

$$p'_{pc} = 678 \text{ psia}$$

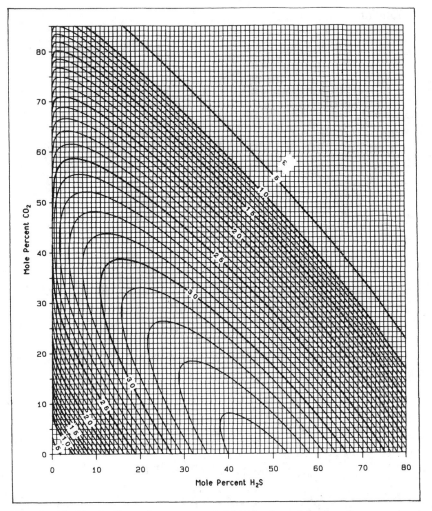

Fig. 3–12. Nonhydrocarbon component adjustment factors for pseudocritical properties of natural gases. (Adapted from Wichert and Aziz, *Hyd. Proc. 51,* 119, 1972.)

Exercises

3–1. Calculate the molar volume of an ideal gas at 100 psia and 90°F.

3–2. A pure gaseous hydrocarbon has a density of 0.103 lb/cu ft at 14.7 psia and 100°F. Chemical analysis shows that there are two hydrogen atoms for each carbon atom in each molecule. What is the formula of this molecule? Assume that the hydrocarbon acts like an ideal gas.

3–3. Compute the composition in weight fraction and the composition in volume fraction of a gas with the following composition.

Component	Composition, mole fraction
Methane	0.870
Ethane	0.049
Propane	0.036
i- Butane	0.025
n-Butane	0.020
	1.000

3–4. Determine the composition in weight fraction and the composition in volume fraction of the gas given below. What assumption did you make?

Component	Composition, mole fraction
Methane	0.6904
Ethane	0.0864
Propane	0.0534
i- Butane	0.0115
n-Butane	0.0233
i- Pentane	0.0093
n-Pentane	0.0085
Hexanes	0.0173
Heptanes plus	0.0999
	1.0000

Properties of heptanes plus	
Specific gravity	0.827
Molecular weight	158 lb/lb mole

3–5. An ideal gas exerts 100 psig in a cylinder at 100°F. What will the pressure be if the temperature is reduced to 32°F?

3–6. A container was found to weigh 80.00 g while evacuated. The container then was filled with oxygen and found to weigh 81.242 g. When filled with water the container weighed 1000.00 g. The temperature and pressure for the procedure were 14.7 psia and 60°F. What is the gas constant in engineering units?

3–7. A piece of sandstone with a bulk volume of 1.3 cc is contained in a 5-cc cell filled with helium at 760 mm Hg. Temperature is maintained constant and the cell is opened to another evacuated cell the same volume. The final pressure in the two vessels is 334.7 mm Hg. What is the porosity of the sandstone?

3–8.   A 20-cu ft tank at 100°F is pressured to 200 psia with a pure paraffin gas. Ten pounds of ethane are added, and the specific gravity of the gas mixture is measured to be 1.68. Assume that the gases act as ideal gases. What was the gas originally in the tank?

3–9.   A 40-cu ft tank is filled with air at 14.7 psia and 80°F. The tank also contains a 15-lb block of dry ice (solid CO_2, specific gravity of 1.53 based on water). After the dry ice sublimes, the temperature is 30°F. What is the pressure? Assume ideal gas behavior.

3–10. A certain machine uses methane as a fuel. The gas is purchased in containers holding 25 lb of methane and costing $3.20 exclusive of the container. If the machine uses 300 cu ft/d at 15 oz/sq in. gauge and 100°F, what is the cost of the methane per day?

3–11. A storage tank has a diameter of 80 ft, a height of 25 ft, and contains oil to a height of 15 ft. The suction pumps, which can handle 15,000 bbl/d, are started, but the safety valves are clogged and a vacuum is drawn in the tank. The roof is flat and rated to withstand 1.0 oz/sq in. before collapsing. The barometric pressure is 30 in Hg. How long will it take the tank to collapse? What is the total force on the roof at collapse? How would more oil in the tank have affected the collapse time? Explain.

3–12. Compute the apparent molecular weight and specific gravity of the gas of Exercise 3–3.

3–13. Calculate the specific gravity of the gas of Exercise 3–4.

3–14. A gas composed of methane and ethane has a specific gravity of 0.75. What is the weight percent and volume percent of the methane in the mixture?

3–15. What is the specific gravity of the gas of Exercise 3–28?

3–16. Calculate the molar volume of an ideal gas with a specific gravity of 0.862 at 3025 psia and 175°F.

3–17. Determine the composition in mole fraction of the following gas.

Component	Composition, weight fraction
Methane	0.880
Ethane	0.043
Propane	0.042
n-Butane	0.035
	1.000

3–18. Determine the partial pressure of each component of the gas in Exercise 3–3 if the total pressure on the gas is 350 psia and the temperature is 90°F.

3–19. What is the specific gravity of the following gas?

Component	Partial pressure, psia
Methane	17.8
Ethane	1.0
Propane	0.4
i- Butane	0.2
n-Butane	0.6
	20.0

3–20. A 2.4-cu ft cylinder of ethane shows a pressure of 1600 psig at 90°F. What is the mass in pounds of the ethane contained in the cylinder? Do not assume ethane is an ideal gas.

3–21. The gas in Exercise 3–3 was produced from a reservoir at 4650 psia and 180°F. What was the density of the gas at reservoir conditions?

3–22. A tank contains methane at 1000 psia and 140°F. Another tank of equal volume contains ethane at 500 psia and 140°F. The two tanks are connected, the gases are allowed to mix, and the temperature is restored to 140°F. Calculate the final pressure, the composition of the mixture, and the partial pressures of the components at final conditions. Do not assume that ideal gas equations apply.

3–23. A cylinder has a volume of 0.5 cu ft and contains a gas at a pressure of 2000 psia and 120°F. The pressure drops to 1000 psia after 0.0923 lb moles of gas are removed. The temperature is constant. The z-factor was 0.90 at 2000 psia. What is the z-factor at 1000 psia?

3–24. A tank with volume of 2.4 cu ft is filled with methane to a pressure of 1500 psia at 104°F. Determine the molecular weight and specific gravity of the gas and the pound moles, pounds, and density of gas in the tank.

3–25. A cylinder with an initial pressure of 14.7 psia and volume of 75,000 cc is held at a constant temperature of 200°F. The volume of the cylinder is reduced by insertion of mercury. The corresponding volumes (cc) and pressures (psia) of the gas inside the cylinder are recorded as follows.

Pressure: 400,   800,   1200,   2000,   2500,   3000,
            4000,   5000
Volume: 2,448,   1,080,   648.6,   350.6,   295.9,   266.4,
          234.6,   206.2

Calculate the ideal volumes for the gas at each pressure and use them to calculate the z-factors. Plot the z-factors against pressure.

3–26. The gas of Exercise 3–3 is in a reservoir at 5580 psig and 226°F. What is its density?

3–27. The gas of Exercise 3–4 is a retrograde gas. However, discovery pressure in reservoir, 7000 psig, is higher than dew point pressure, 6010 psig. Reservoir temperature is 256°F. The reservoir lies under 7040 acres, has an average thickness of 13 ft, has a porosity of 11%, and has a water saturation of 40%. Calculate the mass of gas in the reservoir at initial conditions. Give your answer in lb moles.

3–28. Determine the value of z-factor for the gas given below at 5420 psig and 257°F.

Component	Composition, mole fraction
Hydrogen sulfide	0.100
Carbon dioxide	0.050
Nitrogen	0.021
Methane	0.703
Ethane	0.062
Propane	0.037
n-Butane	0.027
	1.000

3–29. Repeat Exercise 3–16. Do not assume that the gas is an ideal gas.

3–30. A gas from a dry gas reservoir has a specific gravity of 0.697. Reservoir temperature is 90°F and reservoir pressure is 455 psia. Calculate the molar volume of this gas.

3–31. The reservoir gas in a wet gas reservoir has a specific gravity of 1.295, a hydrogen sulfide concentration of 20.9 mole percent, and a carbon dioxide concentration of 44.7 mole percent. Determine a value of z-factor for use at reservoir conditions of 5720 psig and 268°F.

3–32. A wet gas has composition as given below. Determine a value of z-factor for this gas at reservoir conditions of 5709 psig and 293°F.

Component	Composition, mole percent
Hydrogen sulfide	8.67
Carbon dioxide	1.40
Nitrogen	0.45
Methane	79.81
Ethane	5.28
Propane	1.71
i- Butane	0.55
n-Butane	0.53
i- Pentane	0.24
n-Pentane	0.19
Hexanes	0.24
Heptanes plus	0.93
	100.00

Properties of heptanes plus
Specific gravity   0.759
Molecular weight   132 lb/lb mole

Compare your answer with laboratory measured z-factor of 1.048 at 5709 psig and 293°F for this gas.

3–33. A wet gas has composition as given below. Determine a value of z-factor for this gas at reservoir conditions of 5720 psig and 268°F.

Component	Composition, mole percent
Hydrogen sulfide	20.90
Carbon dioxide	44.69
Nitrogen	1.22
Methane	21.80
Ethane	3.68
Propane	2.05
i- Butane	0.58
n-Butane	1.09
i- Pentane	0.46
n-Pentane	0.56
Hexanes	0.72
Heptanes plus	2.25
	100.00

Properties of heptanes plus
Specific gravity   0.844
Molecular weight   115 lb/lb mole

Compare your answer with the laboratory measured z-factor of 0.914 at 5720 psig and 268°F for this gas.

References

1. Cope, J.Q., Lewis, W.K., and Weber, H.C.: "Generalized Thermodynamic Properties of Higher Hydrocarbon Vapors," *Ind. Eng. Chem.* (1931) *23*, 887–892.
2. Brown, G.G., Katz, D.L., Oberfell, G.G., and Alden, R.C.: *Natural Gasoline and the Volatile Hydrocarbons*, Natural Gasoline Assn. of America, Tulsa (1948), 24–32.
3. van der Waals, J.D.: *Over de Continuiteit van den Gas-en Vloeistoftoestand*, dissertation, Leiden (1873).
4. Brown, G.G., Sounders, M., and Smith, R.L.: "Fundamental Design of High Pressure Equipment Involving Paraffin Hydrocarbons," *Ind. Eng. Chem.* (1932) *24*, 513–515.
5. Kay, W.B.: "Density of Hydrocarbon Gases and Vapors at High Temperature and Pressure," *Ind. Eng. Chem.* (Sept. 1936) *28*, 1014–1019.
6. Standing, M.B. and Katz, D.L.: "Density of Natural Gases," *Trans.*, AIME (1942) *146*, 140–149.
7. Kessler, M.G. and Lee, B.I.: "Improved Prediction of Enthalpy of Fractions," *Hyd. Proc.* (March 1976) *55*, 153–158.
8. Sutton, R.P.: "Compressibility Factors for High-Molecular-Weight Reservoir Gases," paper SPE 14265 presented at the 1985 SPE Technical Conference and Exhibition, Las Vegas, Sept. 22–25.
9. Alexander, R.A.: personal communication, Cawley, Gillespie & Associates, Inc., Fort Worth, TX, 1987.
10. Wichert, E. and Aziz, K.: "Calculate Z's for Sour Gases," *Hyd. Proc.* (May 1972) *51*, 119–122.

4

Other Equations of State for Real Gases

One of the limitations in the use of the compressibility equation of state to describe the behavior of gases is that the compressibility factor is not constant. Therefore, mathematical manipulations cannot be made directly but must be accomplished through graphical or numerical techniques. Most of the other commonly used equations of state were devised so that the coefficients which correct the ideal gas law for non-ideality may be assumed constant. This permits the equations to be used in mathematical calculations involving differentiation or integration.

van der Waals' Equation of State

One of the earliest attempts to represent the behavior of real gases by an equation was that of van der Waals (1873).[1] He proposed the following equation.

$$\left(p + \frac{a}{V_M^2} \right) (V_M - b) = RT \qquad (4-1)$$

This equation differs from the ideal gas equation by the addition of the term a/V_M^2 to pressure and the subtraction of the constant b from molar volume.

The term a/V_M^2 represents an attempt to correct pressure for the forces of *attraction between the molecules*. The actual pressure exerted on the walls of the vessel by real gas is less, by the amount a/V_M^2, than the pressure exerted by an ideal gas.

The constant b is regarded as the correction to the molar volume due to the *volume occupied by the molecules*. Constants a and b are characteristic of the particular gas, whereas R is the universal gas constant.

The van der Waals equation is an improvement over the ideal gas equation. However, it has limited use since it is accurate only at low pressures. The van der Waals equation is presented here to illustrate the semitheoretical basis which most researchers have used to develop equations of state from the basic form of the ideal gas equation.

Equations like the van der Waals equation often are called *two-constant equations of state*, although there are actually three constants: a, b, and R. These equations also are called *cubic equations of state*. The van der Waals equation in cubic form is

$$V_M^3 - (b + RT/p)V_M^2 + (a/p)V_M - ab/p = 0. \quad (4-2)$$

Other Equations of State in the "Spirit of van der Waals"

Soon after the appearance of the van der Waals equation, other researchers began attempts to improve it. These attempts have continued for over one hundred years. Usually a change of the molecular attraction term, a/V_M^2, is proposed. Occasionally a researcher suggests a change to the molecular volume term, b.

All of these equations have some utility; however, none is particularly accurate as pressure and temperature approach the critical point. Also, these equations were formulated for pure substances. Arbitrary *mixing rules* must be used to determine values of constants a and b for gas mixtures.

Clausius (1880) proposed that the molecular attraction term is inversely proportional to temperature.[2]

$$\left[p + \frac{a}{T(V_M + c)^2} \right] (V_M - b) = RT \quad (4-3)$$

The addition of the fourth constant, c, enables better agreement with data. However, mathematical manipulations required in thermodynamic calculations are more diffcult. So Berthelot (1899) removed the constant c, resulting in[3]

$$\left(p + \frac{a}{TV_M^2} \right) (V_M - b) = RT. \quad (4-4)$$

Dieterici (1899) handled the temperature dependence of the molecular attraction term in a different manner.[4]

$$\left[p \, \text{EXP} \left(\frac{a}{V_M RT} \right) \right] (V_M - b) = RT \tag{4-5}$$

Lorentz (1881) addressed the molecular volume term.[5]

$$\left(p + \frac{a}{V_M^2} \right) \left(V_M - \frac{bV_M}{V_M + b} \right) = RT \tag{4-6}$$

Wohl (1927) considered the effect of temperature on the molecular attraction term.[6]

$$\left[p + \frac{a}{TV_M(V_M - b)} - \frac{c}{T^2 V_M^3} \right] (V_M - b) = RT \tag{4-7}$$

The constants a, b, and c in the equations above have different values for different substances. Also, note that the constants have different values for each of the equations.

Equations of State at the Critical Point

Already we have seen that the critical temperature isotherm on a pressure-volume diagram for a pure substance has a horizontal point of inflection as it passes through the critical pressure. The data of Figure 2–10 clearly show this. Thus, for a pure substance at the critical point

$$\left(\frac{\partial p}{\partial V_M} \right)_{T_c} = 0 \text{ and } \left(\frac{\partial^2 p}{\partial V_M^2} \right)_{T_c} = 0 . \tag{4-8}$$

This experimental observation can be used to derive equations for constants a and b in terms of the critical temperature and critical pressure of the substance.

As an example, consider the van der Waals equation of state. At the critical point these three equations apply:

$$\left(p_c + \frac{a}{V_{Mc}^2} \right) (V_{Mc} - b) = RT_c , \tag{4-9}$$

$$\left(\frac{\partial p}{\partial V_M} \right)_{T_c} = - \frac{RT_c}{(V_{Mc} - b)^2} + \frac{2a}{V_{Mc}^3} = 0, \tag{4-10}$$

and

$$\left(\frac{\partial^2 p}{\partial V_M^2}\right)_{T_c} = \frac{2RT_c}{(V_{Mc} - b)^3} - \frac{6a}{V_{Mc}^4} = 0. \qquad (4-11)$$

Combination of these three equations results in

$$a = \frac{27R^2T_c^2}{64p_c} \text{ and } b = \frac{RT_c}{8p_c}. \qquad (4-12)$$

EXAMPLE 4-1: *Calculate the van der Waals constants for 3-methyl-hexane.*

Solution

$$a = \frac{(27)(10.732)^2(963.8)^2}{(64)(408.1)} = 110,600 \text{ psia (cu ft)}^2/\text{(lb mole)}^2$$

$$b = \frac{(10.732)(963.8)}{(8)(408.1)} = 3.168 \text{ cu ft/lb mole}$$

Equations for the constants of the other equations of state can be determined in a similar manner. However, this technique will only work for pure substances. Figure 2-20 shows that mixtures do not exhibit horizontal inflection at the critical point.

The Virial Equation of State

Kammerlingh-Onnes (1901) proposed the virial equation of state.[7]

$$pV_M = RT\left[1 + \frac{B}{V_M} + \frac{C}{V_M^2} + \ldots\right] \qquad (4-13)$$

where B, C, ... are not constants but are functions of temperature and are called the second, third, ... *virial coefficients.*

Each of the equations of state previously presented can be expressed in virial form. The van der Waals equation can be approximated as

$$pV_M = RT\left[1 + \frac{1}{V_M}\left(b - \frac{a}{RT}\right) + \frac{b^2}{V_M^2} + \ldots\right]. \qquad (4-14)$$

Thus the second virial coefficient, according to van der Waals, is[8]

$$\left(b - \frac{a}{RT}\right).$$

None of the equations discussed so far in this chapter adequately represents the properties of gases over the ranges of temperature and pressure of interest to the petroleum engineer. These equations are given here to illustrate the various semitheoretical schemes researchers have used in an attempt to modify the ideal gas equation of state to describe real gas properties.

The Beattie-Bridgeman Equation of State

Beattie and Bridgeman analyzed the behavior of real gases under kinetic theory similar to that presented earlier for an ideal gas.[9] They deduced that van der Waals' constants, a and b, are both dependent on the density of the gas. This follows from the fact that the more molecules present, the more interference between molecules. They chose linear functions for simplicity and proposed replacement of the constant a with $A_o(1 - a/V_M)$ and b with $B_o(1 - b/V_M)$.

Further, Beattie and Bridgeman postulated that "when two slowly moving molecules encounter one another, there is a tendency for them to move under the influence of each other for an appreciable length of time due to the intermolecular forces between them." In effect, molecules moving together decreases the number of aggregates in the system.

This will cause a change in average molecular weight of the gas. As Equation 3–26 shows, this will affect the value of R.

This aggregation of molecules is more likely to occur at low kinetic energies and at high densities of molecules. Thus, the value of R is affected by the reciprocals of temperature and molar volume. Beattie and Bridgeman proposed replacement of R by $R(1 - c/V_M T^3)$. The form of the equation is based on theory; the exponent of three on temperature is empirical.

Beattie and Bridgeman made these substitutions in the Lorenz equation, Equation 4–6, producing

$$p = \frac{RT}{V_M^2}\left(1 - \frac{c}{V_M T^3}\right)\left[V_M + B_o\left(1 - \frac{b}{V_M}\right)\right] - \frac{A_o(1 - a/V_M)}{V_M^2}, (4-15)$$

where the coefficients a, b, c, A_o, and B_o are constants for a given gas. Values of these constants for various gases of interest to the petroleum engineer are given in Table 4–1.

TABLE 4–1

Constants in equations of state

Substance	van der Waals constants		Beattie-Bridgeman constants*					Benedict-Webb-Rubin constants*							
	a†	b+	$A_0 \times 10^{-3}$	a	B_0	b	$c \times 10^{-6}$	A_0	B_0	$C_0 \times 10^{-6}$	a	b	$c \times 10^{-6}$	$a \times 10^3$	$\gamma \times 10^2$
Methane	2.253	0.04278	8.5863	0.2972	0.8950	−0.2542	11.986	6,995.25	0.682401	275.763	2,984.12	0.867325	498.106	511.172	153.961
Ethane	5.489	0.06380	22.174	0.9389	1.506	+0.3068	84.08	15,670.7	1.00554	2,194.27	20,850.2	2.85393	6,413.14	1,000.44	302.790
Propane	8.664	0.08445	44.951	1.173	2.899	+0.6877	112.10	25,915.4	1.55884	6,209.93	57,248.0	5.77355	25,247.8	2,495.77	564.524
Isobutane	12.87	0.1142	62.614	1.7895	3.771	+1.233	280.27	38,587.4	2.20329	10,384.7	117,047	10.8890	55,977.7	4,414.96	872.447
n-Butane	14.47	0.1226	67.102	1.9481	3.944	+1.509	326.98	38,029.6	1.99211	12,130.5	113,705	10.2636	61,925.6	4,526.93	872.447
Isopentane	18.05	0.1417	...	...	...	...	...	4,825.36	2.56386	21,336.7	226,902	17.1441	136,025	6,987.77	1,188.07
n-Pentane	19.01	0.1460	106.570	2.4187	6.311	+2.236	373.70	45,928.8	2.51096	25,917.2	246,148	17.1441	161,306	7,439.92	1,218.86
n-Hexane	24.39	0.1735	...	...	...	...	...	5,443.4	2.84835	40,556.2	429,901	28.0032	296,077	11,553.9	1,711.15
n-Heptane	31.51	0.2654	205.600	3.2144	11.344	+3.072	373.70	66,070.6	3.18782	57,984.0	626,106	38.9917	483,427	17,905.6	2,309.42
Nitrogen	1.390	0.03913	5.0702	0.4192	0.8083	−0.1107	3.92								
Carbon dioxide	3.592	0.04267	18.880	1.1425	1.6781	+1.1590	61.66								
Hydrogen sulfide	4.431	0.04287													
Helium	0.03412	0.02370	0.08146	0.9586	0.2243	0	0.0037								
Water	5.464	0.03049													
Hydrogen	0.2444	0.02661	0.7448	−0.0811	0.3358	−0.6983	0.0471								
Ethylene	4.471	0.05714	23.200	0.7952	1.9473	+0.5762	21.19	12,593.6	0.891980	1,602.28	15,645.5	2.20678	4,133.60	731.661	236.844
Propylene	8.379	0.08272	...	...	...	...	...	23,049.2	1.36263	5,365.97	46,758.6	4.79997	20,083.0	1,873.12	469.25

* T in °R (°F + 459.63); p_g in lb moles/cu ft; p in psia. R = 10.7335 psia-cu ft/(lb mole) (°R). † a in liters²-atm mole². + b in liters/mole.

The Beattie-Bridgeman equation, like most equations of state, is explicit in pressure. Certain calculations require an equation that is explicit in volume. Therefore, Beattie rearranged the Beattie-Bridgeman equation and modified it to give the following form.

$$V_M = \left(\frac{RT}{p} + B \right) (1 - E) - \frac{A}{RT}, \qquad (4-16)$$

where

$$A = A_o \left(1 - \frac{ap}{RT} \right), \qquad (4-17)$$

$$B = B_o \left(1 - \frac{bp}{RT} \right), \qquad (4-18)$$

and

$$E = \frac{cp}{RT^4}. \qquad (4-19)$$

The constants are the same as for the Beattie-Bridgeman equation and the same numerical values may be used. The equation is less accurate than the Beattie-Bridgeman equation, but the agreement with the observed data is satisfactory.

Both the Beattie-Bridgeman equation and the Beattie modification have been used with good accuracy at densities up to 2/3 of the critical density of the gas. Unfortunately, this does not cover the entire range of interest of the petroleum engineer.

The constants for these equations have been derived experimentally from data for pure gases. However, the equation can be used for gas mixtures by combining the constants using the following mixture rules.

$$a = \sum_j y_j a_j$$

$$b = \sum_j y_j b_j$$

$$c = \sum_j y_j c_j$$

$$A_o = (\sum_j y_j A_{oj}^{1/2})^2$$

$$B_o = \sum_j y_j B_{oj} \qquad \qquad (4-20)$$

The development of mixture rules will be discussed in detail in connection with the Benedict-Webb-Rubin equation of state.

EXAMPLE 4-2: *Calculate the molar volume of the gas given below at 100°F and 250 psia. Use the Beattie modification of the Beattie-Bridgeman equation.*

Component	Composition, mole fraction
Methane	0.75
Ethane	0.15
Propane	0.10
	1.00

Solution

First, use Equations 4-20 to compute constants for the gas mixture.

Component	y_j	*a_j	$y_j a_j$	*b_j	$y_j b_j$	*$c_j \times 10^{-6}$	$y_j c_j \times 10^{-6}$
C_1	0.75	0.2972	0.2229	-0.2542	-0.1907	11.986	8.99
C_2	0.15	0.9398	0.1410	0.3068	0.0460	84.08	12.61
C_3	0.10	1.173	0.1173	0.6877	0.0688	112.10	11.21
	1.00	$a=$	0.4812	$b=$	-0.0759	$c=$	32.81

Component	y_j	*$A_{oj} \times 10^{-3}$	$A_{oj}^{1/2}$	$y_j A_{oj}^{1/2}$	*B_{oj}	$y_j B_{oj}$
C_1	0.75	8.5863	92.662	69.497	0.8950	0.671
C_2	0.15	22.174	148.91	22.336	1.506	0.226
C_3	0.10	44.951	212.02	21.202	2.899	0.290
	1.00		$A_o^{1/2} =$	113.035	$B_o =$	1.187

*Constants from Table 4-1.

Second, compute the modified Beattie coefficients.

$$A = A_o \left(1 - \frac{ap}{RT}\right) \qquad \qquad (4-17)$$

$$A = (113.035)^2 \left[1 - \frac{(0.4812)(250)}{(10.732)(560)} \right]$$

$$A = 12.521 \times 10^3$$

$$B = B_o \left(1 - \frac{bp}{RT} \right) \tag{4-18}$$

$$B = (1.187) \left[1 - \frac{(-0.0759)(250)}{(10.732)(560)} \right]$$

$$B = 1.191$$

$$E = \frac{cp}{RT^4} \tag{4-19}$$

$$E = \frac{(32.81 \times 10^6)(250)}{(10.732)(560)^4}$$

$$E = 0.007772$$

Third, use Beattie modification of the Beattie-Bridgeman equation to compute molar volume.

$$V_M = \left(\frac{RT}{p} + B \right) (1 - E) - \frac{A}{RT} \tag{4-16}$$

$$V_M = \left[\frac{(10.732)(560)}{(250)} + 1.191 \right] (1 - 0.007772) - \left[\frac{12.521 \times 10^3}{(10.732)(560)} \right]$$

$$V_M = 23.0 \ \frac{cu \ ft}{lb \ mole}$$

The Benedict-Webb-Rubin Equation of State

Benedict et al. rearranged the Beattie-Bridgeman equation.[10]

$$(p - RTd)/d^2 = B_oRT - A_o - Rc/T^2 -$$

$$(B_obRT - A_oa + RB_oc/T^2)d + RB_obcd^2/T^2, \tag{4-21}$$

where d is the molar density, $1/V_M$.

Benedict et al. used pure component data to study the relationship of $(p - RTd)/d^2$ with temperature. They found that $(p - RTd)/d^2$ at constant values of d could be represented by

$$(p - RTd)/d^2 = RTB - A - C/T^2. \qquad (4\text{--}22)$$

They then deduced the relationships of A, B, and C with molar density, combined them with Equation 4–21, and produced an equation with eight empirical constants.

$$p = \frac{RT}{V_M} + \frac{B_oRT - A_o - C_o/T^2}{V_M^2} + \frac{bRT - a}{V_M^3}$$

$$+ \frac{a\alpha}{V_M^6} + \frac{c}{T^2V_M^3}\left(1 + \frac{\gamma}{V_M^2}\right) \text{EXP}\left(\frac{-\gamma}{V_M^2}\right) \qquad (4\text{--}23)$$

While this equation may be considered a modification of the Beattie-Bridgeman equation of state, it is completely empirical.

The constants B_o, A_o, C_o, a, b, c, α, and γ cannot be obtained by analysis of the equation at the critical point but must be derived for each pure substance from experimental p-V_M-T properties of both gas and liquid. The constants for compounds of interest are given in Table 4–1. In order to apply the equation of state to mixtures, Benedict et al. deduced mixture rules from statistical mechanics.[11]

$$B_o = \sum_{ij} y_i y_j B_{oij}$$

$$A_o = \sum_{ij} y_i y_j A_{oij}$$

$$C_o = \sum_{ij} y_i y_j C_{oij}$$

$$b = \sum_{ijk} y_i y_j y_k b_{ijk}$$

$$a = \sum_{ijk} y_i y_j y_k a_{ijk}$$

$$c = \sum_{ijk} y_i y_j y_k c_{ijk} \qquad (4\text{--}24)$$

Constants with repeated subscripts refer to pure compounds, i.e., $B_{o11} = B_{o1}$, and the order of the subscripts is immaterial, i.e., $B_{o12} = B_{o21}$. The coefficients with dissimilar subscripts, such as B_{o12}, arise from the interaction of unlike molecules and are called *interaction coefficients* or *binary interaction coefficients*.

The interaction coefficients may be derived from p-V_M-T data of binary and ternary mixtures of the compounds of interest when data are available. However, calculating the interaction coefficients from properties of the pure components is convenient, if not completely accurate. The following equations have been assumed.

$$B_{oij} = (B_{oi} + B_{oj})/2$$

$$A_{oij} = (A_{oi}A_{oj})^{1/2}$$

$$C_{oij} = (C_{oi}C_{oj})^{1/2}$$

$$b_{ijk} = (b_i b_j b_k)^{1/3}$$

$$a_{ijk} = (a_i a_j a_k)^{1/3}$$

$$c_{ijk} = (c_i c_j c_k)^{1/3} \tag{4-25}$$

When Equations 4–25 are substituted into Equations 4–24, the following mixture rules result.

$$A_o = (\sum_j y_j A_{oj}^{1/2})^2$$

$$B_o = \sum_j y_j B_{oj}$$

$$C_o = (\sum_j y_j C_{oj}^{1/2})^2 \tag{4-26}$$

$$a = (\sum_j y_j a_j^{1/3})^3$$

$$b = (\sum_j y_j b_j^{1/3})^3$$

$$c = (\sum_j y_j c_j^{1/3})^3$$

Note that the interaction coefficients conveniently have disappeared. Benedict et al. also proposed[11]

$$\alpha = (\sum_j y_j \alpha_j^{1/3})^3 \quad \text{and} \quad \gamma = (\sum_j y_j \gamma_j^{1/2})^2 . \tag{4-27}$$

Other researchers examined the deviations of the Benedict-Webb-Rubin constants with pressure, temperature, and molar volume.[12] Equations with as many as 20 empirical constants were developed. Although these equations are very accurate at interpolation of data, their usefulness is limited by the difficulty of mathematical manipulation in the various thermodynamic equations.

Recent Developments in Equations of State

Equation of state research has returned recently to "the spirit of van der Waals," that is, cubic equations with two constants. Many equations of this form have been proposed.[13] Two popularly accepted equations of state in the petroleum industry, Redlich-Kwong and Peng-Robinson, are cubic equations with two empirical constants. These equations have been used widely to calculate physical properties and vapor-liquid equilibria of hydrocarbon mixtures.

Redlich-Kwong Equation of State

Redlich and Kwong proposed an equation of state which takes into account the temperature dependencies of the molecular attraction term in a manner similar to Clausius.[14]

$$\left[p + \frac{a}{T^{1/2}V_M(V_M + b)} \right] (V_M - b) = RT \qquad (4-28)$$

The advantage over the Clausius equation is that a third empirical constant is not included.

Soave suggested that $a/T^{1/2}$ be replaced with a temperature dependent term, a_T.[15]

$$\left[p + \frac{a_T}{V_M(V_M + b)} \right] (V_M - b) = RT \qquad (4-29)$$

The fact that a_T varies with temperature is not inconvenient since most applications of this equation are at constant temperature. The equation for a_T is

$$a_T = a_c\alpha, \qquad (4-30)$$

where a_c is the value of a_T at the critical temperature and α is a nondimensional temperature-dependent term which has a value of 1.0 at the critical temperature. This modification is often called the *Soave-Redlich-Kwong (SRK) equation of state*.

Setting the first and second derivatives of Equation 4–29 equal to zero at the critical point results in

$$b = 0.08664 \frac{RT_c}{p_c} \text{ and } a_c = 0.42747 \frac{R^2 T_c^2}{p_c}. \quad (4-31)$$

The units of b and a_c depend on the units of the selected value of R. Values of α are obtained from

$$\alpha^{1/2} = 1 + m(1 - T_r^{1/2}), \quad (4-32)$$

where

$$m = 0.480 + 1.574\omega - 0.176\omega^2, \quad (4-33)$$

where ω is the *Pitzer acentric factor*.[16] The Pitzer acentric factor is defined as

$$\omega = - (\log p_{vr} + 1) \text{ at } T_r = 0.7, \quad (4-34)$$

where p_{vr} is the reduced vapor pressure, evaluated at $T_r = 0.7$. Thus the acentric factor is a constant for each pure substance. Values are tabulated in Appendix A.

Peng-Robinson Equation of State

Peng and Robinson proposed a slightly different form of the molecular attraction term.[17]

$$\left[p + \frac{a_T}{V_M(V_M + b) + b(V_M - b)} \right] (V_M - b) = RT \quad (4-35)$$

The term a_T is temperature dependent as in the Soave-Redlich-Kwong equation of state; however, it does not have exactly the same values. The coefficients are calculated as follows.

$$b = 0.07780 \frac{RT_c}{p_c} \text{ and } a_c = 0.45724 \frac{R^2 T_c^2}{p_c}, \quad (4-36)$$

$$a_T = a_c \alpha, \quad (4-30)$$

where

$$\alpha^{1/2} = 1 + m(1 - T_r^{1/2}), \qquad (4-32)$$

and

$$m = 0.37464 + 1.54226\omega - 0.26992\omega^2. \qquad (4-37)$$

Mixing Rules

The following mixture rules are recommended for use with both the Soave-Redlich-Kwong and the Peng-Robinson equations of state.[18]

$$b = \sum_j y_j b_j \text{ and } a_T = \sum_i \sum_j y_i y_j a_{Tij}, \qquad (4-38)$$

where

$$a_{Tij} = (1 - \delta_{ij}) (a_{Ti} a_{Tj})^{1/2}. \qquad (4-39)$$

That is:

$$a_T = \sum_i \sum_j y_i y_j (a_{Ti} a_{Tj})^{1/2} (1 - \delta_{ij}). \qquad (4-40)$$

The terms δ_{ij} are *binary interaction coefficients*, which are assumed to be independent of pressure and temperature. Values of the binary interaction coefficients must be obtained by fitting the equation of state to gas-liquid equilibria data for each binary mixture.[19] The binary interaction coefficients have different values for each binary pair and also take on different values for each equation of state.

EXAMPLE 4–3: *A laboratory cell with volume of 0.008829 cu ft (250.0 cc) contains 0.007357 lb mole (79.28 g) of gas. The composition of the gas is given below. The temperature is to be raised to 709.6°R (250°F). Use the SRK equation of state to calculate the pressure to be expected.*

Component	Composition, mole fraction
Methane	0.6500
Ethane	0.2500
n-Butane	0.1000
	1.0000

Use values of 0.02 for binary interaction coefficient between methane and n-butane, 0.01 between ethane and n-butane, and 0.0 between methane and ethane.

Solution

First, calculate the constants a_j and b_j for each component.

$$b_j = 0.08664 \frac{RT_{cj}}{p_{cj}} \qquad (4-31)$$

$$a_{cj} = 0.42747 \frac{(RT_{cj})^2}{p_{cj}} \qquad (4-31)$$

$$m_j = 0.480 + 1.574\omega - 0.176\omega^2 \qquad (4-33)$$

$$\alpha_j^{1/2} = 1 + m_j(1 - T_{rj}^{1/2}) \qquad (4-32)$$

$$a_{Tj} = a_{cj}\,\alpha_j \qquad (4-30)$$

Component	Critical temperature, °R T_{cj}	Critical pressure, psia p_{cj}	b_j	a_{cj}	Acentric factor ω_j	m_j	α_j	a_{Tj}
C_1	342.9	666.4	0.4784	8,687	0.0104	0.4964	0.6120	5,317
C_2	549.5	706.5	0.7232	21,042	0.0979	0.6324	0.8349	17,569
$n\text{-}C_4$	765.2	550.6	1.2922	52,358	0.1995	0.7870	1.0591	55,453

Second, calculate the "mixture constants," b and a_T.

$$b = \sum_j y_j b_j \qquad (4-38)$$

$b = (0.65)(0.4784) + (0.25)(0.7232) + (0.10)(1.2922) = 0.6210$

$$a_T = \sum_i \sum_j y_i y_j (a_{Ti} a_{Tj})^{1/2} (1 - \delta_{ij}).\qquad\qquad (4\text{--}40)$$

$$\begin{aligned}
a_T = \ & (0.65)(0.65)(5317 \times 5317)^{1/2}(1-0.0) \\
+\ & (0.65)(0.25)(5317 \times 17{,}569)^{1/2}(1-0.0) \\
+\ & (0.65)(0.10)(5317 \times 55{,}453)^{1/2}(1-0.02) \\
+\ & (0.25)(0.65)(17{,}569 \times 5317)^{1/2}(1-0.0) \\
+\ & (0.25)(0.25)(17{,}569 \times 17{,}569)^{1/2}(1-0.0) \\
+\ & (0.25)(0.10)(17{,}569 \times 55{,}453)^{1/2}(1-0.01) \\
+\ & (0.10)(0.65)(55{,}453 \times 5317)^{1/2}(1-0.02) \\
+\ & (0.10)(0.25)(55{,}453 \times 17{,}569)^{1/2}(1-0.01) \\
+\ & (0.10)(0.10)(55{,}453 \times 55{,}453)^{1/2}(1-0.0) = 10{,}773
\end{aligned}$$

Third, calculate pressure.

$V_m = 0.008829 \text{ cu ft}/0.007357 \text{ lb mole} = 1.200 \text{ cu ft/lb mole}$

$$p = \frac{RT}{V_M - b} - \frac{a_T}{V_M(V_M + b)}\qquad\qquad (4\text{--}29)$$

$$p = \frac{(10.732)(709.6)}{(1.200 - 0.6210)} - \frac{10{,}773}{(1.200)(1.200 + 0.6210)}$$

$$p = 8{,}223 \text{ psia}$$

Exercises

4–1. Calculate values of van der Waals constants for neopentane.

4–2. A laboratory cell with volume of 0.007063 cu ft contains 0.03589 lb of methane. Temperature is to be raised to 66°F. Calculate the pressure to be expected. Use van der Waals equation of state. Compare your answer with experimental results of 1500 psia.

4–3. Repeat Exercise 4–2. Use the Beattie-Bridgeman equation of state.

4–4. A laboratory cell with volume of 0.007769 cu ft contains 0.001944 lb moles of the mixture given in the table below. Temperature is to be raised to 80°F. Calculate the pressure to be expected. Use the Beattie-Bridgeman equation of state. Compare your answers with experimental results of 1200 psia ±2%.

Component	Composition, mole fraction
Methane	0.850
Ethane	0.100
Propane	0.050
	1.000

4–5. Use the Beattie modification of the Beattie-Bridgeman equation of state to calculate the molar volume of the mixture of Exercise 4–4 at 80°F and 1200 psia.

4–6. Repeat Exercise 4–2. Use the Benedict-Webb-Rubin equation of state.

4–7. Repeat Exercise 4–4. Use the Benedict-Webb-Rubin equation of state.

4–8. Repeat Exercise 4–2. Use the Soave-Redlich-Kwong equation of state.

4–9. Repeat Exercise 4–4. Use the Soave-Redlich-Kwong equation of state. Use values of zero for the binary interaction coefficients.

4–10. Repeat Exercise 4–2. Use the Peng-Robinson equation of state.

4–11. Repeat Exercise 4–4. Use the Peng-Robinson equation of state. Use values of zero for the binary interaction coefficients.

References

1. van der Waals, J.D.: *Over de Continuiteit van den Gas-en Vloeistof-toestand,* dissertation, Leiden (1873).
2. Clausius, R.: *Ann. Phys. Chem.* (1880) *9*, 337.
3. Berthelot, D.J.: *J. Phys.* (1899) *3*, 263.
4. Dieterici, C.: *Ann. Physik.* (1899) *69*, 685.
5. Lorentz: *Wied. Ann.* (1881) *12*, 127, 660.
6. Wohl, K.: *Z Pysik. Chem.* (1927) *133*, 305.
7. Kammerlingh-Onnes, H.K.: *Commun. Phys. Lab.* (1901) Leiden, Holland, 71.
8. Taylor, H.S. and Glasstone, S., Eds.: *A Treatise on Physical Chemistry, Vol. II, States of Matter,* 3rd Ed., D. van Nostrand and Co., Inc., New York (1951).
9. Beattie, J.A. and Bridgeman, O.C.: "A New Equation of State for Fluids," *Proc. Am. Acad. Arts Sci.* (1929) *63*, 229-308.
10. Benedict, M., Webb, G.B., and Rubin, L.C.: "An Empirical Equation for Thermodynamic Properties of Light Hydrocarbons and Their Mixtures, I. Methane, Ethane, Propane and n-Butane," *J. Chem. Phys.* (April 1940) *8*, 334–345.

11. Benedict, M., Webb, G.B., and Rubin, L.C.: "An Empirical Equation for Thermodynamic Properties of Light Hydrocarbons and Their Mixtures, II. Mixtures of Methane, Ethane, Propane, and n-Butane," *J. Chem. Phys.* (Dec. 1931), *10*, 747-758.
12. Starling, K.E.: *Fluid Thermodynamic Properties for Light Hydrocarbon Systems,* Gulf Publishing Co., Houston (1973).
13. Ahmed, T.H.: "Comparative Study of Eight Equations of State for Predicting Hydrocarbon Volumetric Phase Behavior," *SPE Res. Eng.* (Feb. 1988) *3*, No. 1, 337–348.
14. Redlich, O. and Kwong, J.N.S.: "On the Thermodynamics of Solutions. V—An Equation of State. Fugacities of Gaseous Solutions," *Chem. Reviews* (1949) *44*, 233–244.
15. Soave, G.: "Equilibrium Constants from a Modified Redlich-Kwong Equation of State," *Chem. Eng. Sci.* (1972) *27*, No. 6, 1197–1203.
16. Pitzer, K.S., Lippmann, D.Z., Curl, R.F., Jr., Huggins, C.M., and Peterson, D.E.: "The Volumetric and Thermodynamic Properties of Fluids. II. Compressibility Factor, Vapor Pressure and Entropy of Vaporization," *J. Am. Chem. Soc.* (1955) *77*, No. 13, 3433–3440.
17. Peng, D. and Robinson, D.B.: "A New Two-Constant Equation of State," *I.&E.C. Fundamentals* (1965) *15*, No. 1, 59–64.
18. Edmister, W.C. and Lee, B.I.: *Applied Hydrocarbon Thermodynamics* Volume I, 2nd Ed., Gulf Publishing Co., Houston (1984).
19. Katz, D.L. and Firoozabadi, A.: "Predicting Phase Behavior of Condensate/Crude-Oil Systems Using Methane Interaction Coefficients," *Trans.,* AIME (1978) *265*, 1649–1655.

General References

Phase Behavior, Reprint Series *15*, SPE, Dallas (1981).

Chao, K.C. and Robinson, R.L., Jr., Eds.: *Equations of State in Engineering and Research,* Advances in Chemistry Series *182*, ACS, Washington (1979).

The Five Reservoir Fluids

5

Before reading this chapter, go back to the last pages of Chapter 2 and review multicomponent phase diagrams. Remember that all five of Figures 2–32 through 2–36 are for petroleum mixtures. Examination of these figures shows the wide variety of shapes and sizes of petroleum phase diagrams. Numerous components make up these petroleum mixtures. Diverse chemical species are found in them. And the types and quantities of the components of a particular mixture fix the shape of its phase diagram.

The behavior of a reservoir fluid during production is determined by the shape of its phase diagram and the position of its critical point. Our knowledge of the behavior of two-component mixtures will serve as a guide to the behavior of these multicomponent mixtures; a quick review of the first few pages of the two-component mixtures section of Chapter 2 might be useful. Unmistakable similarity exists between the phase diagrams of two-component and multicomponent mixtures. Chapter 5 begins with a short discussion of the relationship of composition to the shape of the phase diagram as evidenced by two-component mixtures.

However, the major purpose of this chapter is to define and describe the five types of petroleum reservoir fluids. Each will be defined by reference to the shape of its typical phase diagram. Several rules of thumb will be given to assist in determining fluid type from normally available production data. Many of the producing characteristics of each type of fluid will be discussed. Ensuing chapters will address the physical properties of these five reservoir fluids, with emphasis on black oils, dry gases, and wet gases.

Multicomponent Phase Diagrams

Figure 2–37 shows phase diagrams for several mixtures of ethane and n-heptane. These are two-component mixtures; however, the shapes of the phase diagrams can be used in understanding the behavior of multicomponent mixtures.

Mixture 2 on Figure 2–37 illustrates a mixture containing a large quantity of the light component. The phase envelope is relatively small and is located at low temperatures. The critical point is located far down the left-hand side of the phase envelope and is fairly close to the critical point of the pure light component. There is a large area in which retrograde condensation can occur.

As heavy component is added to the mixtures—lines 3 and 4, for instance—the phase envelope increases in size and covers wider ranges of temperature and pressure. The critical point moves up closer to the top of the envelope.

Phase behavior of multicomponent reservoir fluids is similar. Reservoir gases, which are predominately methane, have relatively small phase diagrams with critical temperatures not much higher than the critical temperature of methane. The critical point is far down the left slope of the envelope.

Reservoir liquids contain some methane, normally the lightest component of any significance. Reservoir liquids also contain a wide variety of intermediate and very large molecules. Their phase diagrams are extremely large and cover a wide range of temperature, analogous to mixture 6 of Figure 2–37. However, in naturally occurring petroleum liquids, the critical point does not normally appear to the right of the top of the phase envelope. Only those reservoir liquids which are deficient in intermediate components (often found in south Louisiana) or which have considerable dissolved nitrogen will have critical points to the right of the top of the phase envelope.

The Five Reservoir Fluids

There are five types of reservoir fluids. These are usually called *black oil, volatile oil, retrograde gas, wet gas,* and *dry gas.* The five types of reservoir fluids have been defined because each requires different approaches by reservoir engineers and production engineers.

The petroleum engineer should determine the type of fluid very early in the life of his reservoir. Fluid type is the deciding factor in many of the decisions which must be made regarding the reservoir. The method of fluid sampling, the types and sizes of surface equipment, the calculational procedures for determining oil and gas in place, the techniques of predicting oil and gas reserves, the plan of depletion, and the selection of enhanced recovery method are all dependent on the type of reservoir fluid.

Identification of Fluid Type

Reservoir fluid type can be confirmed only by observation in the laboratory. Yet, readily available production information usually will indicate the type of fluid in the reservoir. Rules of thumb will be given for identification of each of the five fluid types. Three properties are readily available: the *initial producing gas-oil ratio,* the *gravity of the stock-tank liquid,* and the *color of the stock-tank liquid.* Initial producing gas-oil ratio is by far the most important indicator of fluid type. The color of stock-tank liquid alone is not a good indicator of fluid type. However, stock-tank liquid gravity and color are useful in confirming the fluid type indicated by the producing gas-oil ratio.

If all three indicators—initial gas-oil ratio, stock-tank liquid gravity, and stock-tank liquid color—do not fit within the ranges given in the rules of thumb, the rules fail and the reservoir fluid must be observed in the laboratory to determine its type.

Do not attempt to compare fluid types as defined here with the reservoir descriptions as defined by the state regulatory agencies which have jurisdiction over the petroleum industry. The legal and regulatory definitions of oil, crude oil, gas, natural gas, condensate, etc., usually do not bear any relationship to the engineering definitions given here. In fact, the regulatory definitions are often contradictory.

Black Oils

Black oils consist of a wide variety of chemical species including large, heavy, nonvolatile molecules. The phase diagram predictably covers a wide temperature range. The critical point is well up the slope of the phase envelope.

Black Oil Phase Diagram

The phase diagram of a typical black oil is shown in Figure 5–1. The lines within the phase envelope represent constant liquid volume, measured as percent of total volume. These lines are called *iso-vols* or *quality lines.* Note that the iso-vols are spaced fairly evenly within the envelope.

The vertical line $\overline{123}$ indicates the reduction in pressure at constant temperature that occurs in the reservoir during production. The pressure and temperature of the separator located at the surface are indicated too.

When reservoir pressure lies anywhere along line $\overline{12}$, the oil is said to be *undersaturated.* The word *undersaturated* is used in this sense to

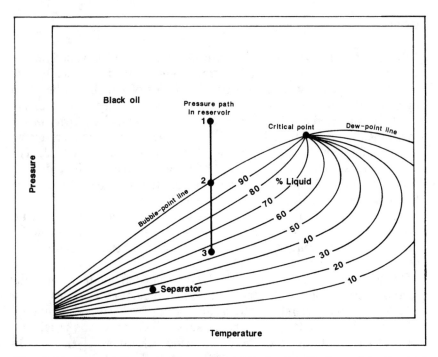

Fig. 5–1. Phase diagram of a typical black oil with line of isothermal reduction of reservoir pressure, 123, and surface separator conditions.

indicate that the oil could dissolve more gas if more gas were present.

If the reservoir pressure is at point 2, the oil is at its bubble point and is said to be *saturated*. The oil contains as much dissolved gas as it can hold. A reduction in pressure will release gas to form a free gas phase in the reservoir.

As reservoir pressure declines along line $\overline{23}$, additional gas is evolved in the reservoir. The volume of gas on a percentage basis is one hundred minus the percent liquid. Water is always present in a petroleum reservoir but is not included in this discussion.

Actually, the oil is "saturated" anywhere along line $\overline{23}$. The bubble point, point 2, is a special case of saturation at which the first bubble of gas forms. Unfortunately, the word "saturated" is often used to mean "bubble point."

Additional gas evolves from the oil as it moves from the reservoir to the surface. This causes some shrinkage of the oil. However, separator conditions lie well within the phase envelope, indicating that a relatively large amount of liquid arrives at the surface.

Comments

The name *black oil* is a misnomer since the color of this type of oil is not always black. This type of reservoir fluid has also been called *low-shrinkage crude oil or ordinary oil.*[1,2]

Field Identification of Black Oils

Black oils are characterized as having initial producing gas-oil ratios of 2000 scf/STB or less.[2] Producing gas-oil ratio will increase during production when reservoir pressure falls below the bubble-point pressure of the oil. The stock-tank oil usually will have a gravity below 45°API. Stock-tank oil gravity will slightly decrease with time until late in the life of the reservoir when it will increase. The stock-tank oil is very dark, indicating the presence of heavy hydrocarbons, often black, sometimes with a greenish cast, or brown.

Laboratory Analysis of Black Oils

Laboratory analysis will indicate an initial oil formation volume factor of 2.0 res bbl/STB or less. Oil formation volume factor is the quantity of reservoir liquid in barrels required to produce one stock-tank barrel. Thus, the volume of oil at point 2 of Figure 5–1 shrinks by one-half or less on its trip to the stock tank.

Laboratory determined composition of heptanes plus will be higher than 20 mole percent, an indication of the large quantity of heavy hydrocarbons in black oils.

Volatile Oils

Volatile oils contain relatively fewer heavy molecules and more intermediates (defined as ethane through hexanes) than black oils.

Volatile Oil Phase Diagram

The phase diagram for a typical volatile oil, Figure 5–2, is somewhat different from the black-oil phase diagram. The temperature range covered by the phase envelope is somewhat smaller, but of more interest is the position of the critical point. The critical temperature is much lower than for a black oil and, in fact, is close to reservoir temperature. Also, the iso-vols are not evenly spaced but are shifted upwards toward the bubble-point line.

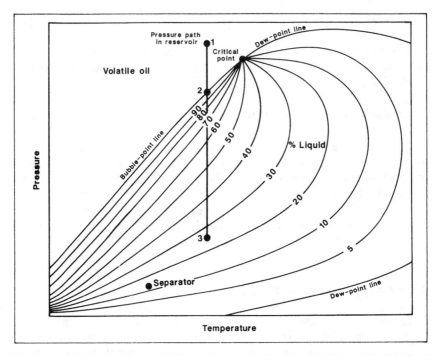

Fig. 5–2. Phase diagram of a typical volatile oil with line of isothermal reduction of reservoir pressure, 123, and surface separator conditions.

The vertical line shows the path taken by the constant-temperature reduction in pressure during production. Notice that a small reduction in pressure below the bubble point, point 2, causes the release of a large amount of gas in the reservoir.

A volatile oil may become as much as 50 percent gas in the reservoir at only a few hundred psi below the bubble-point pressure. Also, an iso-vol with a much lower percent liquid crosses the separator conditions. Hence the name volatile oil.

Comments

Volatile oils also have been called *high-shrinkage crude oils*, and *near-critical oils*.[1,2]

The set of equations known collectively as "material balance equations" which is used for black oils will not work for volatile oils. These equations were derived under the assumption that the gas associated with the reservoir liquid is a dry gas (defined later). This is true for black oils except at low reservoir pressures. However, the gas associated with a volatile oil is very rich, usually a retrograde gas (defined later). This rich gas releases a large quantity of liquid as it moves to the surface. Often over one-half of the stock-tank liquid produced during the life of a volatile oil reservoir entered the wellbore as part of the gas. This situation causes the material balance equations to be invalid for volatile oils.

Field Identification of Volatile Oils

The dividing line between black oils and volatile oils is somewhat arbitrary. The difference depends largely on the point at which the material balance equations begin to have intolerable inaccuracy. The dividing line between volatile oils and retrograde gases is clear. For a fluid to be a volatile oil its critical temperature must be greater than reservoir temperature.

Volatile oils are identified as having initial producing gas-oil ratios between 2000 and 3300 scf/STB.[2] The producing gas-oil ratio increases as production proceeds and reservoir pressure falls below the bubble-point pressure of the oil. The stock-tank oil gravity is usually 40°API or higher and increases during production as reservoir pressure falls below the bubble point. The stock-tank oil is colored (usually brown, orange, or sometimes green).

Laboratory Analysis of Volatile Oils

Laboratory observation of volatile oils will reveal an initial oil formation volume factor greater than 2.0 res bbl/STB. The oil produced at point 2 of Figure 5–2 will shrink by more than one-half, often three-quarters, on the trip to the stock tank. Volatile oils should be produced through three or more stages of surface separation to minimize this shrinkage.

Laboratory determined compositions of volatile oils will have 12.5 to 20 mole percent heptanes plus. The dividing line between volatile oils and retrograde gases of 12.5 mole percent heptanes plus is fairly definite.[2] When the heptanes plus concentration is greater than 12.5 mole percent, the reservoir fluid is almost always liquid and exhibits a bubble point. When the heptanes plus concentration is less than 12.5 mole percent, the reservoir fluid is almost always gas and exhibits a dew point. Any exceptions to this rule normally do not meet the rules of thumb with regard to stock-tank oil gravity and color.

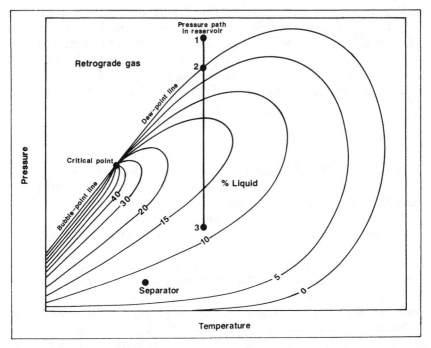

Fig. 5-3. Phase diagram of a typical retrograde gas with line of isothermal reduction of reservoir pressure, 123, and surface separator conditions.

Retrograde Gases

The third type of reservoir fluid we will consider is retrograde gas.

Retrograde Gas Phase Diagram

The phase diagram of a *retrograde gas* is somewhat smaller than that for oils, and the critical point is further down the left side of the envelope. These changes are a result of retrograde gases containing fewer of the heavy hydrocarbons than do the oils.

The phase diagram of a retrograde gas has a critical temperature less than reservoir temperature and a cricondentherm greater than reservoir temperature. See Figure 5-3. Initially, the retrograde gas is totally gas in the reservoir, point 1. As reservoir pressure decreases, the retrograde gas exhibits a dew point, point 2. As pressure is reduced, liquid condenses from the gas to form a free liquid in the reservoir. This liquid will normally not flow and cannot be produced.

The reservoir pressure path on the phase diagram, Figure 5–3, indicates that at some low pressure the liquid begins to revaporize. This occurs in the laboratory; however, it probably does not occur to much extent in the reservoir because during production the overall composition of the reservoir fluid changes.

Field Identification of Retrograde Gases

The lower limit of the initial producing gas-oil ratio for a retrograde gas is approximately 3300 scf/STB.[2] The upper limit is not well defined; values of over 150,000 scf/STB have been observed. Gas-oil ratios this high indicate that the phase diagram is much smaller than shown in Figure 5–3. Gases with high gas-oil ratios have cricondentherms close to reservoir temperature and drop very little retrograde liquid in the reservoir.

As a practical matter, when initial producing gas-oil ratio is above 50,000 scf/STB, the quantity of retrograde liquid in the reservoir is very small and the reservoir fluid can be treated as if it were a wet gas (defined later).

Producing gas-oil ratios for a retrograde gas will increase after production begins when reservoir pressure falls below the dew-point pressure of the gas.

Stock-tank liquid gravities are between 40° and 60°API and increase as reservoir pressure falls below the dew-point pressure.[2] The liquid can be lightly colored, brown, orange, greenish, or water-white.

Laboratory Analysis of Retrograde Gases

Retrograde gases exhibit a dew point when pressure is reduced at reservoir temperature. The heptanes plus fraction is less than 12.5 mole percent. Retrograde behavior will occur at reservoir conditions for gases with less than one percent heptanes plus, but for these gases the quantity of retrograde liquid is negligible.

Comments

Retrograde gases are also called *retrograde gas-condensates, retrograde condensate gases, gas condensates,* or *condensates.*[1,2] The use of the word "condensate" in the name of this reservoir fluid leads to much confusion. Initially, the fluid is gas in the reservoir and exhibits retrograde behavior. Hence, the correct name is *retrograde gas.*

Stock-tank liquid produced from retrograde gas reservoirs often is called *condensate*. The liquid produced in the reservoir is called *condensate* also. A better name is *retrograde liquid*.

An initial producing gas-oil ratio of 3300 to 5000 scf/STB indicates a very rich retrograde gas, one which will condense sufficient liquid to fill 35 percent or more of the reservoir volume. Even this quantity of liquid seldom will flow and normally cannot be produced.

The surface gas is very rich in intermediates and often is processed to remove liquid propane, butanes, pentanes, and heavier hydrocarbons. These liquids often are called *plant liquids*. The gas-oil ratios in the rules of thumb discussed above do not include any of these plant liquids.

Wet Gases

The fourth type of reservoir fluid we will discuss is wet gas.

Wet Gas Phase Diagram

The entire phase diagram of a hydrocarbon mixture of predominately smaller molecules will lie below reservoir temperature. An example of the phase diagram of a *wet gas* is given in Figure 5–4.

A wet gas exists solely as a gas in the reservoir throughout the reduction in reservoir pressure. The pressure path, line $\overline{12}$, does not enter the phase envelope. Thus, no liquid is formed in the reservoir. However, separator conditions lie within the phase envelope, causing some liquid to be formed at the surface.

Comments

The surface liquid normally is called *condensate,* and the reservoir gas sometimes is called *condensate-gas*. This leads to a great deal of confusion between wet gases and retrograde gases.

The word "wet" in wet gas does not mean that the gas is wet with water but refers to the hydrocarbon liquid which condenses at surface conditions. In fact, reservoir gas is normally saturated with water.

Field Identification of Wet Gases

Wet gases produce stock-tank liquids with the same range of gravities as the liquids from retrograde gases. However, the gravity of the stock-tank liquid does not change during the life of the reservoir. The stock-tank liquid is usually water-white. True wet gases have very high

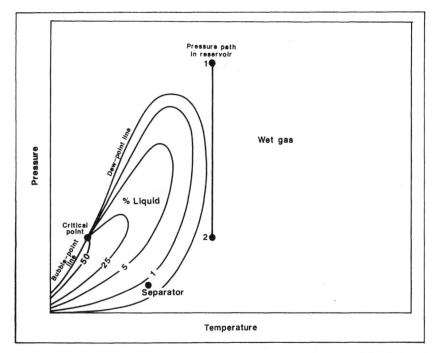

Fig. 5–4. Phase diagram of a typical wet gas with line of isothermal reduction of reservoir pressure, 12., and surface separator conditions.

producing gas-oil ratios. Producing gas-oil ratios will remain constant during the life of a wet gas reservoir.

For engineering purposes, a gas which produces more than 50,000 scf/STB can be treated as if it were a wet gas.

Dry Gases

The fifth type of reservoir fluid we will consider is dry gas.

Dry Gas Phase Diagram

Dry gas is primarily methane with some intermediates. Figure 5–5 shows that the hydrocarbon mixture is solely gas in the reservoir and that normal surface separator conditions fall outside the phase envelope. Thus, no liquid is formed either in the reservoir or at the surface.

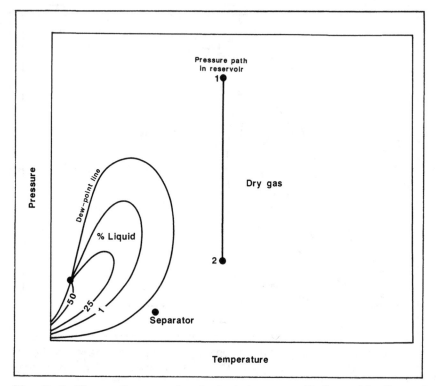

Fig. 5–5. Phase diagram of a typical dry gas with line of isothermal reduction of reservoir pressure, 12 , and surface conditions.

Comments

The word "dry" in dry gas indicates that the gas does not contain enough of the heavier molecules to form hydrocarbon liquid at the surface. Usually some liquid water is condensed at the surface.

A dry gas reservoir often is called simply a *gas* reservoir. This leads to confusion because wet gas reservoirs sometimes are called *gas* reservoirs. Further, a retrograde gas initially exists as *gas* in the reservoir.

A set of equations known collectively as *gas material balance equations* has been devised to determine original gas in place and predict gas reserves. These equations were derived for dry gases and can be used for wet gases, if care is taken in defining the properties of the wet gases. The equations are applicable to retrograde gases only at reservoir pressures above the dew point.

Exercises

5–1. Assume that you have discovered a reservoir which contains a mixture of ethane and n-heptane. Plot on Figure 2–37 the initial reservoir conditions of 1300 psia and 300°F. Connect that point to final reservoir conditions of 400 psia and 300°F. Also plot separator conditions of 100 psia and 150°F.

    a. If the reservoir fluid is the same as mixture 2 on the graph, how should it be classified?

    b. If the reservoir fluid is the same as mixture 3 on the graph, how should it be classified?

    c. If the reservoir fluid is the same as mixture 4 on the graph, how should it be classified?

    d. If the reservoir fluid is the same as mixture 5 on the graph, how should it be classified?

    e. If the reservoir fluid is the same as mixture 6 on the graph, how should it be classified?

NOTE: A reservoir fluid consisting solely of ethane and n-heptane is totally unrealistic, as are the given reservoir conditions. However, the exercise illustrates how changing fluid composition affects the way a fluid is classified.

5–2. The following notices were taken from the *Bryan-College Station Eagle,* Sunday, September 15, 1985. Make an estimate of the type of reservoir fluid in each reservoir.

At a location 5.8 miles south-southeast of Kurten, International Drilling System has finaled a new producer in Brazos County's Buda Field.

The well is designed as the No. 1 Cargill/Chase, flowed ten barrels of oil per day, along with 50,000 CF casinghead gas on an 18/64-in. choke.

Location is in an 814-acre lease in the B.F. Stroud Survey. Bottomed at 10,059 feet, the well will produce from perforations in the Buda Formation, 9,799 to 10,042 feet into the wellbore.

Patterson Petroleum of Houston filed first production figures on a new oil well in the Giddings Field, Washington County. The No. 1 Fischer showed potential to flow 350 barrels of oil per day, along with 750,000 CF casinghead gas on a 20/64-in. choke. Tubing pressure registered 1,100 PSI.

The operator has 160 acres leased, with drillsite in the J. Goocher Survey, 7.5 miles north of Burton.

The well will produce from perforations in the Austin Chalk, 10,222 to 10,392 feet into the wellbore. Total drilling depth was 10,471 feet.

HEA Exploration has filed first production figures on two new oil wells in Brazos County.

The No. 1 A. Varisco Estate indicated ability to flow 368 barrels of oil per day, plus 447,000 CF casinghead gas on a 16/64-in. choke. Tubing pressure registered 1,125 PSI. The well is located two miles southwest of Fountain in a 71-acre lease in the D. Harvey Survey, Northeast Caldwell Field.

What other field data would you like to have to verify your estimates?

5–3. The average gas-oil ratio produced from the Upper Washita-Fredericksburg formation of the Summerland Field is 275 scf/STB. The gravity of the produced oil is 26°API. The color of the stock-tank oil is black. What type of reservoir fluid is in this formation?

5–4. Laboratory analysis of a sample from the Summerland (Upper Washita-Fredericksburg) Field indicates a ratio of volume oil leaving the reservoir to volume of oil arriving at the stock tank of 1.10 res bbl/STB. Does this information confirm your answer to Exercise 5–3? Why or why not?

5–5. One of the wells in the Merit Field, completed in December 1967 in the North Rodessa formation, originally produced 54°API stock-tank liquid at a gas-oil ratio of about 23,000 scf/STB. During July 1969, the well produced 1987 STB of 58°API liquid and 78,946 Mscf of gas. In May 1972, the well was producing liquid at a rate of about 30 STB/d of 59°API liquid and gas at about 2000 Mscf/d. What type of reservoir fluid is this well producing?

5–6. A field in north Louisiana discovered in 1953 and developed by 1956 had an initial producing gas-oil ratio of 2000 scf/STB.[3] The stock-tank liquid was "medium orange" and had a gravity of 51.2°API. Classify this reservoir fluid.

5–7. During the producing history of the field in Exercise 5–6 the stock-tank liquid gravity steadily increased to 63°API, and the producing gas-oil ratio increased to a maximum of 29,000 scf/STB.[4] Does this information confirm your classification? Why or why not?

5-8. Laboratory analysis of a sample from the reservoir of Exercise 5–6 gave the following composition.[3]

Component	Composition, mole fraction
CO_2	0.0218
N_2	0.0167
C_1	0.6051
C_2	0.0752
C_3	0.0474
$C_{4's}$	0.0412
$C_{5's}$	0.0297
$C_{6's}$	0.0138
C_{7+}	0.1491
	1.0000

Properties of heptanes plus	
Specific gravity	0.799
Molecular weight	181 lb/lb mole

The formation volume factor was about 2.6 res bbl/STB. Does this information confirm your classification? Why or why not?

5-9. The Sundance et al. Hoadley 6–2–45–2w5m discovery well in the Hoadley (Lower Cretaceous Glauconite) Field yielded an absolute open flow of 76 MMscf/d with 60 bbl of stock-tank liquid per MMscf of gas. Classify the reservoir fluid type. What other field information would you like to have to confirm your classification?

5-10. The initial reservoir pressure and temperature in the Lower Tuscaloosa reservoir of the East Fork Field were 5043 psig and 263°F. The bubble-point pressure of the 40°API oil plus 1110 scf/STB of gas produced from this field was measured as 3460 psig at 263°F. What type of reservoir fluid is in this reservoir? Is the reservoir oil saturated or undersaturated? How do you know?

5-11. The reported production from the discovery well of the Nancy (Norphlet) Field is given below. How would you classify this reservoir fluid? Why?

Date	Stock-tank Liquid gravity, °API	Monthly Production	
		Oil, STB	Gas, Mscf
9/86	29	4,276	1,165
10/86	28	16,108	5,270
11/86	28	15,232	4,800
12/86	28	15,585	4,960
1/87	28	15,226	4,650
2/87	28	14,147	4,335
3/87	28	15,720	4,707
4/87	28	15,885	4,904
5/87	28	15,434	4,979
6/87	28	12,862	4,339
7/87	28	14,879	4,814
8/87	28	15,192	4,270

5–12. A discovery well produced during testing 46.6°API stock-tank liquid with a gas-oil ratio of 2906 scf/STB. The stock-tank liquid was orange-brown. How would you classify this reservoir fluid?

5–13. The reservoir fluid of Exercise 5–12 was sampled and the composition determined.

Component	Composition, mole fraction
CO_2	0.0042
N_2	0.0035
C_1	0.6349
C_2	0.1132
C_3	0.0591
i-C_4	0.0102
n-C_4	0.0186
i-C_5	0.0087
n-C_5	0.0065
C_6	0.0104
C_{7+}	0.1307
	1.0000

Properties of heptanes plus
Specific gravity	0.819
Molecular weight	212 lb/lb mole

Also, the oil formation volume factor at the bubble point was measured as 2.504 res bbl/STB. Does this information confirm your previous analysis? If so, in what ways?

5–14. The Crown Zellerbach No. 1 was the discovery well in the Hooker (Rodessa) Field. The reported production during the first year of production is given below. How would you classify this reservoir fluid? Why?

Date		Stock-tank Liquid Gravity, °API	Monthly Production		
			Oil, STB	Water, STB	Gas, Mscf
Apr	1984	—	112	—	3,362
May	1984	55	1,810	12,090	54,809
Jun	1984	55	2,519	180	64,104
Jul	1984	55	3,230	240	94,419
Aug	1984	55	3,722	279	119,151
Sep	1984	54	2,780	248	100,235
Oct	1984	55	3,137	270	113,359
Nov	1984	56	2,291	210	80,083
Dec	1984	56	2,108	217	71,412
Jan	1985	56	1,799	203	60,279
Feb	1985	56	1,422	196	57,626
Mar	1985	56	1,861	186	60,330

5–15. The discovery well in the Splunge (Mississippian-Carter) Field was completed in 1973. As of January 1985, the field had 16 wells which had produced a total of 623,000 Mscf, 43 bbl of water, and no oil. How would you classify the fluid in this reservoir?

5–16. The following item was in the December 14, 1987 edition of *Oil & Gas Journal*.

Mobil Exploration Norway Inc. discovered gas and condensate in its 35/11–2 wildcat on Norwegian North Sea Block 35/11. The well, drilled to 13,205 ft in 1,210 ft of water north of Troll gas field, flowed at maximum sustained rates of 18.8 MMcfd of gas and 3,290 b/d of condensate through a 40/64-in. choke.

How would you classify the reservoir fluid? What other information would you request?

5–17. The annual production statistics for the West Oakvale (Sligo) Field are given below.

Date	Stock-tank Liquid Gravity, °API	Annual Production		
		Oil, STB	Water, STB	Gas, Mscf
1982	46	4,646	1,484	463,265
1983	50	2,608	1,177	342,075
1984	47	1,350	1,215	241,048
1985	48	1,430	932	221,020
1986	50	1,662	1,122	267,106
*1987	51	1,110	665	178,951

*through August 1987

Classify this reservoir fluid.

References

1. Clark, N.J.: *Elements of Petroleum Reservoirs,* Henry L. Doherty Series, SPE, Dallas (1960).
2. Moses, P.L.: "Engineering Applications of Phase Behavior of Crude Oil and Condensate Systems," *J. Pet. Tech.* (July 1986) *38,* 715-723.
3. Jacoby, R.H. and Berry, V.J., Jr.: "A Method for Predicting Depletion Performance of a Reservoir Producing Volatile Crude Oil," *Trans.,* AIME (1957) *210, 27*–33.
4. Cordell, J.C. and Ebert, C.K.: "A Case History-Comparison of Predicted and Actual Performance of a Reservoir Producing Volatile Crude Oil," *J. Pet. Tech.* (November 1965) *17,* 1291–1293.

Properties of Dry Gases

6

This chapter describes several properties of dry gases which commonly are used by the petroleum engineer. We will define each property and then give correlations useful for estimating values of the property using normally available information about the gas.

Dry gases will be considered in this chapter. The adjustments necessary to use the correlations for wet gases will be discussed in Chapter 7. Also, comments on retrograde gases will be made in Chapter 7.

Standard Conditions

Since the volume of a gas varies greatly with pressure and temperature, defining the conditions at which gas volume is reported is necessary. This is especially important in the sale of gas. Most states have specified the temperature and pressure which are to be used to report gas volume. These are called *standard conditions*, and the volume of gas measured or calculated at these conditions is called *standard cubic feet*, scf. Standard conditions sometimes are called *base conditions*. The standard temperature used throughout the United States is 60°F. Standard pressure varies as shown in Table 6–1.

This text uses standard conditions of 60°F and 14.65 psia. Any equations given in this book which have coefficients that depend on the values of the standard conditions must be adjusted to fit the applicable conditions.

EXAMPLE 6–1: *Calculate the volume occupied by one pound mole of natural gas at standard conditions.*

Solution

Assume that the gas acts like an ideal gas at standard conditons.

$$V_M = \frac{RT}{p} \qquad\qquad (3-13)$$

$$V_M = \frac{\left(10.732 \ \frac{\text{psia cu ft}}{\text{lb mole } °R}\right)(60 + 459.7)°R}{14.65 \ \text{psia}} = 380.7 \ \frac{\text{scf}}{\text{lb mole}}$$

Note that in this example a value of 459.7° was added to degrees Fahrenheit to obtain degrees Rankin. A less accurate but more commonly used value of 460° will be applied throughout most of the remainder of this book.

The result of this example gives an important conversion factor for use in gas calculations.

TABLE 6–1

Values of standard (base) pressure (from various sources — accuracy not guaranteed).

Alabama	15.025	Montana	15.025
Alaska	14.65	Nebraska	15.025
Arizona	14.65	New Mexico	15.025
Arkansas	14.65	New York	14.65
California	14.73	North Dakota	14.73
Colorado	15.025	Ohio	14.65
Florida	14.65	Oklahoma	14.65
Illinois	14.65	Pennsylvania	14.65
Indiana	14.65	South Dakota	14.73
Kansas	14.65	Texas	14.65
Kentucky	14.65	Utah	15.025
Louisiana	15.025	West Virginia	14.85
Michigan	14.73	Wyoming	15.025
Mississippi	15.025		

EXAMPLE 6–2: *Convert the mass of gas in Example 3–9 into standard cubic feet.*

Solution

$$(37,400 \ \text{lb moles})(380.7 \ \text{scf/lb mole}) = 14.2 \times 10^6 \ \text{scf} = 14.2 \ \text{MMscf}$$

Students are often confused by *three entirely different calculations* which use numbers approximately equal to 14.7.

The *first* is the calculation of standard volume using standard pressure as defined above. The value of standard pressure is arbitrary and is specified by the appropriate political body. Values in the United States range from 14.65 to 15.025 psia.

The *second* is conversion of values of pressure measured in atmospheres to pounds per square inch absolute. The conversion factor is a constant, 14.696 psia/atm.

The *third* calculation is the conversion of pressure units from pounds per square inch read from a gauge (psig) to pounds per square inch absolute (psia). Strictly speaking, the value of barometric pressure in psi read from a barometer at the time the pressure gauge was calibrated should be added to psig to get psia. At sea level, a value of 14.7 psi is usually sufficiently accurate.

Dry Gases

Dry gases are the easiest to deal with because no liquid condenses from the gas as it moves from the reservoir to the surface. The composition of the surface gas is equal to the composition of the gas in the reservoir, and the specific gravity of the surface gas is equal to the specific gravity of the reservoir gas. Thus, a gas sample taken at the surface can be analyzed and the resulting composition or specific gravity used in correlations to determine the properties of the gas in the reservoir.

Gas Formation Volume Factor

The *gas formation volume factor* is defined as the volume of gas at reservoir conditions required to produce one standard cubic foot of gas at the surface. Units vary. Sometimes units of reservoir cubic feet per standard cubic foot, res cu ft/scf, are used. Reservoir cubic feet simply represents the gas volume measured or calculated at reservoir temperature and reservoir pressure. Often the units are reservoir barrels of gas per standard cubic foot, res bbl/scf.

Formation volume factor also is known as *reservoir volume factor*. The reciprocal of the formation volume factor sometimes is called *gas expansion factor*. Unfortunately, the term formation volume factor is used occasionally when gas expansion factor is meant. The engineer must always examine the units to be sure which is intended.

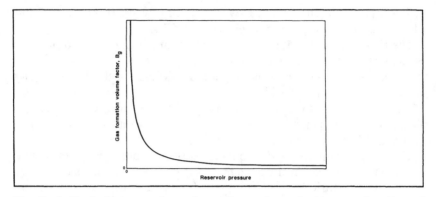

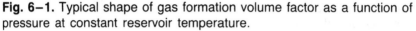

Fig. 6–1. Typical shape of gas formation volume factor as a function of pressure at constant reservoir temperature.

The shape of a plot of gas formation volume factor versus reservoir pressure at constant temperature for a typical dry gas is given in Figure 6–1.

Formation volume factor of a gas may be calculated as the volume occupied by the gas at reservoir temperature and pressure divided by the volume occupied by the same mass of gas at standard conditions.

$$B_g = \frac{V_R}{V_{sc}} \tag{6–1}$$

The volume of n moles of a gas at reservoir conditions may be obtained with the compressibility equation of state.

$$V_R = \frac{znRT}{p}, \tag{3–39}$$

where T and p represent reservoir temperature and pressure.

The volume of the same number of moles of the gas at standard conditions, T_{sc} and p_{sc}, is

$$V_{sc} = \frac{z_{sc}nRT_{sc}}{p_{sc}}. \tag{3–39}$$

Thus, the formation volume factor for the gas is

$$B_g = \frac{V_R}{V_{sc}} = \frac{\dfrac{znRT}{p}}{\dfrac{z_{sc}nRT_{sc}}{p_{sc}}} = \frac{zTp_{sc}}{z_{sc}T_{sc}p} . \qquad (6-1)$$

Since in this book $T_{sc} = 520°R$, $p_{sc} = 14.65$ psia, and for all practical purposes $z_{sc} = 1$, then

$$B_g = \frac{zT(14.65)}{(1.0)(520)p} = \boxed{0.0282 \ \frac{zT}{p} \ \frac{cu\ ft}{scf}} . \qquad (6-2)$$

Also,

$$B_g = \left(0.0282 \ \frac{zT}{p} \ \frac{cu\ ft}{scf}\right)\left(\frac{bbl}{5.615\ cu\ ft}\right) = 0.00502 \ \frac{zT}{p} \ \frac{res\ bbl}{scf} , \qquad (6-3)$$

where temperature must be in degrees Rankine and pressure in psia.

Values of z-factor of natural gases for use in Equations 6–2 or 6–3 may be obtained by methods presented in Chapter 3. If an experimental value of z-factor at reservoir temperature and pressure is available for the gas of interest, it should be used. If an experimental value is unavailable but the composition of the gas is known, the pseudoreduced temperature and pseudoreduced pressure can be computed and z-factor obtained from Figure 3–7. If only the specific gravity of the gas is known, the pseudocritical properties can be obtained from Figure 3–11, and then Figure 3–7 can be used to estimate a z-factor value. The pseudocritical temperature and pseudocritical pressure from either method are adjusted to account for nonhydrocarbon components using Figure 3–12.

EXAMPLE 6–3: *Calculate a value of the formation volume factor of a dry gas with a specific gravity of 0.818 at reservoir temperature of 220°F and reservoir pressure of 2100 psig.*

Solution

First, estimate pseudocritical properties, calculate pseudoreduced properties, and get a value of z-factor.

$T_{pc} = 406°R$ and $p_{pc} = 647$ psia at $\gamma_g = 0.818$, Figure 3–11

$$T_{pr} = \frac{(220 + 460)°R}{406°R} = 1.68 \text{ and } p_{pr} = \frac{(2100 + 14.7) \text{ psia}}{647 \text{ psia}} = 3.27 \quad (3\text{-}43)$$

$z = 0.855$, Figure 3–7

Second, calculate B_g.

$$B_g = 0.00502 \frac{zT}{p} \quad\quad (6\text{-}3)$$

$$B_g = (0.00502) \frac{(0.855)(220 + 460)}{(2100 + 14.7)} = 0.00138 \frac{\text{res bbl}}{\text{scf}}$$

The Coefficient of Isothermal Compressibility of Gas

The *coefficient of isothermal compressibility* is defined as the fractional change of volume as pressure is changed at constant temperature. The defining equations are

$$c_g = -\frac{1}{V}\left(\frac{\partial V}{\partial p}\right)_T \text{ or } c_g = -\frac{1}{V_M}\left(\frac{\partial V_M}{\partial p}\right)_T$$

$$\text{or } c_g = -\frac{1}{v}\left(\frac{\partial v}{\partial p}\right)_T. \quad\quad (6\text{-}4)$$

Units are psi^{-1}. The relationship of c_g to reservoir pressure for a typical dry gas at constant temperature is given in Figure 6–2.

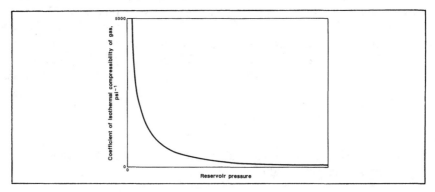

Fig. 6–2. Typical shape of the coefficient of isothermal compressibility of a gas as a function of pressure at constant reservoir temperature.

This coefficient normally is referred to simply as *compressibility* or *gas compressibility*. You must understand that the term compressibility is used to designate the coefficient of isothermal compressibility; whereas, the term compressibility factor refers to z-factor, the coefficient in the compressibility equation of state. Although both are related to the effect of pressure on the volume of a gas, the two are distinctly not equivalent.

The partial derivative rather than the ordinary derivative is used in Equations 6–4 since only one independent variable, pressure, is permitted to vary. The subscript T indicates that temperature is held constant.

EXAMPLE 6–4: *The following table gives volumetric data at 150°F for a natural gas. Determine the coefficient of isothermal compressibility for this gas at 150°F and 1000 psia.*

Pressure, psia	Molar volume, cu ft/lb mole
700	8.5
800	7.4
900	6.5
1000	5.7
1100	5.0
1200	4.6
1300	4.2

Solution

First, plot V_M versus p and determine the slope of the line at 1000 psia, Figure 6–3.

$$\text{slope} = \frac{4.18 - 7.07}{1200 - 800} = -0.00723 \; \frac{\text{cu ft}}{\text{lb mole psi}}$$

Second, compute c_g

$$c_g = -\frac{1}{V_M} \frac{(\partial V_M)}{(\partial p)} \qquad (6\text{--}4)$$

$$c_g = -\left(\frac{1 \text{ lb mole}}{5.7 \text{ cu ft}} \right) \left(\frac{-0.00723 \text{ cu ft}}{\text{lb mole psi}} \right) = 0.0013 \text{ psi}^{-1}$$

$$= 1300 \times 10^{-6} \text{ psi}^{-1}$$

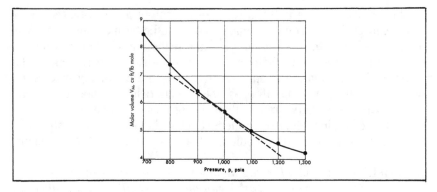

Fig. 6–3. Molar volumes of a natural gas at 150°F., (part of solution to Example 6–4.)

The petroleum engineer often must combine the compressibilities of gas, oil, water, and rock. In order to have the numbers on the same basis, reporting compressibilities on the order of 10^{-6} is convenient. The reciprocal of psi, psi^{-1}, sometimes is called *sip*. A value of $10^{-6} \, psi^{-1}$ is a *microsip*. The answer to Example 6–4 is 1300 microsips.

The Coefficient of Isothermal Compressibility of an Ideal Gas

In order for Equation 6–4 to be useful, it must be combined with an equation which relates volume and pressure so that one of these two variables can be eliminated. An equation of state can be used for this purpose.

The simplest equation of state is that for ideal gases.

$$pV = nRT \text{ or } V = \frac{nRT}{p} \qquad (3-14)$$

We wish to eliminate the term $\partial V / \partial p$ in Equation 6–4, so we derive this term from Equation 3–14 as

$$\left(\frac{\partial V}{\partial p} \right)_T = - \frac{nRT}{p^2} . \qquad (6-5)$$

Combining Equation 6–5 with Equation 6–4 gives

$$c_g = \left(- \frac{1}{V} \right) \left(- \frac{nRT}{p^2} \right)$$

$$c_g = \left(-\frac{p}{nRT}\right)\left(-\frac{nRT}{p^2}\right) = \frac{1}{p}. \qquad (6-6)$$

We recognize that the equation of state for an ideal gas does not describe adequately the behavior of gases at temperatures and pressures normally encountered in petroleum reservoirs. However, Equation 6–6 does illustrate that we can expect the coefficient of isothermal compressibility of a gas to be inversely proportional to pressure. Equation 6–6 can be used to determine the expected order of magnitude of gas compressibility.

EXAMPLE 6–5: *Estimate the coefficient of isothermal compressibility of a gas at 1700 psia. Assume that the gas acts like an ideal gas.*

Solution

$$c_g = \frac{1}{p} \qquad (6-6)$$

$$c_g = \frac{1}{1700 \text{ psia}} = 0.000588 = 588 \times 10^{-6} \text{ psi}^{-1}$$

Therefore, the coefficient of isothermal compressibility of a gas can be expected to be of the order of several hundred microsips.

The Coefficient of Isothermal Compressibility of Real Gases

The compressibility equation is the most commonly used equation of state in the petroleum industry. We will combine this equation with the equation which defines the coefficient of isothermal compressibility. Since z-factor changes as pressure changes, it must be considered to be a variable.

$$V = nRT \frac{z}{p} \qquad (3-39)$$

Thus,

$$\left(\frac{\partial V}{\partial p}\right)_T = nRT \frac{p\left(\frac{\partial z}{\partial p}\right)_T - z}{p^2} \qquad (6-7)$$

and

$$c_g = -\frac{1}{V}\left(\frac{\partial V}{\partial p}\right)_T \qquad (6-4)$$

$$c_g = \left[-\frac{p}{znRT}\right]\left\{\frac{nRT}{p^2}\left[p\left(\frac{\partial z}{\partial p}\right)_T - z\right]\right\}$$

$$c_g = \frac{1}{p} - \frac{1}{z}\left(\frac{\partial z}{\partial p}\right)_T. \qquad (6-8)$$

For the special case of an ideal gas in which z-factor is a constant equal to 1.0, the partial derivative of z-factor with respect to p is equal to zero, and Equation 6–8 reduces to Equation 6–6.

The partial derivative, $(\partial z/\partial p)_T$, is the slope of z-factor plotted against pressure at constant temperature. The slopes of the isotherms of Figures 3–2, 3–3, and 3–4 show that the second term of Equation 6–8 can be significantly large.

At low pressures, the z-factor decreases as pressure increases. Therefore, the partial derivative of z-factor with respect to p is negative, and c_g is larger than in the case of an ideal gas. At high pressures, however, z-factor increases with pressure; the partial derivative of z-factor with respect to p is positive, and c_g is less than in the case of an ideal gas.

EXAMPLE 6–6: *Compute the coefficient of isothermal compressibility of methane at 1000 psia and 68°F.*

Solution

First, use the data of Figure 3–2 to determine z and $\left(\frac{\partial p}{\partial z}\right)_T$

$$z = 0.890, \text{ Figure } 3-2$$

$$\left(\frac{\partial z}{\partial p}\right)_T = -0.000106 \text{ psi}^{-1} = \text{ slope of tangent to } 68°F$$

isotherm at p = 1000 psia, Figure 3–2

Second, calculate c_g

$$c_g = \frac{1}{p} - \frac{1}{z}\left(\frac{\partial z}{\partial p}\right)_T \qquad (6-8)$$

$$c_g = \frac{1}{1000 \text{ psia}} - \left(\frac{1}{0.890}\right)(-0.000106 \text{ psi}^{-1}) = 1,120 \times 10^{-6} \text{ psi}^{-1}$$

Pseudoreduced Compressibility

The Law of Corresponding States can be used to put Equation 6–8 into reduced form. Equation 3–43 can be used to replace the pressure term.

$$p = p_{pc}p_{pr} \tag{3-43}$$

In order to place the partial derivative into reduced form, we must rely on the chain rule.

$$\left(\frac{\partial z}{\partial p}\right) = \left(\frac{\partial p_{pr}}{\partial p}\right)\left(\frac{\partial z}{\partial p_{pr}}\right) \tag{6-9}$$

From Equation 3–43

$$\left(\frac{\partial p_{pr}}{\partial p}\right) = \frac{1}{p_{pc}}, \tag{6-10}$$

so that

$$\left(\frac{\partial z}{\partial p}\right) = \frac{1}{p_{pc}}\left(\frac{\partial z}{\partial p_{pr}}\right) \tag{6-11}$$

Combination of Equations 6–8, 3–43, and 6–11 yields

$$c_g = \frac{1}{p_{pc}p_{pr}} - \frac{1}{zp_{pc}}\left(\frac{\partial z}{\partial p_{pr}}\right)_{T_{pr}} \tag{6-12}$$

or

$$c_g p_{pc} = \frac{1}{p_{pr}} - \frac{1}{z}\left(\frac{\partial z}{\partial p_{pr}}\right)_{T_{pr}}. \tag{6-13}$$

Since the dimensions of c_g are reciprocal pressure, the product of c_g and p_{pc} is dimensionless. This product is called *pseudoreduced compressibility*, c_{pr}.

$$c_{pr} = c_g p_{pc} = \frac{1}{p_{pr}} - \frac{1}{z}\left(\frac{\partial z}{\partial p_{pr}}\right)_{T_{pr}} \tag{6-14}$$

Pseudoreduced compressibility is a function of z-factor and pseudoreduced pressure. Thus, a graph relating z-factor to pseudoreduced pressure, Figure 3–7, Figure 3–8, or Figure 3–9, can be used with Equation 6–14 to calculate values of c_{pr}.

EXAMPLE 6–7: *Calculate the coefficient of isothermal compressibility of the dry gas given in Example 6–3 at 220°F and 2100 psig.*

Solution
First, determine the pseudocritical temperature and pseudocritical pressure.

$$T_{pc} = 406°R, \quad p_{pc} = 647 \text{ psia, from Example 6–3}$$

Second, calculate pseudoreduced temperature and psuedoreduced pressure.

$$T_{pr} = 1.68, \quad p_{pr} = 3.27, \text{ from Example 6–3}$$

Third, determine z and $\left(\dfrac{\partial z}{\partial p_{pr}}\right)_{T_{pr}}$.

$$z = 0.855, \text{ Figure 3–7}$$

$\left(\dfrac{\partial z}{\partial p_{pr}}\right)_{T_{pr}} = $ slope of 1.68 isotherm at p_{pr} of 3.27 = -0.0132, Figure 3–7

Fourth, calculate c_{pr}.

$$c_{pr} = \frac{1}{p_{pr}} - \frac{1}{z}\left(\frac{\partial z}{\partial p_{pr}}\right)_{T_{pr}} \qquad (6\text{–}14)$$

$$c_{pr} = \frac{1}{3.27} - \frac{1}{0.855}(-0.0132) = 0.321$$

Fifth, calculate c_g.

$$c_g = \frac{c_{pr}}{p_{pc}} \qquad (6\text{–}14)$$

$$c_g = \frac{0.321}{647 \text{ psia}} = 497 \times 10^{-6} \text{ psi}^{-1}$$

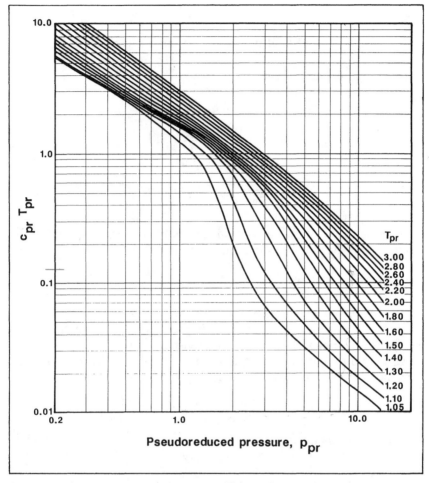

Fig. 6–4. Pseudoreduced compressibilities of natural gases.

Calculations of the type given in the Example 6–7 have been made using an equation which represents the data of Figure 3–7.[1] The results of these calculations are given in Figure 6–4.

Values of pseudocritical temperature and pseudocritical pressure are required in the use of Figure 6–4. Methods presented in Chapter 3, including adjustments for carbon dioxide and hydrogen sulfide, can be used to estimate these properties.

EXAMPLE 6–8: *Repeat Example 6–7 using $c_{pr}T_{pr}$ correlation.*

Solution

$$T_{pr} = 1.68 \text{ and } p_{pr} = 3.27 \text{ from Example } 6-7$$

using appropriate $c_{pr}T_{pr}$ chart.

$$c_{pr}T_{pr} = 0.528, \text{ Figure } 6-4$$

$$c_{pr} = \frac{0.528}{1.68} = 0.314$$

$$c_g = \frac{c_{pr}}{p_{pc}} = \frac{0.314}{647 \text{ psia}} = 486 \times 10^{-6} \text{ psi}^{-1} \qquad (6-14)$$

The accuracy of these calculations has not been tested; however, the results should be suitable for engineering purposes.

The Coefficient of Viscosity of Gas

The *coefficient of viscosity* is a measure of the resistance to flow exerted by a fluid. Usually, *viscosity* is given in units of centipoise. A centipoise is a g mass/100 sec cm. This viscosity term is called *dynamic viscosity* to differentiate it from *kinematic viscosity*, which is defined as dynamic viscosity divided by the density of the fluid.

$$\text{kinematic viscosity } \nu = \frac{\text{dynamic viscosity } \mu}{\text{density } \rho_g} = \frac{\text{centipoise}}{\text{g/cc}} \qquad (6-15)$$

Usually, kinematic viscosity is given in centistokes. The centistoke is defined as centipoise divided by density in g/cc. Thus, the units of centistoke are sq cm/100 sec.

Gas viscosity decreases as reservoir pressure decreases. The molecules are simply further apart at lower pressure and move past each other more easily. The relationship for a dry gas or wet gas is given in Figure 6–5.

The figure also shows the effect of temperature on viscosity. At low pressures, gas viscosity increases as temperature increases. However, at high pressures gas viscosity decreases as temperature increases. The reciprocal of viscosity is called *fluidity*.

Experimental determination of gas viscosity is difficult. Usually, the petroleum engineer must rely on viscosity correlations. We will look at correlations of gas viscosity data which apply to the gases normally encountered in petroleum reservoirs.

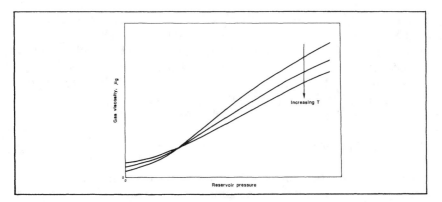

Fig. 6–5. Typical shape of gas viscosity as a function of pressure at three reservoir temperatures.

Viscosity of Pure Hydrocarbon Gases

Figure 6–6 gives the viscosity of ethane.[2] Note the similarity between this figure and the graph of the densities of pure hydrocarbons given in Chapter 2. The dotted line is the saturation line, and the point of maximum temperature on the dotted line indicates the critical point.

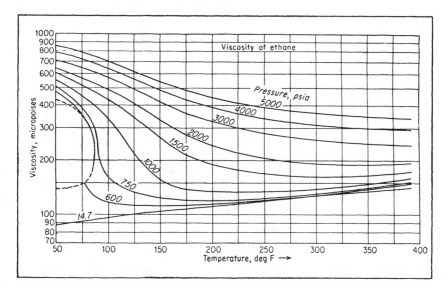

Fig. 6–6. Viscosities of ethane. (From *Handbook of Natural Gas Engineering* by Katz et al. Copyright 1959 by McGraw-Hill Book Co. Used with permission of McGraw-Hill Book Co.)

Note that the viscosity of the saturated liquid is equal to the viscosity of the saturated vapor at the critical point. The isobars above the saturation line give the viscosity of liquid ethane, and the isobars below the saturation line give the viscosity of ethane gas.

The similarity of this graph to the graph showing the density of a pure substance indicates that the Law of Corresponding States should hold for viscosity as well as for volumetric behavior.

Viscosity of Gas Mixtures

The following equation may be used to calculate the viscosity of a mixture of gases when the composition of the gas mixture is known and the viscosities of the components are known at the pressure and temperature of interest.[3]

$$\mu_g = \frac{\sum\limits_j \mu_{gj} y_j M_j^{1/2}}{\sum\limits_j y_j M_j^{1/2}} \tag{6-16}$$

Figure 6–7 may be used to obtain the viscosities of the usual constituents of natural gas at atmospheric pressure for use in Equation 6–16.[4]

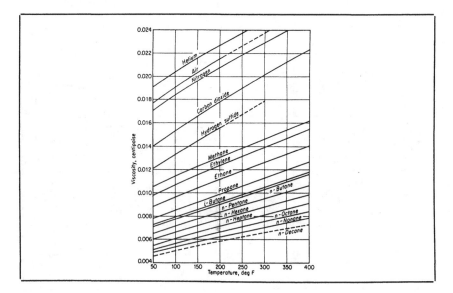

Fig. 6–7. Viscosities of pure gases at atmospheric pressure.

EXAMPLE 6–9: *Calculate the viscosity of the gas mixture given below at 200°F and a pressure of one atmosphere absolute.*

Component	Composition, mole fraction
Methane	0.850
Ethane	0.090
Propane	0.040
n-Butane	0.020
	1.000

Solution

First, determine the viscosities of the individual gases at 200°F and one atmosphere.

$$\mu_{gC1} = 0.0130 \text{ cp}$$
$$\mu_{gC2} = 0.0112 \text{ cp}$$
$$\mu_{gC3} = 0.0098 \text{ cp}$$
$$\mu_{gC4} = 0.0091 \text{ cp, Figure } 6–7$$

Second, calculate viscosity of the gas mixture.

Component	y_j	M_j	$M_j^{1/2}$	$y_jM_j^{1/2}$	μ_{gj}	$\mu_{gj}y_jM_j^{1/2}$
C_1	0.850	16.04	4.00	3.404	0.0130	0.0443
C_2	0.090	30.07	5.48	0.494	0.0112	0.0055
C_3	0.040	44.10	6.64	0.266	0.0098	0.0026
n-C_4	0.020	58.12	7.62	0.152	0.0091	0.0014
	1.000			$\Sigma y_jM_j^{1/2} = 4.316$	$\Sigma\mu_jy_jM_j^{1/2} =$	0.0538

$$\mu_g = \frac{\sum\limits_j \mu_{gj}y_jM_j^{1/2}}{\sum\limits_j y_jM_j^{1/2}} \tag{6–16}$$

$$\mu_g = \frac{0.0538}{4.316} = 0.0125 \text{ cp}$$

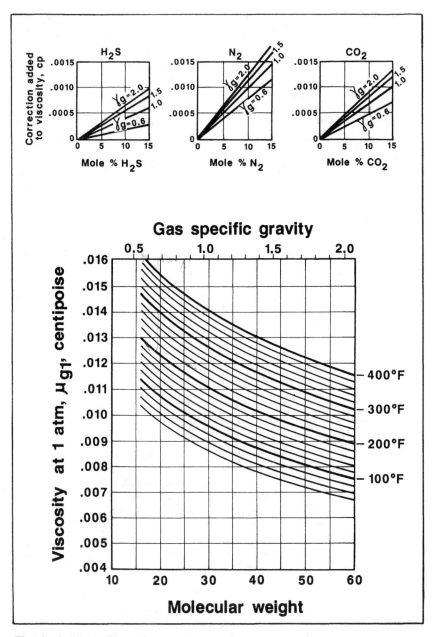

Fig. 6–8. Viscosities of natural gases at atmospheric pressure. (Adapted from Carr et al., *Trans.*, AIME, *201*, 997.)

If the composition of the gas is unavailable, Figure 6–8 can be used to obtain viscosities of mixtures of hydrocarbon gases at one atmosphere pressure.[5] Insert plots show corrections to the value of hydrocarbon viscosity which may be applied to take into account the effect of the presence of hydrogen sulfide, nitrogen, or carbon dioxide. The effect of each of the nonhydrocarbon gases is to increase the viscosity of the gas mixture.

Values of viscosity obtained from Figure 6–8 have been compared with the limited available experimental data. The agreement was quite good for hydrocarbon gases, within less than two percent. The presence of nonhydrocarbon components caused the deviation between Figure 6–8 and experimental viscosities to be somewhat greater than two percent, even after the corrections in the inset figures were applied.

EXAMPLE 6–10: *Use Figure 6–8 to rework Example 6–9.*

Solution

First, calculate the specific gravity of the gas.

$$M_a = \sum_j y_j M_j \qquad (3-35)$$

Component	y_j	M_j	$y_j M_j$
C_1	0.85	16.04	13.63
C_2	0.09	30.07	2.71
C_3	0.04	44.10	1.76
$n\text{-}C_4$	0.02	58.12	1.16
	1.02	$=$	$19.26 = M_a$

$$\gamma_g = \frac{M_a}{29} \qquad (3-38)$$

$$\gamma_g = \frac{19.26}{29} = 0.664$$

Second, determine viscosity.

$$\mu_{g1} = 0.0125 \text{ cp at } 200°F, \text{ Figure } 6-8$$

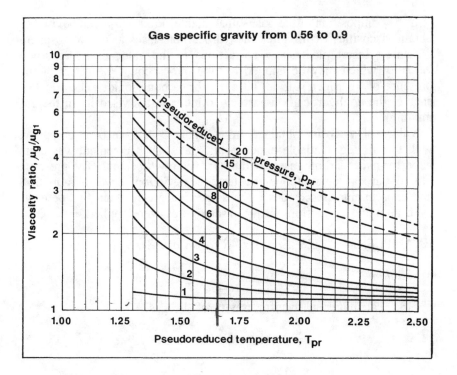

Fig. 6–9. Viscosity ratios for natural gases with specific gravities from 0.56 to 0.9

Gas Viscosity at High Pressure

In most instances, the petroleum engineer is concerned with the viscosity of gases at pressures far removed from one atmosphere, so we must now turn to a method of calculating gas viscosity at high pressure.

The Law of Corresponding States has been used to develop a gas viscosity correlation.[6] This correlation is given as Figures 6–9 through 6–12. Four figures are required to cover the specific gravity range. These figures give a viscosity ratio, μ_g/μ_{g1}, which is multiplied by the viscosity at one atmosphere to determine viscosity at high pressure.

Dry gases, wet gases, and gases separated from black oils normally have specific gravities which lie within the range of Figure 6–9. Values of viscosity obtained from this figure and Figure 6–8 agree remarkably

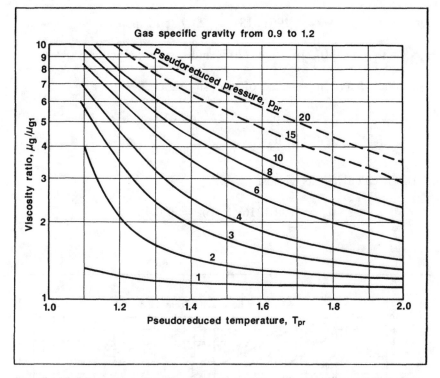

Fig. 6–10. Viscosity ratios for natural gases with specific gravities from 0.9 to 1.2.

well with experimental viscosity data. At low pseudoreduced pressure the agreement is within two percent. At the higher values of pseudoreduced pressure the agreement is within three to four percent.

Retrograde gases and gases associated with volatile oils generally have specific gravities greater than 1.0. Figures 6–10 through 6–12 can be used to obtain viscosity ratios for these gases. Agreement with experimental viscosity data becomes increasingly less accurate as gas specific gravity increases. Figure 6–12 has an accuracy to within 20 percent, with results usually lower than published viscosity data.

EXAMPLE 6–11: *Calculate a value of viscosity for the dry gas of Example 6–3.*

Solution

First, determine a value of gas viscosity at one atmosphere.

$$M_a = 29(0.818) = 23.7 \text{ lb/lb mole} \qquad (3\text{--}38)$$

$\mu_{g1} = 0.01216$ cp at 220°F and one atm, Figure 6–8

Second, determine a value of viscosity ratio.

$$T_{pr} = 1.68, \ p_{pr} = 3.27, \text{ from Exercise 6–3}$$

$$\frac{\mu_g}{\mu_{g1}} = 1.50, \text{ Figure 6–9}$$

Third, calculate gas viscosity.

$$\mu_g = (0.01216)(1.50) = 0.0182 \text{ cp}$$

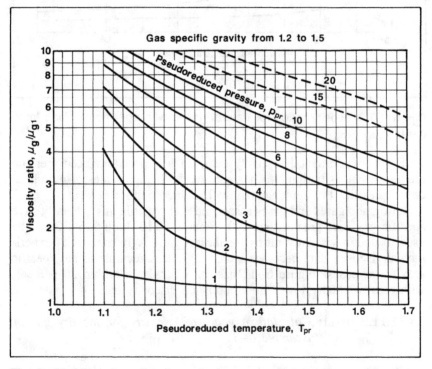

Fig. 6–11. Viscosity ratios for natural gases with specific gravities from 1.2 to 1.5.

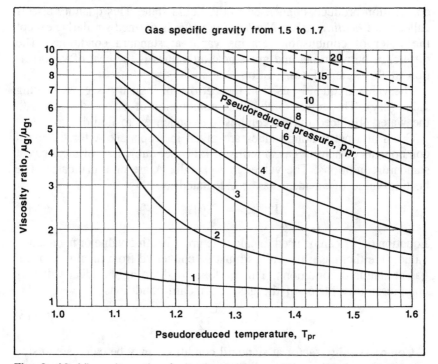

Fig. 6–12. Viscosity ratios for natural gases with specific gravities from 1.5 to 1.7.

Heating Value

The *heating value* of a gas is the quantity of heat produced when the gas is burned completely to carbon dioxide and water. Heating value usually is expressed as British thermal units per standard cubic foot of gas, BTU/scf.

The petroleum industry uses four adjectives to describe heating values: *wet*, *dry*, *gross*, and *net*.[7]

When used with heating values, the words wet and dry refer to the condition of the gas prior to combustion. *Wet* means that the gas is saturated with water vapor, about 1.75 volume percent. And *dry* means that the gas contains no water vapor. Sometimes, the term *bone dry* is used.

The words gross and net refer to the condition of the water of combustion after burning takes place. *Gross heating value* is the heat produced in complete combustion under constant pressure with the combustion products cooled to standard conditions and the water in the combustion products condensed to the liquid state. This quantity also is called *total heating value*. *Net heating value* is defined similarly, except the water of combustion remains vapor at standard conditions. The difference between net and gross heating values is the heat of vaporization of the water of combustion.

The value frequently used in the petroleum industry is gross heating value (dry). This refers to complete combustion of a bone-dry gas with the water produced during combustion condensed to liquid.

Gross and net heating values are given in Appendix A for the usual components of natural gas in terms of BTU per cubic foot of ideal gas at standard conditions. The values in Appendix A are on a dry basis. The heating value of an ideal gas is calculated as

$$L_{c\ ideal} = \sum_j y_j L_{cj} .$$
(6–17)

Equation 6–17 can be used to calculate gross heating value or net heating value. In either case, the values must be converted from ideal gas to real gas at standard conditions. This is done by dividing the ideal value by compressibility factor of the gas at standard conditions.

$$L_c = \frac{L_{c\ ideal}}{z}$$
(6–18)

Compressibility factor at standard conditions may be calculated using z-factors tabulated in Appendix A, where

$$z = 1 - (\sum_j y_j \sqrt{1 - z_j})^2.$$
(6–19)

EXAMPLE 6–12: *Calculate the gross heating value (dry) of a separator gas of composition given below.*

Component	Composition, mole fraction
Carbon dioxide	0.0167
Nitrogen	0.0032
Methane	0.7102
Ethane	0.1574
Propane	0.0751
i- Butane	0.0089
n-Butane	0.0194
i- Pentane	0.0034
n-Pentane	0.0027
Hexanes	0.0027
Heptanes plus	0.0003
	1.0000

Property of heptanes plus	
Molecular weight	103 lb/lb mole

Solution

First, calculate gross heating value of ideal gas and compressibility factor of the gas at standard conditions.

$$L_{c\ ideal} = \sum_j y_j L_{cj} \qquad (6-17)$$

Component	Mole fraction y_j	Gross Heating Value,* BTU/scf L_{cj}	$y_j L_{cj}$	Compressibility Factor Standard Conditions z_j	$y_j\sqrt{1-z_j}$
CO_2	0.0167	0.0	0.0	0.9943	0.00126
N_2	0.0032	0.0	0.0	0.9997	0.00006
C_1	0.7102	1010.0	717.3	0.9980	0.03176
C_2	0.1574	1769.6	278.5	0.9919	0.01417
C_3	0.0751	2516.1	189.0	0.9825	0.00993
i- C_4	0.0089	3251.9	28.9	0.9711	0.00151
n-C_4	0.0194	3262.3	63.3	0.9667	0.00354
i- C_5	0.0034	4000.9	13.6	0.9480	0.00078
n-C_5	0.0027	4008.9	10.8	0.9420	0.00065
C_6	0.0027	4755.9	12.8	0.9100	0.00081
C_{7+}	0.0003	5502.5	1.7	0.8520	0.00012
	1.0000		1315.9 BTU/scf		0.06459

*From Appendix A.

$$z = 1 - \left(\sum_j y_j\sqrt{1 - z_j}\right)^2 \qquad (6-19)$$

$$z = 1 - (0.06459)^2 = 0.9958$$

Second, calculate gross heating value of the gas as a real gas.

$$L_c = \frac{L_{c\ ideal}}{z} \qquad (6-18)$$

$$L_c = (1315.9\ BTU/scf)/0.9958 = 1321\ BTU/scf$$

The change from ideal gas to real gas heating values usually is less than ½ percent and normally is ignored.

Net heating value (dry) may be converted into net heating value (wet) by

$$L_{c(wet)} = (1 - 0.0175) L_{c(dry)} , \qquad (6-20)$$

where 0.0175 is the mole fraction of water vapor in gas when saturated at standard conditions. Gross heating value (dry) may be converted into gross heating value (wet) by

$$L_{c(wet)} = (1 - 0.0175) L_{c(dry)} + 0.9 , \qquad (6-21)$$

where 0.9 accounts for the heat released (BTU/scf) during condensation of the water vapor which was in the gas prior to combustion.

Joule-Thomson Effect

Temperature changes as pressure is reduced when a flowing stream of gas passes through a throttle, i.e., a valve, choke, or perforations in casing. This is called the *Joule-Thomson effect*. The change in temperature is directly related to the attraction of the molecules for each other.

The equation

$$\Delta T = \frac{T \left(\dfrac{\partial V_M}{\partial T}\right)_p - V_M}{C_p} \Delta p \qquad (6-22)$$

gives the change in temperature as pressure is changed.[8] This equation is valid when the pressure change is adiabatic, that is, no heat enters or leaves the system, and when pressure change is small.

The term C_p is the heat capacity of the gas at constant pressure; it is related to the energy required to overcome molecular attraction as the gas expands due to heating at constant pressure.

Substitution of the compressibility equation of state

$$pV_M = zRT \qquad (3-39)$$

into Equation 6-22 results in

$$\Delta T = \frac{\dfrac{V_M T}{z} \left(\dfrac{\partial z}{\partial T}\right)_p}{C_p} \Delta p . \qquad (6-23)$$

The terms V, T, z, and C_p are always positive. Thus, the direction of the change in temperature depends on the signs of both the Δp term and the derivative, $(\partial z/\partial T)_p$.

Petroleum engineers usually deal with situations in which pressure decreases. Flow of gas through restrictions such as a valve or choke or through perforations at the bottom of the well are examples. Consequently, we will examine Equation 6–23 for negative values of Δp.

Examination of Figure 3–7 shows that at moderate pressure, say values of pseudoreduced pressure less than about 6.0, z-factor increases as temperature increases at constant pressure. That is, at pseudoreduced pressures less than about 6.0, the derivative of z-factor with respect to T is positive.

Use of a negative value of Δp and a positive value of $(\partial z/\partial T)_p$ in Equation 6–23 causes ΔT to be negative. That is, temperature *decreases* as pressure decreases.

At high pressure (for instance, at pseudoreduced pressures greater than about 11.0 on Figure 3–7 or Figure 3–8) z-factor decreases as temperature increases. Under these conditions, the derivative of z-factor with respect to T is negative. Thus, Equation 6–23 indicates that temperature *increases* as pressure decreases.

There is a range of pressures in which the isotherms on Figure 3–7 cross. This transition zone occurs between 5000 psia and 7000 psia for natural gases.

At high pressure, temperature *increases* as pressure is decreased. The effect is greater at higher pressures. This situation occurs when high-pressure gas expands through perforations in the casing in a deep gas well. Temperature surveys in the well show "hot spots" where the greatest amount of gas is entering the wellbore.

At pressures below about 5000 psia, temperature *decreases* as pressure is decreased. The greatest temperature decrease occurs when the pressure decrease starts at values between 1500 and 2000 psia. This is observed as gas flows through the surface choke at common surface conditions.

Exercises

6–1. A vessel contains 100 lb moles of a hydrocarbon gas. Calculate the standard cubic feet of gas in the vessel.

6–2. List the standard conditions used in your state. List the standard conditions used in this book. Convert a pressure of 10 atmospheres (absolute) to psia.

6–3. The West Panhandle Gas Field is in the panhandle of Texas. Atmospheric pressure in the area averages about 13.2 psia. What are the standard conditions used to calculate gas quantities produced from this field? A pressure gauge on a wellhead reads 47.0. What is the pressure in psia? What is the pressure in atmospheres (absolute)?

6–4. Calculate the standard cubic feet in the cylinder of Example 3–6.

6–5. Recalculate Exercise 3–27. Give answer in scf.

6–6. Calculate the formation volume factor of the ethane of Example 3–7 at 918 psia and 117°F.

6–7. Calculate the formation volume factor of the gas of Example 3–10 at 3810 psia and 194°F.

6–8. The wells which produce the gas of Exercise 6–7 (Example 3–10) are drilled on 640-acre spacing. How much gas, scf, is contained in the drainage area of one well which has an average formation thickness of 21 feet, average porosity of 18 percent, and average water saturation of 33 percent.

6–9. Calculate the formation volume factor of the gas of Example 3–10 at 5000 psig and 194°F.

6–10. Calculate the formation volume factor of the gas of Exercise 3–28 at 5420 psig and 257°F.

6–11. Calculate the formation volume factor of the gas of Exercise 3–30 at 90°F and 455 psia.

6–12. Compressed into a 6-inch (ID=5.761 inches) pipeline at 1000 psia is 2.4 MMscfd of the separator gas of page 9 of Table 10–1. Assume the temperature is 100°F. Calculate the actual volumetric flow rate in cubic feet per day, the mass flow rate in pounds per hour, and the superficial gas velocity (average linear velocity) in feet per second. Assume the heptanes plus fraction has the properties of n-heptane.

6–13. Determine a value of the coefficient of isothermal compressibility of the gas of Example 6–4 at 1100 psia and 150°F.

6–14. You need an estimate of the coefficient of isothermal compressibility of a gas at 1100 psia and 150°F. Unfortunately, you don't have the correlations in this book with you. What value do you use?

6–15. A modified form of the virial equation of state is

$$V_M = \frac{RT}{p} + B,$$

where B is the second virial coefficient. The second virial coefficient of a certain gas is -2.12 cu ft/lb mole. Derive an equation for the coefficient of isothermal compressibility of the gas in terms of the virial equation. Calculate the coefficient of isothermal compressibility at 200 psia and 100°F. Compare with the compressibility of an ideal gas.

6–16. Compute a value of the coefficient of isothermal compressibility of ethane at 1000 psia and 284°F.

6–17. The molar volumes of a dry gas, measured at 200°F, are given below. What is the coefficient of isothermal compressibility of the gas at 900 psia and 200°F?

Pressure, psia	Molar volume, cu ft/lb mole
600	10.6
700	9.7
800	8.9
900	8.4
1000	8.1
1100	7.9
1200	7.8

Compare your result with the compressibility of an ideal gas at 900 psia and 200°F.

6–18. Determine a value of the coefficient of isothermal compressibility for the gas of Example 3–10 at 3335 psia and 170°F. Use Figure 3–7 and Equation 6–14.

6–19. Calculate the coefficient of isothermal compressibility of the gas of Exercise 3–28 at 5420 psig and 257°F. Use Figure 6–4.

6–20. Calculate the coefficient of isothermal compressibility of a gas with specific gravity of 0.862 at 3025 psia and 175°F. Use Figure 6–4.

6–21. Calculate the coefficient of isothermal compressibility of the gas of Exercise 3–30 at 455 psia and 90°F. Use Figure 6–4.

6–22. Calculate the viscosity at one atmosphere of the dry gas of Example 3–10 at 194°F. Use Equation 6–16 and Figure 6–7.

6–23. Repeat Exercise 6–22. Use Figure 6–8.

6–24. Calculate the viscosity of the gas of Exercise 6–22 at 3810 psia and 194°F.

6–25. Estimate the coefficient of viscosity of the gas of Exercise 3–28 at 5420 psig and 257°F.

6–26. Estimate the coefficient of viscosity of the gas of Exercise 6–20 at 3025 psia and 175°F.

6–27. Estimate the coefficient of viscosity of the gas of Exercise 3–30 at 455 psia and 90°F.

6–28. Calculate the gross heating value (dry) of the gas of Example 3–5.

6–29. Calculate the gross heating value (dry) of the gas of Exercise 3–28.

6–30. Calculate the gross heating value (dry) of the gas of Example 3–10.

References

1. Dranchuk, P.M. and Abou-Kassem, J.H.: "Calculation of z-Factors for Natural Gases Using Equations of State," *J. Can. Pet. Tech.* (July–Sept, 1975) *14*, No. 3, 34-36.
2. Smith, A.S. and Brown, G.G.: "Correlating Fluid Viscosity," *Ind. Eng. Chem.* (June 1943) *35*, 705–711.
3. Herning, F. and Zipperer, L.: "Calculations of the Viscosity of Technical Gas Mixtures from the Viscosity of Individual Gases," *Gas. and Wasserfach* (1936) *79*, 49–69.
4. Katz, D.L., et al.: *Handbook of Natural Gas Engineering,* McGraw-Hill Book Co., Inc., New York City (1959).
5. Carr, N.L., Kobayashi, R., and Burrows, D.B.: "Viscosity of Hydrocarbon Gases Under Pressure," *Trans.,* AIME (1954) *201*, 264–272.
6. Lee, A.J., Gonzales, M.H., and Eakin, B.E.: "The Viscosity of Natural Gases,: *Trans.,* AIME (1966) *237*, 997–1000.
7. McClanahan, D.N.: "Gas Heating Value: What It Is, How to Measure, Calculate," *Oil and Gas J.* (Feb. 20, 1967) 84–89.
8. Taylor, H.S. and Glasstone, S.: *A Treatise on Physical Chemistry Vol. II—States of Matter*, 3rd Ed., D. van Nostrand Co. Inc., New York City (1951).

7

Properties of Wet Gases

The key to the analysis of the properties of a wet gas is that the properties of the surface gas are not the same as the properties of the reservoir gas. Liquid condenses from the reservoir gas as it moves from reservoir conditions to surface conditions. Thus the composition of the surface gas is quite different from the composition of the reservoir gas, having considerably less of the intermediate and heavy components.

Wet gases are usually separated in two-stage separation systems such as Figure 13–1. At the surface, the well stream is separated into stock-tank liquid (condensate), separator gas, and stock-tank gas. All three of these fluids must be included in the recombination calculation.

If a three-stage separator system is used, all three gases must be included in the recombination calculation.

Recombination of Surface Fluids—Compositions Known

The surface liquid and gases must be recombined in order to determine the properties of the gas in the reservoir. This recombination is done by calculation if the compositions of the surface fluids are known.

Surface Composition–Known

The composition of the reservoir gas can be calculated given the compositions of the stock-tank liquid, separator gas, and stock-tank vent gas. The producing gas-oil ratios must be known also. The calculation simulates laboratory recombination of the liquid and gases in quantities indicated by the gas-oil ratios.

Once the composition of the reservoir gas has been calculated, its physical properties can be calculated as for a dry gas.

The first step is conversion of gas-oil ratios in scf/STB to lb mole gas/lb mole stock-tank liquid. To accomplish this, the density and apparent molecular weight of the stock-tank liquid must be calculated. Liquid density calculations are discussed in Chapter 11. Apparent molecular weight of a liquid is calculated from composition exactly as for a gas, Equation 3–35.

These gas-oil ratios in terms of lb moles are used to combine the compositions of the separator gas, stock-tank gas, and stock-tank oil in the proper ratios. An example will best illustrate the procedure.

EXAMPLE 7–1: *A wet gas produces through a separator at 300 psia and 73°F to a stock tank at 76°F. The separator produces 69,551 scf/STB and the stock tank vents 366 scf/STB. The stock-tank liquid gravity is 55.9°API. Compositions are given below. Calculate the composition of the reservoir gas.*

Component	Composition, separator gas, mole fraction	Composition, stock-tank gas, mole fraction	Composition, stock-tank liquid, mole fraction
C_1	0.8372	0.3190	0.0018
C_2	0.0960	0.1949	0.0063
C_3	0.0455	0.2532	0.0295
i-C_4	0.0060	0.0548	0.0177
n-C_4	0.0087	0.0909	0.0403
i-C_5	0.0028	0.0362	0.0417
n-C_5	0.0022	0.0303	0.0435
C_6	0.0014	0.0191	0.0999
C_{7+}	0.0002	0.0016	0.7193
	1.0000	1.0000	1.0000

Properties of heptanes plus of the stock-tank liquid
Specific gravity　　　　　　　　　　　　　　　　　　　　　　　　　0.794
Molecular weight　　　　　　　　　　　　　　　　　113 lb/lb mole

Solution

First, calculate the apparent molecular weight and density of the stock-tank liquid.

M_{STO} = 100.9 lb/lb mole, Equation 3–35　 pg.103

ρ_{STO} = 47.11 lb/cu ft, from 55.9°API, Equations 8–2, 8–1

Second, convert gas-oil ratios to lb mole gas/lb mole stock-tank liquid.

Table for Example 7-1

Component	Composition, separator gas, mole fraction	lb mole j SP gas	Composition, stock-tank gas, mole fraction	lb mole j ST gas	Composition, stock-tank oil mole fraction	lb mole j res gas	Composition, recombined gas, mole fraction
	lb mole j SP gas	lb mole STO	lb mole j ST gas	lb mole STO	lb mole j STO	lb mole STO	lb mole j res gas
	y_{SP}	$69.69\,y_{SP}$	y_{ST}	$0.3667\,y_{ST}$	x_{ST}	$69.69\,y_{SP}$ $+\,0.3667\,y_{ST} + x_{ST}$	y_{res}
C_1	0.8372	58.344	0.3190	0.117	0.0018	58.463	0.8228
C_2	0.0960	6.690	0.1949	0.072	0.0063	6.768	0.0952
C_3	0.0455	3.171	0.2532	0.093	0.0295	3.294	0.0464
i-C_4	0.0060	0.418	0.0548	0.020	0.0177	0.456	0.0064
n-C_4	0.0087	0.606	0.0909	0.033	0.0403	0.679	0.0096
i-C_5	0.0028	0.195	0.0362	0.013	0.0417	0.250	0.0035
n-C_5	0.0022	0.153	0.0303	0.011	0.0435	0.208	0.0029
C_6	0.0014	0.098	0.0191	0.007	0.0999	0.205	0.0029
C_{7+}	0.0002	0.014	0.0016	0.001	0.7193	0.734	0.0103
	1.0000	69.690	1.0000	0.367	1.0000	71.057	1.0000

$$\left(69,551\frac{\text{scf SP gas}}{\text{STB}}\right)\left(\frac{\text{lb mole}}{380.7 \text{ scf}}\right)\left(\frac{\text{bbl}}{5.615 \text{ cu ft}}\right)\left(\frac{\text{cu ft STO}}{47.11 \text{ lb STO}}\right)$$

$$\left(\frac{100.9 \text{ lb STO}}{\text{lb mole STO}}\right) = 69.69 \ \frac{\text{lb mole SP gas}}{\text{lb mole STO}}$$

$$\left(366 \ \frac{\text{scf ST gas}}{\text{STB}}\right)\left(\frac{\text{lb mole}}{380.7 \text{ scf}}\right)\left(\frac{\text{bbl}}{5.615 \text{ cu ft}}\right)\left(\frac{\text{cu ft STO}}{47.11 \text{ lb STO}}\right)$$

$$\left(\frac{100.9 \text{ lb STO}}{\text{lb mole STO}}\right) = 0.3667 \ \frac{\text{lb mole ST gas}}{\text{lb mole STO}}$$

Third, calculate mole fraction of the recombined reservoir gas. See Table for Example 7–1.

The composition of the reservoir gas as obtained in Example 7–1 is used in correlations previously given to estimate z-factors, gas viscosities, and other properties.

Separator Compositions Known

Often samples of gas and liquid are taken from the primary separator. The compositions of these two fluids can be recombined in the manner of Example 7–1. Sometimes the separator gas-oil ratio is given in scf/STB. If so, it must be converted to scf/SP bbl by dividing by the ratio of volume of separator liquid to volume of stock-tank liquid which is given normally as a part of the compositional analysis.

EXAMPLE 7–2: *Separator gas and separator liquid are sampled from a separator operating at 300 psia and 73°F. The producing gas-oil ratio of the separator at the time of sampling was 69,551 scf/STB. Laboratory analysis gave the following compositions. Also the laboratory reported a separator/stock-tank volume ratio of 1.216 bbl SP liquid at separator conditions per STB at standard conditions.*

Component	Composition, separator gas, mole fraction	Composition, separator liquid, mole fraction
C_1	0.8372	0.0869
C_2	0.0960	0.0569
C_3	0.0455	0.0896
i-C_4	0.0060	0.0276
n-C_4	0.0087	0.0539
i-C_5	0.0028	0.0402
n-C_5	0.0022	0.0400
C_6	0.0014	0.0782
C_{7+}	0.0002	0.5267
	1.0000	1.0000

Properties of heptanes plus of separator liquid
Specific gravity   0.794
Molecular weight   113 lb/lb mole

You may assume the heptanes plus fraction of separator gas has a molecular weight of 103 lb/lb mole.

Solution

First, calculate the separator gas-oil ratio in terms of separator liquid.

$$\left(69{,}551 \frac{\text{scf SP gas}}{\text{STB}}\right) \left(\frac{\text{STB}}{1.216 \text{ SP bbl}}\right) = 57{,}197 \frac{\text{scf SP gas}}{\text{SP bbl}}$$

Second, calculate the apparent molecular weight and density of the separator liquid (at separator conditions).

$M_a = 83.8$ lb/lb mole, Equation 3–35

$\rho_o = 43.97$ lb/cu ft at 300 psia and 73°F, procedure of Chapter 11

Third, convert gas-oil ratio to lb mole SP gas/lb mole SP liquid.

$$\left(57{,}197 \frac{\text{scf SP gas}}{\text{SP bbl}}\right) \left(\frac{\text{lb mole}}{380.7 \text{ scf}}\right) \left(\frac{\text{bbl}}{5.615 \text{ cu ft}}\right) \left(\frac{\text{cu ft SP liq}}{43.97 \text{ lb SP liq}}\right)$$

$$\left(\frac{83.8 \text{ lb SP liq}}{\text{lb mole SP liq}}\right) = 50.99 \frac{\text{lb mole SP gas}}{\text{lb mole SP liq}}$$

Fourth, calculate the composition of the recombined reservoir gas.

Component	Separator gas, mole fraction		Separator liquid, mole fraction		Recombined gas, mole fraction
	$\dfrac{\text{lb mole } j \text{ SP gas}}{\text{lb mole SP gas}}$	$\dfrac{\text{lb mole } j \text{ SP gas}}{\text{lb mole SP liq}}$	$\dfrac{\text{lb mole } j \text{ SP liq}}{\text{lb mole SP liq}}$	$\dfrac{\text{lb mole } j \text{ in well stream}}{\text{lb mole SP liq}}$	$\dfrac{\text{lb mole } j \text{ in well stream}}{\text{lb mole well stream}}$
	y_j	$50.99 y_j$	x_j	$x_j + 50.99 y_j$	z_j
C_1	0.8372	42.689	0.0869	42.776	0.8228
C_2	0.0960	4.895	0.0569	4.952	0.0952
C_3	0.0455	2.320	0.0896	2.410	0.0464
i- C_4	0.0060	0.306	0.0276	0.334	0.0064
n- C_4	0.0087	0.444	0.0539	0.498	0.0096
i- C_5	0.0028	0.143	0.0402	0.183	0.0035
n- C_5	0.0022	0.112	0.0400	0.152	0.0029
C_6	0.0014	0.071	0.0782	0.149	0.0029
C_{7+}	0.0002	0.010	0.5267	0.537	0.0103
	1.0000	50.990	1.0000	51.990	1.0000

Recombination of Surface Fluids—Compositions Unknown

If compositional analysis is unavailable, the engineer must rely on production data to estimate the specific gravity of the reservoir gas. Other properties of the reservoir gas are estimated using specific gravity.

Next we will look at two methods of estimating the specific gravity of the reservoir gas from production data. In the first case, the properties and quantities of all surface gas streams are known. In the second case, only the properties of the gas from the primary separator are known.

Separator Gas and Stock-Tank Vent Gas Properties Known[1]

The surface gas is represented by a weighted average of the specific gravities of the separator gas and stock-tank gas.

$$\gamma_g = \frac{R_{SP}\gamma_{gSP} + R_{ST}\gamma_{gST}}{R_{SP} + R_{ST}} \qquad (7-1)$$

The producing gas-oil ratio is

$$R = R_{SP} + R_{ST} . \qquad (7-2)$$

For a three-stage separator system the equations are

$$\gamma_g = \frac{R_{SP1}\gamma_{gSP1} + R_{SP2}\gamma_{gSP2} + R_{ST}\gamma_{gST}}{R_{SP1} + R_{SP2} + R_{ST}} \qquad (7-3)$$

and

$$R = R_{SP1} + R_{SP2} + R_{ST} . \qquad (7-4)$$

Then, on the basis of one stock-tank barrel, the mass of the reservoir gas in *pounds* is

$$
m_R = \frac{\left(R \frac{scf}{STB}\right)\left(29\,\gamma_g \frac{lb\,gas}{lb\,mole\,gas}\right)}{\left(380.7 \frac{scf}{lb\,mole\,gas}\right)} + \left(62.37\gamma_{STO} \frac{lb\,oil}{cu\,ft\,oil}\right)\left(5.615 \frac{cu\,ft\,oil}{STB}\right),
$$

$$
m_R = 0.0762R\gamma_g + 350.2\gamma_{STO}, \frac{lb\,of\,gas}{STB}. \tag{7-5}
$$

Again, on the basis of one stock-tank barrel, the mass of reservoir gas in *pound moles* is

$$
n_R = \frac{R \frac{scf}{STB}}{380.7 \frac{scf}{lb\,mole\,gas}} + \frac{350.2\gamma_{STO} \frac{lb\,oil}{STB}}{M_{STO} \frac{lb\,oil}{lb\,mole\,oil}},
$$

$$
n_R = 0.00263R + 350.2\gamma_{STO}/M_{STO}, \frac{lb\,mole\,of\,gas}{STB}. \tag{7-6}
$$

The molecular weight of the reservoir gas is m_R/n_R, and the specific gravity of the reservoir gas is molecular weight divided by 29.

$$
\gamma_{gR} = \frac{R\gamma_g + 4600\gamma_{STO}}{R + 133,300\gamma_{STO}/M_o} \tag{7-7}
$$

The term $133,300\,\gamma_{STO}/M_{STO}$ in the denominator represents the volume in standard cubic feet of gas that the stock-tank liquid would occupy if it were vaporized. This is sometimes called the *gaseous equivalent* of the stock-tank liquid.

If the molecular weight of the stock-tank liquid is not known, it can be estimated as[2]

$$M_{STO} = \frac{5954}{°API - 8.8} = \frac{42.43\gamma_{STO}}{1.008 - \gamma_{STO}} \tag{7-8}$$

where γ_{STO} is the specific gravity of the stock-tank liquid.

EXAMPLE 7-3: *A wet gas produces through a separator at 300 psia and 73°F to a stock tank at 76°F. The separator produces 0.679 specific gravity gas at 69,551 scf/STB and the stock tank vents 1.283 specific gravity gas at 366 scf/STB. The stock-tank liquid gravity is 55.9°API. Calculate the specific gravity of the reservoir gas.*

Solution

First, calculate the specific gravity of the surface gas.

$$\gamma_g = \frac{R_{SP}\gamma_{gSP} + R_{ST}\gamma_{gST}}{R_{SP} + R_{ST}} \tag{7-1}$$

$$\gamma_g = \frac{(69,551)(0.679) + (366)(1.283)}{69,551 + 366} = 0.682$$

Second, calculate the molecular weight of the stock-tank liquid.

$$M_{STO} = \frac{5954}{°API - 8.8} \tag{7-8}$$

$$M_{STO} = \frac{5954}{55.9 - 8.8} = 126 \text{ lb/lb mole}$$

Third, calculate reservoir gas specific gravity.

$$\gamma_{gR} = \frac{R\gamma_g + 4600\gamma_{STO}}{R + 133,300\gamma_{STO}/M_{STO}} \qquad (7\text{-}7)$$

$$\gamma_{gR} = \frac{(69,917)(0.682) + (4600)(0.755)}{69,917 + (133,300)(0.755)/(126)}$$

$$\gamma_{gR} = 0.723$$

A quicker but less accurate method of obtaining the specific gravity of the gas in the reservoir is given in Figure 7–1.[3] Separator gas and stock-tank gas must be added to obtain gas volume for calculating condensate production rate. Also, the weighted average of separator and stock-tank specific gravities must be used as surface gas specific gravity. The inset in Figure 7–1 gives a relationship between stock-tank oil gravity and molecular weight. The graph is not as accurate as Equation 7–8.

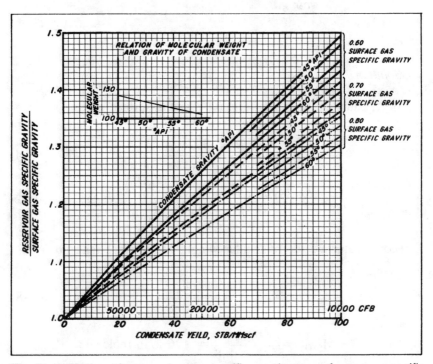

Fig. 7–1. Ratios of reservoir-gas specific gravity to surface-gas specific gravity. (Standing, *Volumetric and Phase Behavior of Oil Field Hydrocarbon Systems*, SPE, Dallas, 1951. Copyright 1951 SPE-AIME.

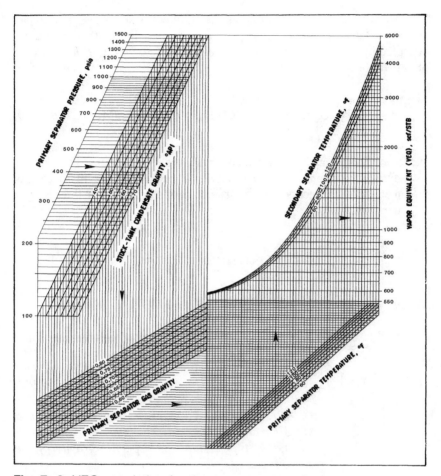

Fig. 7–2. VEQ correlation for three-stage separation.

Properties of Stock-Tank Gas Unknown

Very often the quantity and specific gravity of the stock-tank vent are not measured. Sometimes, in a three-stage system, the quantity of second separator gas is unknown. A correlation is available for use in these situations.[2]

Equation 7–7 has been rewritten so that only the properties and quantities of the primary separator gas and the stock-tank liquid are required.

$$\gamma_{gR} = \frac{R_{SP1}\gamma_{gSP1} + 4600\gamma_{STO} + AGP}{R_{SP1} + VEQ} \qquad (7\text{–}9)$$

The *equivalent volume*, VEQ, is the volume of second-separator gas and stock-tank gas in scf/STB *plus* the volume in scf that would be occupied by a barrel of stock-tank liquid if it were gas. For *three stages* of separation,

$$VEQ = R_{SP2} + R_{ST} + 133,300\gamma_{STO}/M_{STO}. \quad (7-10)$$

Values of VEQ can be obtained from Figure 7–2.

The *additional gas produced*, AGP, is related to the mass of gas produced from the second separator and the stock tank. For three stages of separation,

$$AGP = R_{SP2}\gamma_{gSP2} + R_{ST}\gamma_{gST}. \quad (7-11)$$

Values of AGP can be obtained from Figure 7-3.

EXAMPLE 7–4: *A wet gas produces through a three-stage separator system. The primary separator produces 61,015 scf/STB of 0.669 specific gravity gas and the second-stage separator produces 1,002 scf/STB of 0.988 specific gravity gas. Stock-tank oil gravity is 60.7°API. The primary separator operates at 900 psia and 73°F, and the second separator operates at 75 psia and 74°F. Calculate the specific gravity of the reservoir gas.*

Solution

First, determine a value of VEQ.

VEQ = 1600 scf/STB at p_{SP1} = 900 psia, γ_{SP1} = 0.669,

γ_{STO} = 60.7°API, T_{SP1} = 73°F, T_{SP2} = 74°F, Figure 7–2.

Second, determine a value of AGP.

AGP = 756 scf-gravity/STB, Figure 7–3.

Third, calculate reservoir gas specific gravity

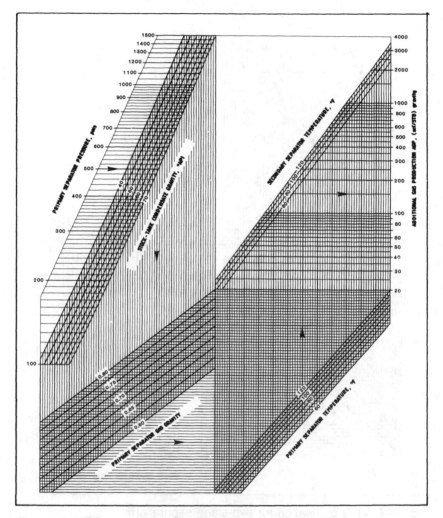

Fig. 7–3. AGP correlation for three-stage separation.

$$\gamma_{gR} = \frac{R_{SP1}\gamma_{gSP1} + 4600\gamma_o + AGP}{R_{SP1} + VEQ} \qquad (7\text{–}9)$$

$$\gamma_{gR} = \frac{(61,015)\ (0.669)\ +\ (4600)\ (0.736)\ +756}{61,015\ +\ 1600} = 0.718$$

NOTE: The second separator gas-oil ratio is not used since the correlations of Figures 7–2 and 7–3 include it.

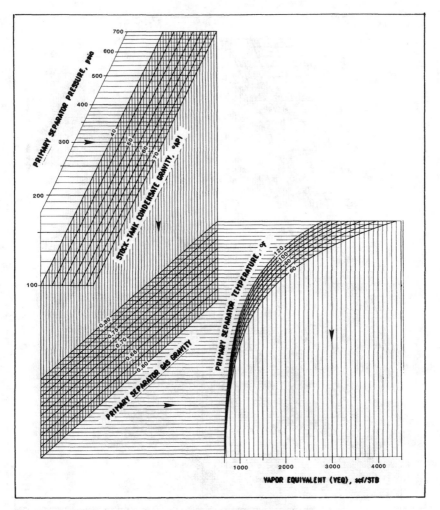

Fig. 7–4. VEQ correlation for two-stage separation.

Equation 7–9 can be used with Figures 7–4 and 7–5 for *two-stage* separation. VEQ in Figure 7–4 accounts for stock-tank gas and liquid and AGP in Figure 7–5 accounts for stock-tank gas.

Values of specific gravities of reservoir gases calculated with Equation 7–9 and the appropriate figures will be within 2% of laboratory determined values.

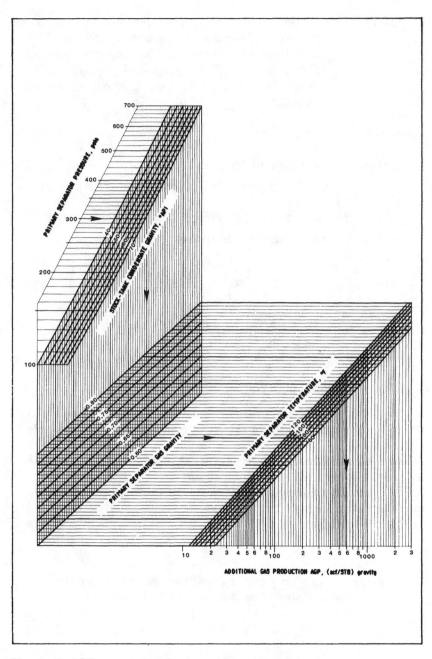

Fig. 7–5. AGP correlation for two-stage separation.

EXAMPLE 7–5: *A wet gas produces through a two-stage separator system. The separator produces 69,551 scf/STB of 0.679 specific gravity gas. The separator operates at 300 psia and 73°F. Stock-tank liquid gravity is 56.0°API. Calculate the specific gravity of the reservoir gas.*

Solution

First, determine values of VEQ and AGP.

VEQ = 1040 scf/STB at p_{SP} = 300 psia, γ_{STO} = 56.0°API, T_{SP} = 73°F, Figure 7–4.

AGP = 330 scf-gravity/STB, Figure 7–5

Second, calculate reservoir gas specific gravity

$$\gamma_{gR} = \frac{R_{SP1}\gamma_{gSP1} + 4600\gamma_o + AGP}{R_{SP1} + VEQ} \tag{7-9}$$

$$\gamma_{gR} = \frac{(69,551)\,(0.679) + (4600)\,(0.755) + 330}{69,551 + 1,040} = 0.723$$

Formation Volume Factor of Wet Gas

The equations for the formation volume factor of gas, Equations 6–2 and 6–3, only apply to dry gases. These equations are not applicable to wet gases.

However, it is important to be able to calculate the reservoir volume of wet gas associated with quantities of surface liquid and gas.

The *formation volume factor of a wet gas* is defined as the volume of reservoir gas required to produce one stock-tank barrel of liquid at the surface. By definition

$$B_{wg} = \frac{\text{volume of reservoir gas at reservoir pressure and temperature}}{\text{volume of stock-tank liquid at standard conditions}}. \tag{7-12}$$

Usually the units are barrels of a gas at reservoir conditions per barrel of stock-tank liquid.

Two methods for estimating formation volume factors for wet gases will be discussed. Each of the methods requires a different set of starting data.

Surface Compositions Known

If the compositions of the produced gases and liquid are known and the producing gas-oil ratios are available, the composition of the reservoir gas may be calculated as in Example 7–1. The results of such a recombination calculation can be used to calculate the formation volume factor. The volume of gas in the reservoir and the volume of stock-tank liquid must be calculated.

The volume of the gas in the reservoir may be calculated using the compressibility equation of state. This calculation is based on 1 lb mole of gas using Equation 3–39. The composition of the gas in the reservoir calculated by the recombination method can be used to compute the pseudocritical properties so that the compressibility factor may be obtained in the same manner as illustrated in Examples 3–10 and 3–12.

The volume of stock-tank liquid which condenses during production of 1 lb mole of reservoir gas can be computed with values which result from the recombination calculation. The molecular weight divided by the stock-tank liquid density (both calculated in the first step of Example 7–1) is the volume of one lb mole of stock-tank liquid. The sum of column 7 of the third step of Example 7–1 gives the pound moles of reservoir gas per pound mole of stock-tank liquid.

The first of these numbers divided by the second gives the volume of stock-tank liquid which comes from 1 lb mole of reservoir gas. This number divided into the molar volume of the reservoir gas gives the formation volume factor.

EXAMPLE 7–6: *Continue Example 7–1 by calculating wet gas forma-tion volume factor at reservoir conditions of 2360 psig and 204°F.*

Solution

First, calculate the pseudocritical properties of the reservoir gas.

Component	Composition, mole fraction	Critical temperature, °R		Critical pressure, psia	
	y_j	T_{cj}	$y_j T_{cj}$	p_{cj}	$y_j p_{cj}$
C_1	0.8228	342.91	282.15	666.4	548.31
C_2	0.0952	549.50	52.31	706.5	67.26
C_3	0.0464	665.64	30.89	616.0	28.58
i- C_4	0.0064	734.04	4.70	527.0	3.38
n-C_4	0.0096	765.20	7.35	550.6	5.29
i- C_5	0.0035	828.68	2.90	490.4	1.72
n-C_5	0.0029	845.38	2.45	488.6	1.42
C_6	0.0029	913.18	2.65	436.9	1.27
C_{7+}	0.0103	*1068	11.00	*455	4.69
	1.0000		$T_{pc} = 396.4°R$		$p_{pc} = 661.9$ psia

*Pseudocritical properties for C_{7+} from Figure 3–10.

Second, determine z-factor and calculate the molar volume of the reservoir gas.

$$T_{pr} = \frac{T}{T_{pc}} = \frac{664}{396.4°R} = 1.68 \qquad (3-43)$$

$$p_{pr} = \frac{p}{p_{pc}} = \frac{2374.7 \text{ psia}}{661.9 \text{ psi}} = 3.59 \qquad (3-43)$$

$$z = 0.849, \text{ Figure } 3-7$$

$$V_M = \frac{zRT}{p} = \frac{(0.849)\left[10.732 \dfrac{\text{psia cu ft}}{\text{lb mole °R}}\right](664°R)}{2374.7 \text{ psia}}$$

$$= 2.548 \text{ cu ft/lb mole} \qquad (3-39)$$

Third, calculate the volume of stock-tank liquid condensed from one lb mole of reservoir gas.

$$\text{molar volume of stock-tank liquid} = \frac{M_{STO}}{\rho_{STO}}$$

$$= \frac{100.9 \text{ lb/lb mole}}{47.11 \text{ lb/cu ft}} = 2.142 \ \frac{\text{cu ft STO}}{\text{lb mole STO}} , \text{ values from Example 7-1.}$$

volume of STO from one lb mole reservoir gas =

$$\frac{2.142 \text{ cu ft STO/lb mole STO}}{71.057 \text{ lb mole res gas/lb mole STO}} = 0.03014 \ \frac{\text{cu ft STO}}{\text{lb mole res gas}},$$

values from Example 7-1.

Fourth, calculate wet gas formation volume factor.

$$B_{wg} = \frac{2.548 \ \dfrac{\text{cu ft res gas}}{\text{lb mole res gas}}}{0.03014 \ \dfrac{\text{cu ft STO}}{\text{lb mole res gas}}} = 84.5 \ \frac{\text{bbl res gas}}{\text{STB}}$$

The recombination method gives results as accurate as laboratory results if accurate values of compositions and gas-oil ratios are available.

Compositions Unknown

Very often the gas compositions are unknown. Usually the volume of stock-tank gas is not known. Under these circumstances an accurate value of the formation volume factor of a wet gas can be estimated using equivalent volume, VEQ. Only the primary separator gas-oil ratio is needed. The second separator and stock-tank gas-oil ratios are ignored; the VEQ correlation includes these gases.

The sum of the primary separator gas-oil ratio and VEQ is the standard cubic feet of reservoir wet gas required to produce one barrel of stock-tank liquid.[2]

volume of reservoir wet gas $= R_{SP1} + VEQ$, scf/STB   (7-13)

EXAMPLE 7-7: *Continue Example 7-4 by calculating the quantity of reservoir wet gas required to produce 61,015 scf/STB of primary separator gas.*

Solution

volume of reservoir wet gas $= R_{SPl} + VEQ$, scf/STB (7–13)

volume of reservoir wet gas $= 61,015$ scf/STB $+ 1600$ scf/STB $= 62,615$ scf/STB

The volume of wet gas in standard cubic feet can be converted to reservoir conditions through use of Equation 3–39. The mass in pound moles of wet gas per stock-tank barrel of surface liquid can be calculated as

$$n = \frac{(R_{SPl} + VEQ) \text{ scf/STB}}{380.7 \text{ scf/lb mole gas}} .$$
(7–14)

Substitution into Equation 3–39 gives

$$B_{wg} = V_R = \frac{z(R_{SPl} + VEQ)RT}{380.7p}$$

$$= 0.0282z(R_{SPl} + VEQ)T/p, \text{ res cu ft gas/STB}$$
(7–15)

or

$$B_{wg} = 0.00502z(R_{SPl} + VEQ)T/p, \text{ res bbl gas/STB} .$$
(7–16)

The z-factor can be obtained using the specific gravity of the reservoir wet gas calculated as in Examples 7–4 or 7–5.

EXAMPLE 7–8: *Continue Example 7–4 by calculating wet gas formation volume factor at reservoir conditions of 2360 psig and 204°F.*

Solution

First, determine a value of z-factor.

$$T_{pc} = 382°R, \; p_{pc} = 661 \text{ psia at } \gamma_g = 0.718, \text{ Figure } 3\text{--}11.$$

$$T_{pr} = \frac{664°R}{382°R} = 1.74, \; p_{pr} = \frac{2374.7 \text{ psia}}{661 \text{ psi}} = 3.59 \qquad (3\text{--}43)$$

$$z = 0.876, \text{ Figure } 3\text{--}7$$

Second, calculate wet gas formation volume factor.

$$B_{wg} = 0.00502z(R_{SPl} + VEQ)T/p \qquad (7\text{--}16)$$

$$B_{wg} = \frac{(0.00502) \, (0.876) \, (61,015 + 1600) \, (664)}{(2374.7)}$$

$$= 77.0 \; \frac{\text{res bbl gas}}{\text{STB}}$$

Plant Products

Frequently, processing surface gas to remove and liquefy the intermediate hydrocarbons is economically feasible. These liquids often are called *plant products*. The quantities of liquid products which can be obtained usually are determined in gallons of liquid per thousand standard cubic feet of gas processed, gal/Mscf, or GPM.

The composition of the gas must be known in order to make these calculations.

The units of mole fraction are pound moles of component j per pound mole of gas. Mole fraction can be converted into gal/Mscf as follows.

$$GPM_j = \left(y_j \frac{\text{lb mole j}}{\text{lb mole gas}} \right) \left(\frac{\text{lb mole gas}}{380.7 \text{ scf}} \right) \left(\frac{1000 \text{ scf}}{\text{Mscf}} \right) \left(M_j \frac{\text{lb j}}{\text{lb mole j}} \right)$$

$$\left(\frac{\text{cu ft liq}}{\rho_{oj} \text{ lb j}} \right) \left(\frac{7.481 \text{ gal}}{\text{cu ft liq}} \right)$$

$$= 19.65 \; \frac{y_j M_j}{\rho_{oj}} \; \frac{\text{gal}}{\text{Mscf}}, \qquad (7\text{--}17)$$

where ρ_{oj} is the density of component j, lb/cu ft, as a liquid at standard conditions, or

$$GPM_j = 0.3151 \frac{y_j M_j}{\gamma_{oj}} \frac{gal}{Mscf} , \qquad (7-18)$$

where γ_{oj} is the specific gravity of the component as a liquid at standard conditions. These data are available in Appendix A.

EXAMPLE 7-9: *Determine the maximum available liquid products from a gas of composition given below.*

Components	Composition, mole fraction
Carbon dioxide	0.0167
Nitrogen	0.0032
Methane	0.7102
Ethane	0.1574
Propane	0.0751
i- Butane	0.0089
n-Butane	0.0194
i- Pentane	0.0034
n-Pentane	0.0027
Hexanes	0.0027
Heptanes plus	0.0003
	1.0000

Note: When heptanes plus concentration is
this low, the molecular weight and specific
gravity cannot be measured. Use 103 lb/lb
mole and 0.7.

Solution

$$GPM_j = 0.3151 \frac{y_j M_j}{\gamma_{oj}} \frac{gal}{Mscf} , \qquad (7-18)$$

Component	Composition, mole fraction y_j	Molecular weight M_j	Liquid specific gravity γ_{oj}	Liquid content $\dfrac{0.3151 y_j M_j}{\gamma_{oj}}$
CO_2	0.0167			
N_2	0.0032			
C_1	0.7102			
C_2	0.1574	30.07	0.3562	4.187
C_3	0.0751	44.10	0.5070	2.058
i- C_4	0.0089	58.12	0.5629	0.290
n-C_4	0.0194	58.12	0.5840	0.608
i- C_5	0.0034	72.15	0.6247	0.124
n-C_5	0.0027	72.15	0.6311	0.097
C_6	0.0027	86.18	0.6638	0.110
C_{7+}	0.0003	103.00	0.7000	0.014
	1.0000			7.488 GPM

Complete recovery of these products is not feasible. A general rule of thumb is 5 to 25 percent of the ethane, 80 to 90 percent of the propane, 95 percent or more of the butanes, and 100 percent of the heavier components can be recovered in a relatively simple "plant."

Retrograde Gases

The discussion above for wet gases applies to retrograde gases as long as the reservoir pressure is above the dew-point pressure of the retrograde gas. At reservoir pressures below the dew point, none of the recombination calculations given in this chapter are valid for retrograde gases. Special laboratory analyses (not discussed in this book) are required for engineering of retrograde gas reservoirs.

A reasonably accurate procedure for estimating the dew-point pressure of a retrograde gas is given in Appendix B.

The surface gas from a retrograde gas reservoir is rich in intermediate molecules. Processing it through a "plant" to produce additional liquids is usually economically attractive. The calculation of "plant liquids" is the same as described previously.

Exercises

7–1. A wet gas is produced through a two-stage separator system. Separator conditions are 580 psia and 95°F. The stock tank is at 80°F. Gas-oil ratios are 98,835 scf/STB and 408 scf/STB. Stock-tank liquid is 55.4°API. Compositions of the surface streams are given below. Calculate the composition of the reservoir gas.

Component	Composition, separator gas, mole fraction	Composition, stock-tank gas, mole fraction	Composition, stock-tank liquid, mole fraction
C_1	0.8814	0.4795	0.0026
C_2	0.0730	0.1769	0.0054
C_3	0.0297	0.1755	0.0194
i-C_4	0.0054	0.0490	0.0148
n-C_4	0.0045	0.0466	0.0193
i-C_5	0.0023	0.0293	0.0312
i-C_5	0.0017	0.0222	0.0293
C_6	0.0016	0.0190	0.0902
C_{7+}	0.0004	0.0020	0.7878
	1.0000	1.0000	1.0000

Properties of heptanes plus of the stock-tank liquid
Specific gravity   0.779
Molecular weight   131 lb/lb mole

7–2. A wet gas is produced through a two-stage separator system. The 555 psia-89°F separator produces 170,516 scf/STB, and the 70°F stock tank produces 492 scf/STB. The stock-tank liquid has a gravity of 64.7°API. The compositions of the surface production are given below. Calculate the composition of the reservoir gas.

Component	Composition, separator gas, mole fraction	Composition, stock-tank gas, mole fraction	Composition, stock-tank liquid, mole fraction
C_1	0.8943	0.4615	0.0026
C_2	0.0518	0.1218	0.0042
C_3	0.0313	0.1845	0.0234
i-C_4	0.0043	0.0373	0.0134
n-C_4	0.0103	0.1020	0.0499
i-C_5	0.0028	0.0327	0.0424
n-C_5	0.0032	0.0384	0.0629
C_6	0.0018	0.0206	0.1244
C_{7+}	0.0002	0.0012	0.6768
	1.0000	1.0000	1.0000

Properties of heptanes plus of the stock-tank liquid
Specific gravity   0.754
Molecular weight   111 lb/lb mole

7–3. The compositions of gas and liquid samples taken from a separator are given below. The separator was stabilized at 945 scf/STB. Laboratory measurement indicated a separator/stock-tank volume ratio of 1.052 SP bbl/STB. The density of the separator liquid, calculated with procedures given in Chapter 11, is 49.8 lb/cu ft at separator conditions of 115 psia and 100°F.

Component	Composition, separator gas, mole fraction	Composition, separator liquid, mole fraction
C_1	0.7833	0.0307
C_2	0.0965	0.0191
C_3	0.0663	0.0423
i-C_4	0.0098	0.0147
n-C_4	0.0270	0.0537
i-C_5	0.0063	0.0310
n-C_5	0.0064	0.0373
C_6	0.0037	0.0622
C_{7+}	0.0007	0.7090
	1.0000	1.0000

Properties of heptanes plus of separator liquid
Specific gravity   0.850
Molecular weight   202 lb/lb mole

You may assume heptanes plus of separator gas has a molecular weight of 103 lb/lb mole.

Producing gas-oil ratio remained constant prior to sampling, so you may assume that the reservoir fluid is single phase. Calculate the composition of the reservoir fluid. What type of reservoir fluid is this?

7-4. Samples of gas and liquid are taken from a first stage separator operating at 500 psia and 75°F. The separator gas-oil ratio is constant at 2347 scf/SP bbl. The compositions of the samples are given in the table below. The density of the separator liquid at separator conditions, calculated with procedures given in Chapter 11, is 47.7 lb/cu ft. Calculate the composition of the reservoir fluid. What type of reservoir fluid is this?

Component	Composition, separator gas, mole fraction	Composition, separator liquid, mole fraction
C_1	0.8739	0.1437
C_2	0.0776	0.0674
C_3	0.0332	0.0902
i- C_4	0.0033	0.0191
n-C_4	0.0077	0.0587
i- C_5	0.0014	0.0228
n-C_5	0.0018	0.0380
C_6	0.0009	0.0513
C_{7+}	0.0002	0.5088
	1.0000	1.0000

Properties of heptanes plus of separator liquid
Specific gravity   0.840
Molecular weight   207 lb/lb mole

7-5. A reservoir containing retrograde gas has a reservoir pressure higher than the dew-point pressure of the gas as evidenced by constant producing gas-oil ratio. The gas is produced through three stages of separation. Surface production rates, specific gravities, and conditions are given below.

	Gas rate, scf/STB	Gas specific gravity	Temperature, °F	Pressure, psia
Primary separator	4187	0.666	105	650
Second separator	334	0.946	95	70
Stock tank	88	1.546	90	14.7
	4609			

Stock-tank oil gravity is 55.2°API. Calculate the specific gravity of the reservoir gas.

7-6. A wet gas produces 99,835 scf/STB through a 580 psia-90°F separator and 408 scf/STB through a 80°F stock tank. Specific gravities of the separator gas and stock-tank gas are 0.644 and 1.099, and the stock-tank liquid is 55.4°API. Calculate the specific gravity of the reservoir gas.

7-7. A wet gas reservoir produces 170,516 scf/STB of 0.646 specific gravity separator gas, 492 scf/STB of 1.184 specific gravity stock-tank gas, and 64.7°API stock-tank liquid. Calculate the specific gravity of the reservoir gas.

7-8. A retrograde gas is produced through a three-stage separation system. The table below gives producing rates and specific gravities of the produced gases. Stock-tank liquid gravity is 54.8°API. Calculate the specific gravity of the reservoir gas.

	Gas produced, scf/STB	Gas specific gravity
Primary separator	8885	0.676
Second separator	514	1.015
Stock tank	128	1.568
	9527	

7-9. A wet gas produces through a two-stage separator system. Separator gas rate is 99,835 scf/STB with specific gravity of 0.644. The separator operates at 580 psia and 90°F. Stock-tank liquid gravity is 55.4°API. Calculate the specific gravity of the reservoir gas.

7-10. A wet gas reservoir produces 170,516 scf/STB of 0.646 specific gravity separator gas and 64.7°API stock-tank liquid. The separator operates at 555 psia and 89°F. The separator system is two-stage. Calculate the specific gravity of the reservoir gas.

7-11. A reservoir containing retrograde gas has a reservoir pressure higher than the dew-point pressure of the gas as evidenced by constant producing gas-oil ratio. The gas is produced through three stages of separation. The primary separator produces 4187 scf/STB of 0.666 specific gravity gas. Stock-tank oil gravity is 55.2°API. The primary separator is operated at 650 psia and 105°F. The second separator is at 95°F. Calculate the specific gravity of the reservoir gas.

7–12. A retrograde gas is produced through a three-stage separator system. The primary separator at 625 psia and 75°F produces 8885 scf/STB of 0.676 specific gravity gas. The second separator operates at 60 psia and 75°F. Stock-tank liquid gravity is 54.8°API. Estimate the specific gravity of the gas in the reservoir.

7–13. Calculate the wet gas formation volume factor of the gas of Exercise 7–1 when reservoir conditions are 6500 psia and 225°F.

7–14. Calculate the wet gas formation volume factor of the gas of Exercise 7–2 when reservoir conditions are 2000 psia and 156°F.

7–15. Continue Exercise 7–11. Calculate the quantity of reservoir gas which must be produced to result in one barrel of liquid in the stock tank. Give your answer in scf of reservoir gas/STB.

7–16. Continue Exercise 7–11. Calculate the quantity of reservoir gas which must be produced to result in one barrel of liquid in the stock tank. Give your answer in barrels of reservoir gas/STB when reservoir conditions are 5900 psig and 249°F.

7–17. Continue Exercise 7–12. Calculate the quantity of reservoir gas which must be produced to create one barrel of stock-tank liquid. Give your answer in scf of reservoir gas/STB.

7–18. Continue Exercise 7–12. Calculate the quantity of reservoir gas which must be produced to create one barrel of stock-tank liquid. Reservoir conditions are 5025 psia and 222°F. Give your answer in barrels of reservoir gas/STB.

7–19. Continue Exercise 7–12. Calculate the quantity of primary separator gas resulting from the production of one million scf of reservoir gas.

7–20. Continue Exercises 7–2 and 7–14. Calculate the quantity of reservoir gas in scf required to produce one barrel of stock-tank liquid.

7–21. Calculate the stock-tank liquid production in STB/MMscf surface gas and STB/MMscf reservoir gas for the wet gas of Exercise 7–1.

7–22. Calculate the total liquids content of the separator gas of Exercise 7-1.

7–23. Calculate the gross heating value (dry) of the separator gas of Exercise 7-1.

7–24. Calculate the total liquids content of the separator gas of Exercise 7-2.

7–25. Calculate the total liquids content of the separator gas of Exercise 7–3.

References

1. Craft, B.C. and Hawkins, M.F.: *Applied Petroleum Reservoir Engineering,* Prentice-Hall, Inc., Englewood Cliffs, NJ (1959).

2. Gold, D.K., McCain, W.D., Jr., and Jennings, J.W.: "An Improved Method for the Determination of the Reservoir-Gas Specific Gravity for Retrograde Gases," *J. Pet. Tech.* (July 1989) *41,* 747-752.

3. Standing, M.B.: *Volumetric and Phase Behavior of Oil Field Hydrocarbon Systems,* Reinhold Publishing Corp., New York City (1952).

8

Properties of Black Oils—Definitions

We now turn to black oils. We consider those physical properties which are required for the reservoir engineering calculations known as material balance calculations. These properties are *formation volume factor of oil, solution gas-oil ratio, total formation volume factor, coefficient of isothermal compressibility,* and *oil viscosity*. Also, *interfacial tension* is discussed.

These properties are defined in this chapter. The physical processes involved in the way black oil properties change as reservoir pressure is reduced at constant temperature are explained. Later chapters address methods of determining values of these properties using field data, laboratory fluid studies, and correlations.

The subscript o is used to indicate a liquid property since the petroleum engineer often uses the word *oil* to describe the liquids with which he deals.

Specific Gravity of a Liquid

Liquid specific gravity, γ_o, is defined as the ratio of the density of the liquid to the density of water, both taken at the same temperature and pressure.

$$\gamma_o = \frac{\rho_o}{\rho_w} \qquad (8-1)$$

Specific gravity appears to be nondimensional since the units of the density of the liquid are the same as the units of the density of water; however, this is not strictly true. Actually, in the English system the units are

$$\gamma_o = \frac{\rho_o}{\rho_w} = \frac{\text{lb oil/cu ft oil}}{\text{lb water/cu ft water}} . \qquad (8-1)$$

Sometimes specific gravity is given as sp. gr. 60°/60°, which means that the densities of both the liquid and the water were measured at 60°F and atmospheric pressure.

The petroleum industry also uses another gravity term called *API gravity* which is defined as

$$°API = \frac{141.5}{\gamma_o} - 131.5 , \qquad (8-2)$$

where γ_o is the specific gravity at 60°/60°. This equation was devised so that hydrometers could be constructed with linear scales.

EXAMPLE 8–1: *The density of a stock-tank oil at 60°F is 51.25 lb/cu ft. Calculate the specific gravity and gravity in °API.*

Solution

First, calculate the specific gravity

$$\gamma_o = \frac{\rho_o}{\rho_w} \qquad (8-1)$$

$$\gamma_o = \frac{51.25 \text{ lb/cu ft}}{62.37 \text{ lb/cu ft}} = 0.8217$$

Second, calculate gravity in °API

$$°API = \frac{141.5}{\gamma_o} - 131.5 \qquad (8-2)$$

$$°API = \frac{141.5}{0.8217} - 131.5 = 40.7°API$$

Formation Volume Factor of Oil

The volume of oil that enters the stock tank at the surface is less than the volume of oil which flows into the wellbore from the reservoir. This change in oil volume which accompanies the change from reservoir conditions to surface conditions is due to *three* factors.

The most important factor is the evolution of gas from the oil as pressure is decreased from reservoir pressure to surface pressure. This causes a rather large decrease in volume of the oil when there is a significant amount of dissolved gas.

$$B_{ob} \doteq \frac{\vartheta_{sto} + 0.01357 R_s \gamma_g}{\vartheta_{bo}}$$

The reduction in pressure also causes the remaining oil to expand slightly, but this is somewhat offset by the contraction of the oil due to the reduction of temperature.

The change in oil volume due to these three factors is expressed in terms of the *formation volume factor of oil*. Oil formation volume factor is defined as the volume of reservoir oil required to produce one barrel of oil in the stock tank. Since the reservoir oil includes dissolved gas,

$$P > P_b : \quad B_o = B_{ob} e^{C_o(P_r - P_b)}$$

$B_o =$

$$\frac{\text{volume of oil + dissolved gas leaving reservoir at reservoir conditions}}{\text{volume of oil entering stock tank at standard conditions}} . \quad (8-3)$$

The units are barrels of oil at reservoir conditions per barrel of stock-tank oil, res bbl/STB. The volume of stock-tank oil is always reported at 60°F, regardless of the temperature of the stock tank. Thus, stock-tank liquid volume, like surface gas volume, is reported at standard conditions.

EXAMPLE 8–2: *A sample of reservoir liquid with volume of 400 cc under reservoir conditions was passed through a separator and into a stock tank at atmospheric pressure and 60°F. The liquid volume in the stock tank was 274 cc. A total of 1.21 scf of gas was released. Calculate the oil formation volume factor.*

Solution

$$B_o = \frac{400 \text{ res cc}}{274 \text{ ST cc}} = 1.46 \frac{\text{res bbl}}{\text{STB}} \quad (8-3)$$

Another way to express formation volume factor of oil is that it is the volume of reservoir occupied by one STB plus the gas in solution at reservoir temperature and pressure.

The relationship of formation volume factor of oil to reservoir pressure for a typical black oil is given in Figure 8–1.

This figure shows the initial reservoir pressure to be above the bubble-point pressure of the oil. As reservoir pressure is decreased from initial pressure to bubble-point pressure, the formation volume factor increases slightly because of the expansion of the liquid in the reservoir.

A reduction in reservoir pressure below bubble-point pressure results in the evolution of gas in the pore spaces of the reservoir. The liquid remaining in the reservoir has less gas in solution and, consequently, a smaller formation volume factor.

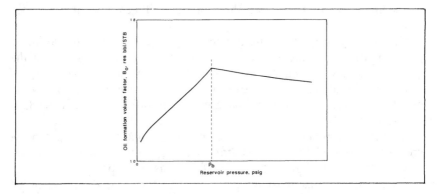

Fig. 8–1. Typical shape of formation volume factor of a black oil as a function of pressure at constant reservoir temperature.

If the reservoir pressure could be reduced to atmospheric, the value of the formation volume factor would nearly equal 1.0 res bbl/STB. A reduction in temperature to 60°F is necessary to bring the formation volume factor to exactly 1.0 res bbl/STB.

The reciprocal of the formation volume factor is called the *shrinkage factor*.

$$b_o = \frac{1}{B_o} \qquad (8-4)$$

The formation volume factor may be multiplied by the volume of stock-tank oil to find the volume of reservoir oil required to produce that volume of stock-tank oil. The shrinkage factor can be multiplied by the reservoir volume to find the corresponding stock-tank volume. Both terms are in use, but petroleum engineers have adopted universally the formation volume factor.

Formation volume factor also is called *reservoir volume factor*.

Since the method of processing the produced fluids has an effect on the volume of stock-tank oil, the value of the formation volume factor will depend on the method of surface processing. However, the effect is small for black oils.

Solution Gas-Oil Ratio

We often refer to the solubility of natural gas in crude oil as if we are dealing with a two-component system. Although it is convenient to discuss dissolved gas in this manner, in fact, the gas and oil are both multicomponent mixtures, and the quantities of gas and oil are established by gas-liquid equilibrium.

The quantity of gas-forming molecules (light molecules) in the liquid phase at reservoir temperature is limited only by the pressure and the quantity of light molecules present. A black oil is said to be *saturated* when a slight decrease in pressure will allow release of some gas. The bubble-point pressure is a special case of saturation at which the first release of gas occurs.

On the other hand, when the black oil is above its bubble-point pressure, it is said to be *undersaturated*. An undersaturated oil could dissolve more gas (light molecules) if the gas were present.

The quantity of gas dissolved in an oil at reservoir conditions is called *solution gas-oil ratio*. Solution gas-oil ratio is the amount of gas that evolves from the oil as the oil is transported from the reservoir to surface conditions. This ratio is defined in terms of the quantities of gas and oil which appear at the surface during production.

$$R_s = \frac{\text{volume of gas produced at surface at standard conditions}}{\text{volume of oil entering stock tank at standard conditions}} \quad (8-5)$$

The surface volumes of both gas and liquid are referred to standard conditions so that the units are standard cubic feet per stock-tank barrel, scf/STB.

Solution gas-oil ratio is also called *dissolved gas-oil ratio* and occasionally *gas solubility*.

Figure 8–2 shows the way the solution gas-oil ratio of a typical black oil changes as reservoir pressure is reduced at constant temperature.

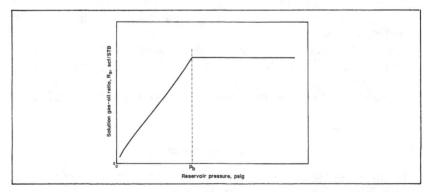

Fig. 8–2. Typical shape of solution gas-oil ratio of a black oil as a function of pressure at constant reservoir temperature.

The line is horizontal at pressures above the bubble-point pressure because at these pressures no gas is evolved in the pore space and the entire liquid mixture is produced into the wellbore. When reservoir pressure is reduced below bubble-point pressure, gas evolves in the reservoir, leaving less gas dissolved in the liquid.

EXAMPLE 8–3: *Calculate the solution gas-oil ratio of the reservoir liquid of Example 8–2.*

Solution

$$R_s = \frac{1.21 \text{ scf}}{(274 \text{ ST cc})(6.2898 \times 10^{-6} \text{ bbl/cc})} = 702 \frac{\text{scf}}{\text{STB}} \quad (8-5)$$

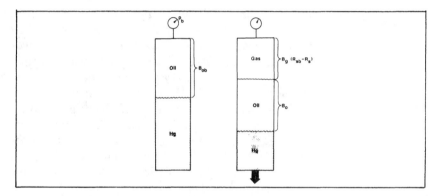

Fig. 8–3. Volume change as pressure is reduced below the bubble point at constant reservoir temperature.

Total Formation Volume Factor

Figure 8–3 shows the volume occupied by one barrel of stock-tank oil plus its dissolved gas at bubble-point pressure. The figure also shows the volume occupied by the same mass of material after an increase in cell volume has caused a reduction in pressure. The volume of oil has decreased; however, the total volume has increased.

The volume of oil at the lower pressure is B_o. The quantity of gas evolved is the quantity in solution at the bubble point, R_{sb}, minus the quantity remaining in solution at the lower pressure, R_s. The evolved gas is called *free gas*. It is converted to reservoir conditions by multiplying by the formation volume factor of gas, B_g.

This total volume is the *total formation volume factor*.

$$B_t = B_o + B_g(R_{sb} - R_s) \qquad\qquad (8-6)$$

The gas formation volume factor must be expressed in units of res bbl/scf, and total formation volume factor has units of res bbl/STB.

Figure 8-4 gives a comparison of total formation volume factor with the formation volume factor of oil. The two formation volume factors are identical at pressures above the bubble-point pressure since no gas is released into the reservoir at these pressures.

The difference between the two factors at pressures below the bubble-point pressure represents the volume of gas released in the reservoir. The volume of this gas is $B_g(R_{sb} - R_s)$ res bbl of gas/STB.

Total formation volume factor is also called *two-phase formation volume factor*.

EXAMPLE 8-4: *Exactly one stock-tank barrel was placed in a laboratory cell. 768 scf of gas was added. Cell temperature was raised to 220°F, the cell was agitated to attain equilibrium between gas and liquid, and pressure was raised until the final bubble of gas disappeared. At that point cell volume was 1.474 barrels and pressure was 2620 psig. Pressure in the cell was reduced to 2253 psig by increasing total cell volume to 1.569 barrels. At that point the oil volume in the cell was 1.418 barrels and the gas volume in the cell was 0.151 barrels. Calculate the total formation volume factor at 2253 psig.*

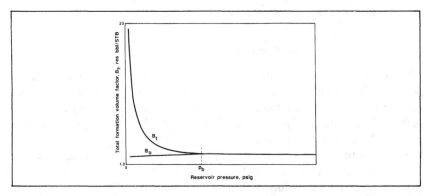

Fig. 8-4. Typical shape of total formation volume factor of a black oil as a function of pressure at constant reservoir temperature compared to shape of black oil formation volume factor at same conditions.

Solution

$$B_t = 1.569 \text{ res bbl/STB}$$

The Coefficient of Isothermal Compressibility of Oil

At pressures above the bubble point, the *coefficient of isothermal compressibility of oil* is defined exactly as the coefficient of isothermal compressibility of a gas. At pressures below the bubble point an additional term must be added to the definition to account for the volume of gas which evolves.

As with gases, the coefficient of isothermal compressibility of oil usually is called *compressibility* or, in this case, *oil compressibility*.

Pressures Above the Bubble-Point Pressure

The definition of the coefficient of isothermal compressibility at pressures above the bubble point is

$$c_o = -\frac{1}{V}\left(\frac{\partial V}{\partial p}\right)_T \text{ or } c_o = -\frac{1}{V_M}\left(\frac{\partial V_M}{\partial p}\right)_T$$

$$\text{or } c_o = -\frac{1}{v}\left(\frac{\partial v}{\partial p}\right)_T . \tag{8-7}$$

These equations simply give the fractional change in volume of a liquid as pressure is changed at constant temperature. The partial derivative is used rather than the ordinary derivative because only one independent variable, pressure, is permitted to vary. Remember that the subscript T indicates that temperature is held constant.

The relationship of oil compressibility to pressure for a typical black oil at constant temperature is shown in Figure 8–5. Black oil compressibility is virtually constant except at pressures near the bubble point. Values rarely exceed 35×10^{-6} psi^{-1}. Equations 8–7 apply only at pressures above the bubble-point pressure, so the line on Figure 8–5 ends at the bubble point.

Equations 8–7 can be written as

$$c_o = -\left(\frac{\partial \ln V}{\partial p}\right)_T \text{ or } c_o = -\left(\frac{\partial \ln V_M}{\partial p}\right)_T$$

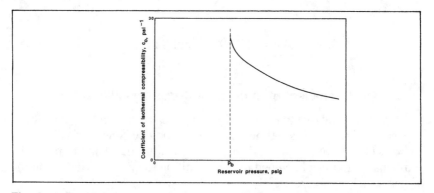

Fig. 8–5. Typical shape of the coefficient of isothermal compressibility of oil as a function of pressure at constant reservoir temperature at pressures above the bubble point.

$$\text{or} \quad c_o = - \left(\frac{\partial \ln v}{\partial p} \right)_T . \qquad (8\text{–}8)$$

Direct substitution of formation volume factor of oil into the first of Equations 8–7 results in

$$c_o = - \frac{1}{B_o} \left(\frac{\partial B_o}{\partial p} \right)_T . \qquad (8\text{–}9)$$

Equation 8–7 can be integrated if c_o is assumed to remain constant as pressure changes.

$$c_o \int_{p_1}^{p_2} dp = - \int_{v_1}^{v_2} \frac{dv}{v} \qquad (8\text{–}10)$$

results in

$$c_o (p_2 - p_1) = - \ln \frac{v_2}{v_1}$$

or

$$c_o (p_2 - p_1) = - \ln \frac{V_2}{V_1} \qquad (8\text{–}11)$$

The resulting equation is usually rearranged so that changes in specific volume can be calculated for known changes in pressure.

$$v_2 = v_1 \; EXP \left[c_o(p_1 - p_2) \right] \qquad (8-12)$$

EXAMPLE 8–5: *A sample of reservoir oil was placed in a laboratory cell at 5000 psig and 220°F. The volume was 59.55 cc. Pressure was reduced to 4000 psig by increasing the oil volume to 60.37 cc. Calculate the coefficient of isothermal compressibility for this oil at cell conditions.*

Solution

$$c_o \; (p_2 - p_1) = - \; \ln \frac{V_2}{V_1} \qquad (8-11)$$

$$c_o = - \; \frac{\ln(60.37 \; cc/59.55 \; cc)}{(4014.7 - 5014.7) \; psia} = 13.68 \times 10^{-6} \; psi^{-1}$$

The definition of oil compressibility can be written in terms of oil density. We will start with

$$c_o = - \; \frac{1}{v} \left(\frac{\partial v}{\partial p} \right)_T . \qquad (8-7)$$

Remember that by definition

$$v = \frac{1}{\rho_o} . \qquad (8-13)$$

The partial derivative of this equation with respect to pressure results in

$$\left(\frac{\partial v}{\partial p} \right)_T = - \; \frac{1}{\rho_o^2} \left(\frac{\partial \rho_o}{\partial p} \right)_T . \qquad (8-14)$$

Substitution of Equations 8–13 and 8–14 into Equation 8–7 gives

$$c_o = - \left[\frac{1}{1/\rho_o} \right] \left[- \; \frac{1}{\rho_o^2} \left(\frac{\partial \rho_o}{\partial p} \right)_T \right] . \qquad (8-15)$$

Thus,

$$c_o = \frac{1}{\rho_o} \left(\frac{\partial \rho_o}{\partial p} \right)_T. \qquad (8-16)$$

Equation 8–16 can be integrated under the assumption that c_o remains constant as pressure changes. Our future use of this equation will be related to bubble-point pressure, so we will use a lower limit of p_b.

$$c_o \int_{p_b}^{p} dp = \int_{\rho_{ob}}^{\rho_o} \frac{d\rho_o}{\rho_o} \qquad (8-17)$$

results in

$$c_o (p - p_b) = \ln \frac{\rho_o}{\rho_{ob}} \qquad (8-18)$$

or

$$\rho_o = \rho_{ob} \, EXP \left[c_o (p - p_b) \right]. \qquad (8-19)$$

Equation 8–19 is used to compute the density of an oil at pressures above the bubble point. The density at the bubble point is the starting point.

Pressures Below the Bubble-Point Pressure

When reservoir pressure is below bubble-point pressure, the situation is much different. As Figure 8–6 shows, the volume of the reservoir liquid decreases as pressure is reduced. However, the reservoir volume occupied by the mass that was originally liquid increases due to the evolution of gas. The change in liquid volume may be represented by

$$\left(\frac{\partial B_o}{\partial p} \right)_T. \qquad (8-20)$$

The change in the amount of dissolved gas is

$$\left(\frac{\partial R_s}{\partial p} \right)_T \qquad (8-21)$$

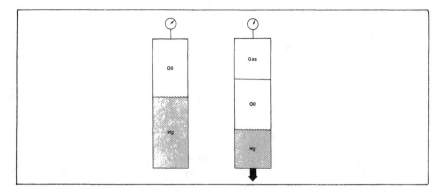

Fig. 8–6. Illustration of the coefficient of isothermal compressibility of oil at pressures below the bubble point at constant reservoir temperature.

and so, the change in volume of free gas is

$$- \left(\frac{\partial R_s}{\partial p} \right)_T . \qquad (8\text{--}22)$$

Thus, at reservoir pressures below the bubble point, the total change in volume is the sum of the change in liquid volume and the change in free gas volume.

$$\left[\left(\frac{\partial B_o}{\partial p} \right)_T - B_g \left(\frac{\partial R_s}{\partial p} \right)_T \right] , \qquad (8\text{--}23)$$

where B_g is inserted to convert the volume of evolved gas to reservoir conditions.

Consequently, the fractional change in volume as pressure changes is[1]

use points on either side

$$P < P_b$$

$$c_o = - \frac{1}{B_o} \left[\left(\frac{\partial B_o}{\partial p} \right)_T - B_g \left(\frac{\partial R_s}{\partial p} \right)_T \right] . \qquad (8\text{--}24)$$

This is consistent with Equation 8–9 since the derivative of R_s with respect to pressure is zero at pressures above the bubble point.

The complete graph of compressibility as a function of reservoir pressure is given in Figure 8–7. There is a discontinuity at the bubble point. The evolution of the first bubble of gas causes a large shift in the value of compressibility. Equation 8–7 applies at pressures above the bubble point, and Equation 8–24 applies at pressures below the bubble point.

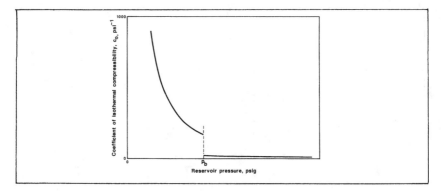

Fig. 8–7. Typical shape of the coefficient of isothermal compressibility of oil as a function of pressure at constant reservoir temperature.

Coefficient of Viscosity of Oil

The *coefficient of viscosity* is a measure of the resistance to flow exerted by a fluid. *Viscosity* appears as a coefficient in many equations that describe fluid flow.

Viscosity of oil usually has units of centipoise, although other units are in use. A discussion of the units of viscosity may be found in Chapter 6.

Viscosity, like other physical properties of liquids, is affected by both pressure and temperature. An increase in temperature causes a decrease in viscosity. A decrease in pressure causes a decrease in viscosity, provided that the only effect of pressure is to compress the liquid. In addition, in the case of reservoir liquids, there is a third parameter which affects viscosity. A decrease in the amount of gas in solution in the liquid causes an increase in viscosity, and, of course, the amount of gas in solution is a direct function of pressure.

The viscosity of a liquid is related directly to the type and size of the molecules which make up the liquid. The variation of liquid viscosity with molecular structure is not known with exactness; however, the viscosities of liquids which are members of a homologous series are known to vary in a regular manner, as do most other physical properties. For example, pure paraffin hydrocarbons exhibit a regular increase in viscosity as the size and complexity of the hydrocarbon molecules increase.

Figure 8–8 shows the relationship of the viscosity of a reservoir oil to pressure at constant temperature. At pressures above bubble point, the viscosity of the oil in a reservoir decreases almost linearly as pressure decreases. At lower pressures the molecules are further apart and therefore move past each other more easily.

However, as reservoir pressure decreases below the bubble point, the liquid changes composition. The gas that evolves takes the smaller molecules from the liquid, leaving the remaining reservoir liquid with relatively more molecules with large complex shapes. This changing liquid composition causes large increases in viscosity of the oil in the reservoir as pressure decreases below the bubble point.

As a black oil reservoir is depleted, not only does production decrease due to the decrease in the pressure available to drive the oil to the well and due to the competition of the free gas for space to flow, but also because the viscosity of the oil has increased. A tenfold increase in oil viscosity between the bubble point and low reservoir pressure is not uncommon.

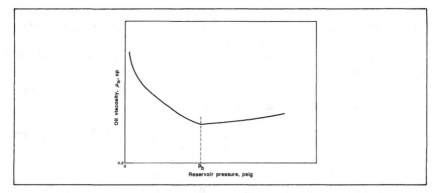

Fig. 8–8. Typical shape of oil viscosity as a function of pressure at constant reservoir temperature.

The Coefficient of Isobaric Thermal Expansion of a Liquid

The *coefficient of isobaric thermal expansion* is defined as the fractional change in volume of a liquid as temperature changes under constant pressure.

$$\beta = \frac{1}{V} \left(\frac{\partial V}{\partial T} \right)_p \tag{8–25}$$

The definition in terms of density follows from Equation 8–25 as

$$\beta = - \frac{1}{\rho_o} \left(\frac{\partial \rho_o}{\partial T} \right)_p. \qquad (8\text{–}26)$$

The *thermal expansion coefficient* usually is assumed to be constant over a limited range of temperatures. Rigorous integration of Equation 8–25 can be performed in the same manner as the integration of Equation 8–7. However, for small changes in temperature, Equation 8–25 can be approximated by

$$V_2 = V_1 \left[1 + \beta(T_2 - T_1) \right] \qquad (8\text{–}27)$$

and Equation 8–26 by

$$\rho_{o2} = \rho_{o1} \left[1 - \beta(T_2 - T_1) \right]. \qquad (8\text{–}28)$$

The petroleum engineer rarely uses this liquid property since petroleum reservoirs normally are operated at constant temperature.

There is also a physical property called *thermal expansion*. This is not defined as above but is simply the ratio of the volume of oil at high temperature to the volume of the same oil at low temperature, with both volumes measured at the same pressure.

$$\text{thermal expansion} = \frac{\text{oil volume at pressure and high temperature}}{\text{oil volume at pressure and low temperature}} \qquad (8\text{–}29)$$

When a value of thermal expansion is reported, it must include the pressure and temperature range for which it is valid. Thermal expansion as defined here must not be used interchangeably with the coefficient of isobaric thermal expansion defined above.

EXAMPLE 8–6: *A sample of reservoir oil was placed in a laboratory cell at 5000 psig and 76°F. The volume was 54.74 cc. Temperature was increased to 220°F and pressure was held constant by increasing cell volume to 59.55 cc. Calculate the coefficient of isobaric thermal expansion and calculate the thermal expansion.*

Solution

First, calculate the coefficient of isobaric thermal expansion.

$$V_2 = V_1 \left[1 + \beta(T_2 - T_1)\right] \tag{8–27}$$

$$B = \frac{(59.55 \text{ cc}/54.74 \text{ cc}) - 1}{(220 - 76)°F} = 610 \times 10^{-6} \, °R^{-1}$$

Second, calculate thermal expansion

$$\text{Thermal expansion} = \frac{59.55 \text{ cc}}{54.74 \text{ cc}} = 1.088 \tag{8–29}$$

Interfacial Tension

There is an imbalance of molecular forces at the interface between two phases. This is caused by physical attraction between molecules. This imbalance of forces is known as *interfacial tension*.

A molecule in a liquid is uniformly attracted to the surrounding molecules. This is represented schematically by the sizes of the arrows on the molecules of Figure 8–9.

A molecule at the surface is attracted more strongly from below because the molecules of the gas are separated much more widely, and the attraction is inversely proportional to the distance between molecules. This imbalance of forces creates a membrane-like surface. It causes a liquid to tend toward a minimum surface area. For instance, a drop of water falling through air tends to be spherical since a sphere has the minimum surface-to-volume ratio.

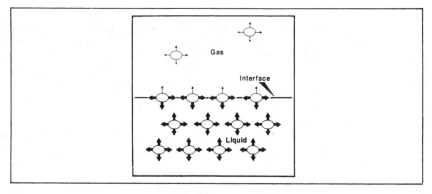

Fig. 8–9. Illustration of intermolecular forces as they affect interfacial tension.

The attraction between molecules is inversely proportional to the square of the distance between them, as previously stated. Also, the attraction is directly proportional to the mass of the molecules. Thus, the interface between two liquids will exhibit interfacial tension due to the differences in mass of the molecules of the two liquids.

Interfacial tension can be thought of as the force required to prevent destruction of the surface. The units are in terms of the force holding the surface together in dynes acting along one centimeter of length (dynes/cm).

The creation of this surface requires work. The work in ergs required to create one square centimeter of surface is called *boundary energy* (erg/sq cm). Interfacial tension and boundary energy are equal. Remember that work equals force times distance, i.e., an erg equals a dyne cm.

Often, the term *surface tension* is used to describe interfacial tension between gas and liquid. Regardless of the terminology, the physical forces which cause the *boundary* or *surface* or *interface* are the same. And the terms can be interchanged.

Volatile Oils

All of the properties discussed in this chapter are defined in exactly the same way for volatile oils as for black oils.

Formation volume factors and solution gas-oil ratios normally are not measured for volatile oils. These quantities are used primarily in material balance calculations which do not apply to volatile oils. If these quantities were measured for volatile oils, they would have the shapes indicated in Figures 8–10 and 8–11. The large decreases in both curves

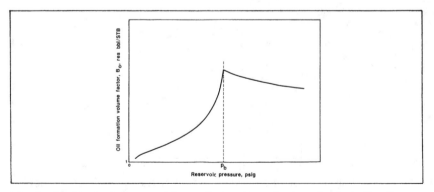

Fig. 8–10. Typical shape of formation volume factor of a volatile oil as a function of pressure at constant reservoir temperature.

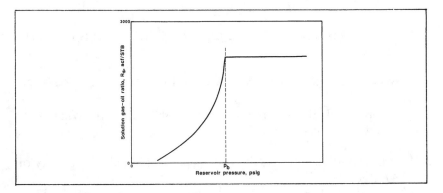

Fig. 8–11. Typical shape of solution gas-oil ratio of a volatile oil as a function of pressure at constant reservoir temperature.

at pressures immediately below bubble point are due to the evolution of large quantities of gas in the reservoir at pressures just below the bubble point. This is indicated by the close spacing of the iso-vol lines just below the bubble-point line on Figure 5–2.

Volatile oil reservoirs are engineered through *compositional material balance* calculations. A special laboratory study (not discussed in this text) is required.

The coefficient of isothermal compressibility is important in the study of volatile oil reservoirs. Values of compressibility are higher for volatile oils than for black oils. Values from $20 \times 10^{-6} \text{psi}^{-1}$ to $60 \times 10^{-6} \text{psi}^{-1}$ are common at pressures above the bubble point. The relationship of compressibility to pressure for volatile oils is the same as given in Figure 8–7. The discontinuity at the bubble point is greater for volatile oils than for black oils.

The viscosities of volatile oils behave as indicated in Figure 8–8. The viscosities of volatile oils are much lower than the viscosities of black oils. Values of 0.1 cp are common at the bubble point and values above 0.2 cp are rare. There is usually a tenfold increase in viscosity between bubble point and low pressure. Volatile oil viscosity is affected by pressure above the bubble point more strongly than is black oil viscosity.

Exercises

8–1. A stock-tank oil has a specific gravity of 0.875. What is its density in lb/cu ft?

8–2. What is the gravity in °API of the oil of Exercise 8–1?

8–3. A stock-tank liquid has a density of 46.4 lb/cu ft. What is its specific gravity?

8–4. What is the specific gravity of a stock-tank liquid with gravity of 47.3°API?

8–5. How many pounds does a barrel of 35.2°API oil weigh?

8–6. What is the gravity in °API of an oil with a density of 56.4 lb/cu ft?

8–7. A liquid sample from a black oil reservoir had a volume of 227.0 cc in a laboratory cell at reservoir temperature and bubble-point pressure. The liquid was expelled through laboratory equipment which is the equivalent of the field separator-stock tank system. The oil volume collected in the stock tank was 167.4 cc. The separator produced 0.537 scf of gas, and the stock tank produced 0.059 scf of gas. Calculate the formation volume factor of the oil and the solution gas-oil ratio.

8–8. You have just discovered a petroleum reservoir. Initial testing of the well produced 76 barrels of 18.2°API stock-tank oil and 14.1 Mscf of gas in 24 hours. What is the solution gas-oil ratio? At what pressures does this solution gas-oil ratio apply?

8–9. A black oil reservoir has just been discovered. Reservoir pressure appears to be above the bubble-point pressure of the oil. Measured at reservoir conditions, 86.3 barrels per day enter the wellbore. The oil is processed through a separator into a stock tank. The stock tank accumulates 57.9 barrels of 44.2°API oil each day. The separator produces 43,150 scf/d of 0.724 gravity gas, and the stock tank vents 7240 scf/d of 1.333 gravity gas. What is the formation volume factor of the oil? What is the solution gas-oil ratio? Which of the previous answers applies only at the bubble point?

$$L_0 @ 2400 psig$$

8–10. You have a laboratory analysis of a reservoir sample from an oil well producing 41.5°API stock-tank oil at 941 scf/STB. The sample was obtained from the reservoir at 184°F and 3463 psig. The results are given below.

Pressure psig	Oil formation volume gas-oil factor B_o res bbl/STB	Solution gas-oil ratio R_s scf/STB	Gas formation volume factor B_g bbl/scf
5000	1.498	941	
4500	1.507	941	
4000	1.517	941	
3500	1.527	941	
3400	1.530	941	
3300	1.532	941	
3200	1.534	941	
3100	1.537	941	
3054 = p_b	1.538	941	0.000866
2700	1.484	819	0.000974
2400	1.441	732	0.001090
2100	1.401	646	0.001252
1800	1.361	562	0.001475
1500	1.323	481	0.001795
1200	1.287	400	0.002285
900	1.252	321	0.003108
600	1.215	240	0.004760
300	1.168	137	0.009683

Plot oil formation volume factor and solution gas-oil ratio against pressure. Compare the shapes of your plots with Figures 8–1 and 8–2. Compare the shapes of the B_o graph and the R_s graph. Save your graphs; you will need them in Exercise 8–17.

8–11. Determine the value of total formation volume factor of the black oil of Exercise 8–10 at 2400 psig.

8–12. Determine the value of total formation volume factor of the black oil of Exercise 8–10 at 3500 psig.

8–13. A sample of the reservoir oil of Exercise 8–8 was placed in a laboratory cell at 300°F. Pressure was changed by increasing cell volume as follows.

Pressure, psig	Oil volume, cc
5000	219.80
4500	220.55
4000	221.33
3500	222.17
3000	223.07
2500	223.99
2200	224.57
1972 $= p_b$	225.05

Tabulate values of coefficient of isothermal compressibility for use in the pressure ranges indicated in the table. Does the trend in the results agree with the shape of Figure 8–5?

8–14. Early in its life a well produced 0.766 gravity gas at 933 scf/STB and 43.7°API stock-tank oil. A sample of reservoir oil was placed in a laboratory cell at reservoir temperature of 186°F. Pressure was varied and oil volume measured as follows.

Pressure, psig	Oil volume, cc
5000	192.10
4500	193.39
4000	194.79
3500	196.32
3300	196.99
3100	197.70
3025 $= p_b$	198.00

Tabulate and graph values of coefficient of isothermal compressibility of this oil for the pressure ranges indicated. Does your graph look like Figure 8–5?

8–15. A sample of a volatile oil was placed in a laboratory cell at reservoir temperature of 209°F. The pressure-volume relationship of the liquid was measured.

Pressure, psig	Oil volume, cc
6000	172.91
5500	174.88
5000	177.02
4500	179.48
4000	182.31
3500	185.66
3000	189.69
2600	193.78
2400	196.28
2300	197.70
2200	199.21
2100	200.92
2000	202.81
1974 = p_b	203.35

Calculate and plot the coefficient of isothermal compressibility against pressure. Does your graph look like Figure 8–5?

8–16. Determine the value of the coefficient of isothermal compressibility of the black oil of Exercise 8–10 for use between 3500 and 4000 psig.

8–17. Determine the value of the coefficient of isothermal compressibility of the black oil of Exercise 8–10 at 2400 psig.

8–18. Initial pressure is above the bubble-point pressure of a black oil in a reservoir. As production begins, reservoir pressure decreases. How do the following fluid properties change? Select "increases," "decreases," or "remains constant." Fill in the blanks.

Formation volume factor of oil   _____
Total formation volume factor   _____
Solution gas-oil ratio   _____
Oil viscosity   _____
Coefficient of isothermal compressibility of oil   _____

8–19. Reservoir pressure of Exercise 8–18 has now decreased to the bubble-point pressure of the black oil. Further production will cause changes in reservoir fluid properties. Select "increases," "decreases," or "remains constant."

Formation volume factor of oil   _____
Total formation volume factor   _____
Solution gas-oil ratio   _____
Oil viscosity   _____
Coefficient of isothermal compressibility of oil   _____
Formation volume factor of the free gas   _____

Reference

1. Martin, J.C.: "Simplified Equations of Flow in Gas Drive Reservoirs and the Theoretical Foundation of Multiphase Pressure Buildup Analyses," *Trans.* AIME (1959) *216*, 309–311.

9

Properties of Black Oils—
Field Data

An examination of information readily available at the surface, early in the life of the reservoir, usually can lead to an understanding of the type of fluid in the reservoir. The rules of thumb given in Chapter 5 with regard to initial producing gas-oil ratio, stock-tank oil gravity, and stock-tank oil color can be used to determine the type of fluid.

We will see in this chapter that field data can also be used to determine some properties of the reservoir fluid—in particular, bubble-point pressure and solution gas-oil ratio at the bubble point. These two properties are insufficient to permit prediction of reservoir behavior, but they can be used to check the results of reservoir fluid studies or correlations.

Black Oil Reservoirs—Initial Reservoir Pressure

The pressure in a black oil reservoir at discovery will be either higher than the bubble-point pressure of the oil or equal to the bubble-point pressure.

Discovery pressure, also called *initial reservoir pressure,* will not be below the bubble-point pressure of the reservoir oil. If reservoir pressure were below the bubble-point pressure of the liquid, gas would have formed and migrated to the top of the reservoir.

This zone of gas, called a *gas cap,* lies in the pore space above a zone of oil. When a gas cap is present, the initial reservoir pressure, measured at the interface between the gas cap and the oil zone, is equal to the bubble-point pressure of the oil.

Possibly, a reservoir could be discovered with the oil at its bubble point with a gas zone so small that it would be difficult to identify. Normally, if no gas cap is present, the initial reservoir pressure is above the bubble-point pressure of the oil.

Black Oil Reservoirs—Gas Production Trends

Consider a black oil at a reservoir pressure above the bubble point. Production into the wellbore will consist solely of liquid. During the trip to the surface and through surface equipment to the stock tank, the *dissolved gas* will come out of solution. This gas will appear as separator gas and stock-tank vent gas. Typical surface facilities for a black oil are shown in Figure 13–1.

The quantity of gas produced with each barrel of stock-tank oil will remain constant as long as reservoir pressure is above bubble-point pressure. The usual methods of keeping track of surface gas and stock-tank oil will cause some variance in the producing gas-oil ratio, but the trend will be constant.

As production proceeds, reservoir pressure will decrease. When bubble-point pressure is reached, gas forms in the pore space. At first this *free gas* does not move. However, as pressure continues to decrease, the volume of gas in the pore space increases, and the gas begins to move to the wellbore. At this point the producing gas-oil ratio at the surface increases since it consists of both *dissolved gas* and *free gas*.

Due to its lower viscosity and other factors, gas flows through the reservoir more readily than oil. Producing gas-oil ratio will increase throughout most of the remaining life of the reservoir. Just before abandonment, when reservoir pressure is very low, producing gas-oil ratio will decline.

Black Oil Reservoirs—Pressure Trends

Reservoir pressure decreases as liquid is removed from the reservoir. At pressures above the bubble point, the oil, water, and reservoir rock must expand to fill the void created by the removal of liquid. The rock and remaining liquids are not very compressible. So a large decrease in pressure is necessary to allow the rock and remaining liquids to expand enough to replace a relatively small amount of oil produced. Thus, as long as reservoir pressure is above the bubble point, pressure decreases rapidly during production.

At pressures below the bubble point, gas forms in the pore space. This free gas occupies considerably more space as a gas than it did as a liquid. Also, the gas readily expands as pressure decreases further. The forming and expanding gas replaces most of the void created by production. Reservoir pressure does not decrease as rapidly as it does when pressure is above the bubble point. Excessive production of this gas is detrimental to maintenance of reservoir pressure.

If the reservoir is connected to a large aquifer, water encroachment will offset the produced fluids, and these pressure trends will be obscured.

Fluid Properties from Production—Pressure History

Figure 9–1 gives production pressure history data for a typical black oil reservoir. The reservoir in this case does not have a gas cap and is not connected to an aquifer. The conventional method of plotting these data is against time. However, the trends we want to see are more apparent when the data are plotted against cumulative production, as in Figure 9–2.

The plot of reservoir pressure exhibits two different slopes. The pressure at which the slope changes is the bubble-point pressure of the reservoir oil.

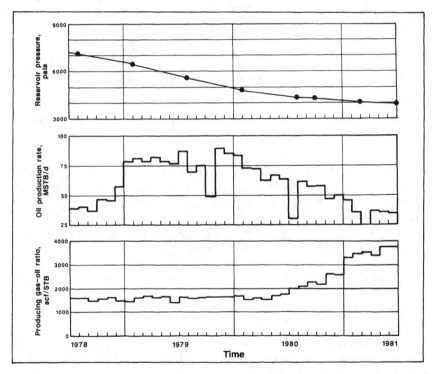

Fig. 9–1. Production-pressure history of a black oil reservoir plotted against time.

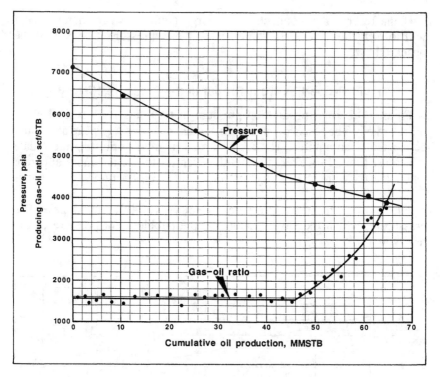

Fig. 9–2. Production-pressure history of Figure 9–1 plotted against cumulative oil production.

Although the producing gas-oil ratio data show some scatter, the initial trend is constant. The producing gas-oil ratio will start to increase very soon after bubble-point pressure is reached in the reservoir. Figure 9–2 shows that producing gas-oil ratio starts its increase at about the same cumulative production as the pressure line changes slope. This gives a check on the estimate of bubble-point pressure.

EXAMPLE 9–1: *Estimate bubble-point pressure from the pressure-production history of Figure 9–2.*

Solution

The pressure data on Figure 9–2 appear to describe two straight lines which cross at about 4550 psia.

$$p_b = 4550 \text{ psia}$$

NOTE: Laboratory analysis of a sample of this reservoir oil gave a bubble-point pressure of 4565 psia.

No dissolved gas is lost from the oil in the reservoir at pressures above bubble point. Thus the producing gas-oil ratio represents the amount of dissolved gas. The horizontal trend of producing gas-oil ratio is equal to the solution gas-oil ratio. This is valid for pressures from initial reservoir pressure to bubble-point pressure.

EXAMPLE 9–2: *Estimate the solution gas-oil ratio at pressures above the bubble point for the reservoir oil of Figure 9–2. It is known that the gas includes both separator and stock-tank gas.*

Solution
The early producing gas-oil ratio appears to have a horizontal trend of about 1570 scf/STB.
NOTE: Laboratory analysis of a sample of this reservoir oil gave a solution gas-oil ratio at the bubble point of 1535 scf/STB.

The producing gas-oil ratio data of Figure 9–2 start to trend upward somewhat later than the point at which the bubble point is reached. This is not unusual.

The first gas that comes out of solution in the reservoir does not flow. Sufficient gas volume must be present in the reservoir in order for gas flow paths to form. Thus, the producing gas-oil ratio decreases slightly at reservoir pressures immediately below bubble point. Then producing gas-oil ratio increases as the quantity of gas in the reservoir increases.

This slight decrease in producing gas-oil ratio usually is obscured by the scatter in the data. However, the slight delay in response is often visible on plots such as Figure 9–2.

Adjustment of Surface Gas Data

Records of the volume of gas produced from the separators are normally available. Very often, the gas from the stock tank is vented. Since it is not sold, its volume is not measured. Separator gas-oil ratio in standard cubic feet of separator gas per barrel of stock-tank oil will show the same trend as in Figure 9–2. However, if the vent gas from the stock tank is ignored, the estimate of solution gas-oil ratio above the bubble point can be low by 10 percent to 20 percent or more.

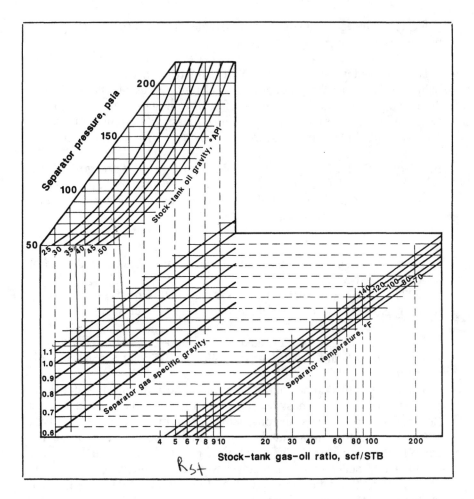

Fig. 9-3. Stock-tank gas-oil ratio correlation.

$$R_{sb} = R_s + R_{st}$$

Figure 9-3 may be used to estimate to volume of stock-tank gas given other surface data.[1]

Solution gas-oil ratio at pressures above the bubble point is calculated by adding the estimate of stock-tank gas-oil ratio from Figure 9-3 to the producing separator gas-oil ratio.

EXAMPLE 9–3: *A reservoir is producing 40.7°API stock-tank oil and 0.786 specific gravity separator gas. Separator gas-oil ratio appears to be constant at 676 scf/STB. Separator conditions are 100 psig and 75°F. Estimate solution gas-oil ratio at the bubble point for this black oil.*

Solution

$$R_{ST} = 80 \text{ scf/STB, Figure 9–3}$$

$$R_{sb} = 676 + 80 = 756 \text{ scf/STB}$$

Calculations of the type given in Example 9–3 have been made for over 500 black oil reservoirs and compared with laboratory results.[1] The average absolute error of R_{sb} was 3 percent.

After reservoir pressure drops below bubble-point pressure, separator gas-oil ratio will increase. However, stock-tank gas-oil ratio will remain nearly constant through the life of the reservoir if separator conditions remain the same.

Volatile Oils

The production data from a volatile oil reservoir exhibit the same trends as those from a black oil, but reservoir pressure above the bubble point for a volatile oil does not decrease as rapidly as for a black oil. And the change in slope of the plot of pressure against cumulative production for a volatile oil is not as sharp as it is for a black oil. The bubble-point pressure is often hard to identify.

Retrograde Gases

Producing gas-oil ratio from a retrograde gas reservoir is also constant during early production and then increases. When reservoir pressure is above the dew point, the gas carries a constant quantity of components which will liquefy at surface conditions. However, at reservoir pressures below the dew point some of these components condense in the reservoir. This liquid does not flow to any appreciable degree. Thus, the produced gas carries fewer condensable components to the surface and producing gas-oil ratio increases. Dew point pressure cannot be determined from a plot of pressure versus cumulative production; the change in slope at the dew point is rarely apparent.

Exercises

9–1. The following pressure-production history is available for an oil well. The producing gas-oil ratio is calculated with sales-gas volumes, i.e., stock-tank gas is vented and not measured. The separator operates at 67°F and 100 psig; the specific gravity of the separator gas is 0.788, and the gravity of the stock-tank oil is 46.0° API. Estimate the bubble-point pressure and the solution gas-oil ratio at the bubble point for this black oil. Compare your answers with laboratory data which indicate a bubble-point pressure of 1928 psig and a solution gas-oil ratio at the bubble point of 623 scf/STB.

Cumulative oil production, STB	Producing gas-oil ratio, scf/STB	Reservoir pressure, psia
11,800	723	–
60,800	493	–
96,000	–	5,931
145,100	518	–
255,800	486	–
379,800	499	–
465,800	–	4,770
513,600	483	–
663,400	530	–
816,200	537	–
961,000	530	–
1,007,000	–	2,652
1,098,300	491	–
1,221,000	514	–
1,354,200	613	–
1,503,600	753	–
1,557,500	–	1,829
1,649,900	848	–
1,801,800	871	–
1,950,000	764	–
1,997,600	–	1,640
2,065,700	820	–

9–2. An oil field with several wells has the following production history. Producing gas-oil ratio is really separator gas-oil ratio; stock-tank gas is vented and not metered. Separator conditions are

30 psig and 120°F. Separator gas specific gravity is 0.888, and stock-tank oil gravity is 36.1°API. Estimate solution gas-oil ratio at the bubble point. Compare your answer with laboratory results of 318 scf/STB.

Cumulative oil production, MSTB	Producing gas-oil ratio, scf/STB
260	310
455	322
761	286
1,024	305
1,319	317
1,699	402
1,910	434
2,201	515
2,400	575
2,607	610
2,800	661
2,959	602
3,111	723
3,231	769
3,360	848
3,480	911
3,591	974
3,676	990
3,799	1,092
3,890	1,044
3,992	1,143
4,039	1,170
4,110	1,318
4,205	1,271
4,275	1,407
4,317	1,485
4,390	1,494
4,435	1,373
4,500	1,467
4,540	1,305
4,600	1,446

9–3. The pressure-production history for the oil field of Exercise 9–2 is given below. Estimate bubble-point pressure. Compare your answer with laboratory results of 1441 psig.

Reservoir pressure, psia	Cumulative oil production, MSTB
3,122	110
2,982	370
1,830	1,050
1,275	1,400
965	3,060
1,095	3,160
933	3,360
495	4,620

9–4. An oil well is producing 930 scf/STB of 0.742 specific gravity separator gas and 40.8°API stock-tank oil. Separator conditions are 100 psig and 90°F. The well has just been completed; reservoir pressure is believed to be above bubble-point pressure. Estimate a value of solution gas-oil ratio for use at and above bubble-point pressure.

9–5. Estimate the solution gas-oil ratio of the black oil of Exercise 5–11. The reported gas production is separator gas with a specific gravity of 0.7. The separator is a heater/treater operating at 100 psig and 140°F.

Reference

1. Rollins, J.B., McCain, W.D., Jr., and Creeger, J.T.: "Estimation of Solution Gas-Oil Ratio of Black Oils," *J. Pet. Tech.* (Jan. 1990) *42*, 92-94.

10

Properties of Black Oils—Reservoir Fluid Studies Flash vap, diff vap

A black oil *reservoir fluid study* consists of a series of laboratory procedures designed to provide values of the physical properties required in the calculation method known as *material balance calculations*. There are five main procedures in the black oil reservoir fluid study. These procedures are performed with samples of reservoir liquid.

The following physical properties can be determined from the results of a black oil reservoir fluid study:

> bubble-point pressure,
> formation volume factor of oil,
> solution gas-oil ratio,
> total formation volume factor,
> coefficient of isothermal compressibility of oil, and
> oil viscosity,

as functions of pressure as pressure decreases from initial reservoir pressure through the bubble point to very low pressure. Also determined from the study are

> z-factor,
> formation volume factor of gas, and
> gas viscosity

of the gas evolved in the reservoir at pressures below the bubble point.

In addition, quantities and properties of

> separator gas,
> stock-tank gas, and
> stock-tank oil

at various separator pressures are determined.

The five major procedures are

> composition measurement,
> flash vaporization,

differential vaporization,
separator tests, and
oil viscosity measurement.
The results of these procedures are called a reservoir fluid study. Often
the term *PVT study* is used.

Collection of Reservoir Oil Samples

A sample which is representative of the liquid originally in the
reservoir must be obtained for the laboratory work. Sampling procedures
are discussed in detail elsewhere; we will not cover them here.[1,2]
Obtaining a representative sample of the reservoir liquid requires great
care in both conditioning the well and in the sampling technique.

Samples can be obtained in two ways. In one method the well is shut
in, and the liquid at the bottom of the wellbore is sampled. This is called
a *bottom-hole sample* or a *subsurface sample*.

In the other method, production rate is carefully controlled, and
separator gas and separator liquid are sampled. These are called
separator samples or *surface samples*. The gas and liquid are recombined at the producing ratio in order to obtain a sample representative of
the reservoir liquid.

An oil reservoir must be sampled before reservoir pressure drops
below the bubble-point pressure of the reservoir liquid. At reservoir
pressures below the bubble point, no sampling method will give a sample
representative of the original reservoir mixture. Bottom-hole samples
generally will contain less gas than the original liquid since some gas has
evolved. Separator samples will be recombined at the wrong ratio
because free gas is produced from the reservoir with the reservoir liquid.

Reservoir Fluid Study

First, we will examine the five major tests performed during a
reservoir fluid study. Later we will see how to convert the results into
fluid properties of interest to the petroleum engineer.

Compositions

Determining the composition of every one of the hundreds of different
chemical species present in a black oil is impossible. Even determining
the composition of a major fraction of the crude is difficult. In every case
the compositions of the light components are determined, and all of the
heavier components are grouped together in a *plus component*. The plus
component consists of hundreds of different chemical species.

The usual analysis gives compositions, in mole fraction or mole
percent, of the light hydrocarbons from methane through hexanes. The

fraction reported as hexanes contains various isomers of hexane as well as some of the lighter naphthenes. The remaining components are lumped together as *heptanes plus*. The apparent molecular weight and specific gravity of the heptanes plus fraction is measured in an attempt to characterize its properties. A typical analysis is shown on page 2 of the example reservoir fluid study given in Table 10–1.

Table 10–1

Typical black oil reservoir fluid study (courtesy Core Laboratories, Inc.)

Page __1__ of __15__

File __RFL 76000__

Company__Good Oil Company__   Date Sampled_____

Well__Oil Well No. 4__   County__Samson__

Field__Productive__   State__Texas__

FORMATION CHARACTERISTICS

Formation Name	Cretaceous
Date First Well Completed	_____, 19___
Original Reservoir Pressure	4100 PSIG @ 8692 Ft.
Original Produced Gas-Oil Ratio	600 SCF/Bbl
Production Rate	300 Bbl/Day
Separator Pressure and Temperature	200 PSIG. 75 °F.
Oil Gravity at 60° F.	°API
Datum	8000 Ft. Subsea
Original Gas Cap	No

WELL CHARACTERISTICS

Elevation	610 Ft.
Total Depth	8943 Ft.
Producing Interval	8684-8700 Ft.
Tubing Size and Depth	2-7/8 In. to 8600 Ft.
Productivity Index	1.1 Bbl/D/PSI @ 300 Bbl/Day
Last Reservoir Pressure	3954* PSIG @ 8500 Ft.
Date	_____, 19___
Reservoir Temperature	217* °F. @ 8500 Ft.
Status of Well	Shut in 72 hours
Pressure Gauge	Amerada
Normal Production Rate	300 Bbl/Day
Gas-Oil Ratio	600 SCF/Bbl
Separator Pressure and Temperature	200 PSIG, 75 °F.
Base Pressure	14.65 PSIA
Well Making Water	None % Cut

SAMPLING CONDITIONS

Sampled at	8500 Ft.
Status of Well	Shut in 72 hours
Gas-Oil Ratio	SCF/Bbl
Separator Pressure and Temperature	PSIG, °F.
Tubing Pressure	1400 PSIG
Casing Pressure	PSIG
Sampled by	
Type Sampler	Wofford

REMARKS:

* Pressure and temperature extrapolated to the mid-point of the producing interval = 4010 PSIG and 220°F.

Table 10-1 (cont.)

Page __2__ of __15__

File __RFL 76000__

Company __Good Oil Company__   Formation __Cretaceous__

Well __Oil Well No. 4__   County __Samson__

Field __Productive__   State __Texas__

HYDROCARBON ANALYSIS OF __Reservoir Fluid__ SAMPLE

COMPONENT	MOL PERCENT	WEIGHT PERCENT	DENSITY @ 60° F. GRAMS PER CUBIC CENTIMETER	° API @ 60° F.	MOLECULAR WEIGHT
Hydrogen Sulfide	Nil	Nil			
Carbon Dioxide	0.91	0.43			
Nitrogen	0.16	0.05			
Methane	36.47	6.24			
Ethane	9.67	3.10			
Propane	6.95	3.27			
iso-Butane	1.44	0.89			
n-Butane	3.93	2.44			
iso-Pentane	1.44	1.11			
n-Pentane	1.41	1.09			
Hexanes	4.33	3.97			
Heptanes plus	33.29	77.41	0.8515	34.5	218
	100.00	100.00			

Table 10-1 (cont.)

Page __3__ of __15__
File __RFL 76000__
Well __Oil Well No. A__

VOLUMETRIC DATA OF __Reservoir Fluid__ SAMPLE

1. Saturation pressure (bubble-point pressure)   __2620__ PSIG @ __220__ °F.

2. Specific volume at saturation pressure: ft³/lb   __0.02441__ @ __220__ °F.

3. Thermal expansion of saturated oil @ __5000__ PSI = $\dfrac{V @ \;220\; °F}{V @ \;76\; °F}$ = __1.08790__

4. Compressibility of saturated oil @ reservoir temperature: Vol/Vol/PSI:

$$\text{From } \underline{5000} \text{ PSI to } \underline{4000} \text{ PSI} = \underline{13.48} \times 10^{-6}$$

$$\text{From } \underline{4000} \text{ PSI to } \underline{3000} \text{ PSI} = \underline{15.88} \times 10^{-6}$$

$$\text{From } \underline{3000} \text{ PSI to } \underline{2620} \text{ PSI} = \underline{18.75} \times 10^{-6}$$

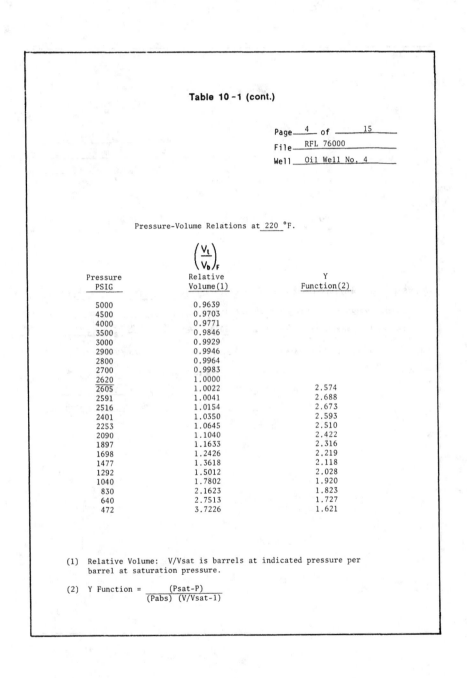

Table 10 -1 (cont.)

Pressure-Volume Relations at _220_ °F.

$$\left(\frac{V_t}{V_b}\right)_F$$

Pressure PSIG	Relative Volume(1)	Y Function(2)
5000	0.9639	
4500	0.9703	
4000	0.9771	
3500	0.9846	
3000	0.9929	
2900	0.9946	
2800	0.9964	
2700	0.9983	
2620	1.0000	
2605	1.0022	2.574
2591	1.0041	2.688
2516	1.0154	2.673
2401	1.0350	2.593
2253	1.0645	2.510
2090	1.1040	2.422
1897	1.1633	2.316
1698	1.2426	2.219
1477	1.3618	2.118
1292	1.5012	2.028
1040	1.7802	1.920
830	2.1623	1.823
640	2.7513	1.727
472	3.7226	1.621

(1) Relative Volume: V/Vsat is barrels at indicated pressure per barrel at saturation pressure.

(2) Y Function = $\dfrac{(Psat-P)}{(Pabs) \ (V/Vsat-1)}$

Table 10-1 (cont.)

Page 5 of 15
File REL 76000
Well Oil Well No. 4

Differential Vaporization at 220 °F.

Pressure PSIG	R_{sD} Solution Gas/Oil Ratio(1)	B_{oD} Relative Oil Volume(2)	B_{tD} Relative Total Volume(3)	Oil Density gm/cc	Deviation Factor Z	Gas Formation Volume Factor(4)	Incremental Gas Gravity
2620	854 = R_{sDb}	1.600 = B_{oDb}	1.600	0.6562			
2350	763	1.554	1.665	0.6655	0.846	0.00685	0.825
2100	684	1.515	1.748	0.6731	0.851	0.00771	0.818
1850	612	1.479	1.859	0.6808	0.859	0.00882	0.797
1600	544	1.445	2.016	0.6889	0.872	0.01034	0.791
1350	479	1.412	2.244	0.6969	0.887	0.01245	0.794
1100	416	1.382	2.593	0.7044	0.903	0.01552	0.809
850	354	1.351	3.169	0.7121	0.922	0.02042	0.831
600	292	1.320	4.254	0.7198	0.941	0.02931	0.881
350	223	1.283	6.975	0.7291	0.965	0.05065	0.988
159	157	1.244	14.693	0.7382	0.984	0.10834	1.213
0	0	1.075		0.7892			2.039

@ 60°F. = 1.000

Gravity of residual oil = 35.1°API @ 60°F.

(1) Cubic feet of gas at 14.65 psia and 60°F. per barrel of residual oil at 60°F.
(2) Barrels of oil at indicated pressure and temperature per barrel of residual oil at 60°F.
(3) Barrels of oil plus liberated gas at indicated pressure and temperature per barrel of residual oil at 60°F.
(4) Cubic feet of gas at indicated pressure and temperature per cubic foot at 14.65 psia and 60°F.

Table 10-1 (cont.)

Page __6__ of __15__
File __RFL 76000__
Well __Oil Well No. 4__

Viscosity Data at __220__ °F.

Pressure PSIG	Oil Viscosity Centipoise	Calculated Gas Viscosity Centipoise	Oil/Gas Viscosity Ratio
5000	0.450		
4500	0.434		
4000	0.418		
3500	0.401		
3000	0.385		
2800	0.379		
2620	0.373		
2350	0.396	0.0191	20.S
2100	0.417	0.0180	23.2
1850	0.442	0.0169	26.2
1600	0.469	0.0160	29.4
1350	0.502	0.0151	33.2
1100	0.542	0.0143	37.9
850	0.592	0.0135	43.9
600	0.654	0.0126	51.S
350	0.738	0.0121	60.9
159	0.855	0.0114	75.3
0	1.286	0.0093	137.9

Table 10-1 (cont.)

Page___7___of_____15_____

File_____RFL 76000_____

Well_____Oil Well No. 4_____

SEPARATOR TESTS OF___Reservoir Fluid___SAMPLE

SEPARATOR PRESSURE, PSI GAUGE	SEPARATOR TEMPERATURE, °F	GAS/OIL RATIO (1)	R_{sSb} GAS/OIL RATIO (2)	STOCK TANK GRAVITY, °API @ 60° F.	B_{oSb} FORMATION VOLUME FACTOR (3)	SEPARATOR VOLUME FACTOR (4)	SPECIFIC GRAVITY OF FLASHED GAS
50	75	715	737			1.031	0.840
to							
0	75	41	41	40.5	1.481	1.007	1.338
			778				
100	75	637	676			1.062	0.786
to							
0	75	91	92	40.7	1.474	1.007	1.363
			768				
200	75	542	602			1.112	0.732
to							
0	75	177	178	40.4	1.483	1.007	1.329
			780				
300	75	478	549			1.148	0.704
to							
0	75	245	246	40.1	1.495	1.007	1.286
			795				

(1) Gas/Oil Ratio in cubic feet of gas @ 60° F. and___14.65___PSI absolute per barrel of oil @ indicated pressure and temperature.

(2) Gas/Oil Ratio in cubic feet of gas @ 60° F. and___14.65___PSI absolute per barrel of stock tank oil @ 60° F.

(3) Formation Volume Factor is barrels of saturated oil @ ___2620___PSI gauge and___220___° F. per barrel of stock tank oil @ 60° F.

(4) Separator Volume Factor is barrels of oil @ indicated pressure and temperature per barrel of stock tank oil @ 60° F.

Table 10-1 (cont.)

Page___8___of___15___

File___RFL 76000___

Company___Good Oil Company___   Formation___Cretaceous___

Well___Oil Well No. 4___   County___Samson___

Field___Productive___   State___Texas___

HYDROCARBON ANALYSIS OF___Separator___ GAS SAMPLE

COMPONENT	MOL PERCENT	G P M
Hydrogen Sulfide	Nil	
Carbon Dioxide	1.62	
Nitrogen	0.30	
Methane	67.00	
Ethane	16.04	4.265
Propane	8.95	2.449
iso-Butane	1.29	0.420
n-Butane	2.91	0.912
iso-Pentane	0.53	0.193
n-Pentane	0.41	0.155
Hexanes	0.44	0.178
Heptanes plus	0.49	0.221
	100.00	8.793

Calculated gas gravity (air = 1.000) =   0.840

Calculated gross heating value = 1405 **BTU**
per cubic foot of dry gas at 14.65 psia at 60° F.

Collected at   50   **psig and**   75   ° F. in the laboratory.

Table 10- 1 (cont.)

Company Good Oil Company   Formation Cretaceous

Well Oil Well No. 4   County Samson

Field Productive   State Texas

HYDROCARBON ANALYSIS OF Separator **GAS SAMPLE**

COMPONENT	MOL PERCENT	G P M
Hydrogen Sulfide	Nil	
Carbon Dioxide	1.67	
Nitrogen	0.32	
Methane	71.08	
Ethane	15.52	4.127
Propane	7.36	2.014
iso-Butane	0.92	0.299
n-Butane	1.98	0.621
iso-Pentane	0.33	0.120
n-Pentane	0.26	0.094
Hexanes	0.27	0.110
Heptanes plus	0.29	0.131
	100.00	7.516

Calculated gas gravity (air = 1.000) = 0.786

Calculated gross heating value = 1321 BTU
per cubic foot of dry gas at 14.65 psia at 60° F.

Collected at 100 psig and 75 ° F. in the laboratory.

Table 10−1 (cont.)

Page 10 of 15

File RFL 76000

Company Good Oil Company   Formation Cretaceous

Well Oil Well No. 4   County Samson

Field Productive   State Texas

HYDROCARBON ANALYSIS OF Separator GAS SAMPLE

COMPONENT	MOL PERCENT	G P M
Hydrogen Sulfide	Nil	
Carbon Dioxide	1.68	
Nitrogen	0.36	
Methane	76.23	
Ethane	13.94	3.707
Propane	5.31	1.453
iso-Butane	0.57	0.185
n-Butane	1.21	0.379
iso-Pentane	0.20	0.073
n-Pentane	0.16	0.058
Hexanes	0.16	0.065
Heptanes plus	0.18	0.081
	100.00	6.001

Calculated gas gravity (air = 1.000) = 0.732

Calculated gross heating value = 1236 BTU
per cubic foot of dry gas at 14.65 psia at 60° F.

Collected at 200 psig and 75 ° F. in the laboratory.

Table 10-1 (cont.)

Company **Good Oil Company** Formation **Cretaceous**

Well **Oil Well No. 4** County **Samson**

Field **Productive** State **Texas**

HYDROCARBON ANALYSIS OF Separator GAS SAMPLE

COMPONENT	MOL PERCENT	G P M
Hydrogen Sulfide	Nil	
Carbon Dioxide	1.65	
Nitrogen	0.39	
Methane	79.42	
Ethane	12.48	3.318
Propane	4.21	1.152
iso-Butane	0.43	0.140
n-Butane	0.90	0.282
iso-Pentane	0.15	0.055
n-Pentane	0.12	0.043
Hexanes	0.12	0.049
Heptanes plus	0.13	0.059
	100.00	5.098

Calculated gas gravity (air = 1.000) = 0.704

Calculated gross heating value = 1192 **BTU**
per cubic foot of dry gas at 14.65 psia at 60° F.

Collected at 300 **psig** and 75 ° **F.** in the laboratory.

Core Laboratories, Inc.

Manager
Reservoir Fluid Analysis

The analysis can be carried somewhat further. Table 10–2 gives an analysis of a heptanes plus fraction carried out to a carbon number of 30.

When surface samples are used the compositions of both the gas and liquid are measured. The composition of the well stream is calculated in the manner of Example 7–2.

TABLE 10–2

Heptanes plus analysis of a typical separator liquid sample

Component	Weight percent	Mole percent
Hexanes	0.06	0.12
Methylcyclopentane	0.39	0.81
Benzene	0.10	0.21
Cyclohexane	0.51	1.06
Heptanes	4.89	8.54
Methylcyclohexane	1.58	2.82
Toluene	0.68	1.29
Octanes	6.07	9.30
Ethylbenzene	0.30	0.49
Meta & Para Xylenes	0.87	1.43
Orthoxylene	0.40	0.66
Nonanes	7.00	9.57
iso-Propyl Benzene	0.31	0.45
n-Propyl Benzene	0.51	0.74
1,2,4-Trimethylbenzene	1.17	1.70
Decanes	6.43	7.91
Undecanes	7.93	8.90
Dodecanes	6.76	6.95
Tridecanes	6.78	6.44
Tetradecanes	6.47	5.71
Pentadecanes	5.86	4.83
Hexadecanes	4.80	3.71
Heptadecanes	3.84	2.78
Octadecanes	3.36	2.31
Nonadecanes	2.90	1.88
Eicosanes	2.26	1.40
Heneicosanes	2.03	1.19
Docosanes	1.82	1.02
Tricosanes	1.54	0.83
Tetracosanes	1.31	0.68
Pentacosanes	1.13	0.56
Hexacosanes	1.32	0.63
Heptacosanes	1.08	0.50
Octacosanes	1.14	0.50
Nonacosanes	0.98	0.42
Triacontanes plus	5.42	1.66
	100.00	100.00

Average molecular weight: 175 lb/lb mole

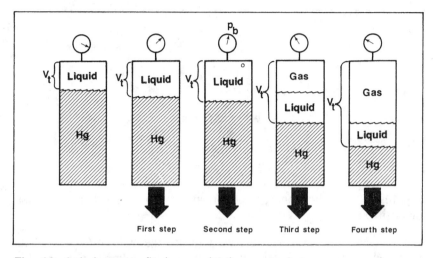

Fig. 10–1. Laboratory flash vaporization procedure.

Flash Vaporization

A sample of the reservoir liquid is placed in a laboratory cell. Pressure is adjusted to a value equal to or greater than initial reservoir pressure. Temperature is set at reservoir temperature. Pressure is reduced by increasing the volume in increments. The procedure is illustrated in Figure 10–1.

The cell is agitated regularly to ensure that the contents are at equilibrium. No reservoir gas or liquid is removed from the cell.

At each step, pressure and volume of the reservoir fluid are measured. The volume is termed total volume, V_t, since at pressures below the bubble point the volume includes both gas and liquid.

This procedure is called *flash vaporization, flash liberation, pressure-volume relations, constant composition expansion,* or *flash expansion.*

Pressure is plotted against total volume, as in Figure 10–2. The plot reproduces part of an isotherm of a pressure-volume diagram. The shape is similar to that shown in Figure 2–20.

The pressure at which the slope changes is the bubble-point pressure of the mixture. The volume at this point is the volume of the bubble-point liquid. Often it is given the symbol V_{sat}. The volume of the bubble-point liquid can be divided by the mass of reservoir fluid in the cell to obtain a value of specific volume at the bubble point. Specific volume at the bubble point also is measured during other tests and is used as a check on the quality of the data.

All values of total volume, V_t, are divided by volume at the bubble point, and the data are reported as relative volume. Sometimes the

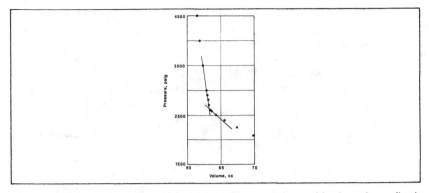

Fig. 10-2. Determination of bubble-point pressure with data from flash vaporization. ⎵here slope changes

symbol V/V_{sat} is used; however, we will use the symbol $(V_t/V_b)_F$. The symbol $(V_t/V_b)_F$ means total volume divided by volume at the bubble point for a flash vaporization.

The results of a flash vaporization are given on page 4 of the example black oil reservoir fluid study, Table 10-1.

EXAMPLE 10-1: *The data from a flash vaporization on a black oil at 220°F are given below. Determine the bubble-point pressure and prepare a table of pressure and relative volume for the reservoir fluid study.*

rel. vol

@ 2900 psig

Pressure, psig	Total volume, cc
5000	61.030
4500	61.435
4000	61.866
3500	62.341
3000	62.866
2900	62.974
2800	63.088
2700	63.208
2605	63.455
2591	63.576
2516	64.291
2401	65.532
2253	67.400
2090	69.901
1897	73.655
1698	78.676
1477	86.224
1292	95.050
1040	112.715
830	136.908
640	174.201
472	235.700

Solution

First, plot pressure against total volume, determine the point at which the two lines cross, and read bubble-point pressure and volume at the bubble point.

$$p_b = 2620 \text{ psig}, \quad V_b = 63.316 \text{ cc} \quad \text{(see Figure 10–2)}$$

Second, divide all volumes in the data set by V_b and add p_b to the table. For instance, at 2516 psig,

$$\text{relative volume} = \frac{64.291 \text{ cc}}{63.316 \text{ cc}} = 1.0154$$

$$= V_t / V_b$$

See page 4 of Table 10–1.

Differential Vaporization

The sample of reservoir liquid in the laboratory cell is brought to bubble-point pressure, and temperature is set at reservoir temperature.

Pressure is reduced by increasing cell volume, and the cell is agitated to ensure equilibrium between the gas and liquid. Then, all the gas is expelled from the cell while pressure in the cell is held constant by reducing cell volume.

The gas is collected, and its quantity and specific gravity are measured. The volume of liquid remaining in the cell, V_o, is measured. This process is shown in Figure 10–3.

The process is repeated in steps until atmospheric pressure is reached. Then temperature is reduced to 60°F, and the volume of remaining liquid is measured. This is called *residual oil from differential vaporization* or *residual oil*.

This process is called *differential vaporization, differential liberation,* or *differential expansion*.

Each of the values of volume of cell liquid, V_o, is divided by the volume of the residual oil. The result is called *relative oil volume* and is given the symbol B_{oD}.

The volume of gas removed during each step is measured both at cell conditions and at standard conditions. The z-factor is calculated as

$$z = \frac{V_R p_R T_{sc}}{V_{sc} p_{sc} T_R}, \tag{6–1}$$

where the subscript R refers to conditions in the cell. Formation volume factors of the gas removed are calculated with these z-factors using Equation 6–2.

$$B_g = 0.0282 \, \frac{zT}{p} \, \frac{cu\ ft}{scf} \tag{6-2}$$

psia

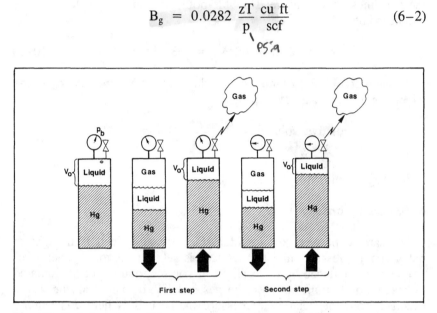

Fig. 10–3. Laboratory differential vaporization procedure.

The total volume of gas removed during the entire process is the amount of gas in solution at the bubble point. This total volume is divided by the volume of residual oil, and the units are converted to standard cubic feet per barrel of residual oil. The symbol R_{sDb} represents standard cubic feet of gas removed per barrel of residual oil.

The gas remaining in solution at any lower pressure is calculated by subtracting the sum of the gas removed down to and including the pressure of interest from the total volume of gas removed. The result is divided by the volume of residual oil, converted to scf/residual bbl, and reported as R_{sD}.

Relative total volume at any pressure is calculated as

$$B_{tD} = B_{oD} + B_g(R_{sDb} - R_{sD}). \tag{10-1}$$

The results of a differential vaporization are shown on page 5 of Table 10–1.

EXAMPLE 10–2: *The data from a differential vaporization on a black oil at 220°F are given below. Prepare a table of solution gas-oil ratios, relative oil volumes, and relative total volumes by this differential process. Also include z-factors and formation volume factors of the increments of gas removed.*

[handwritten: 1100 psig]

Pressure psig	Gas removed,* cc	Gas removed,** scf	Oil volume, in cell CC	Incremental gas specific gravity
2620	–	–	63.316	–
2350	4.396	0.02265	61.496	0.825
2100	4.292	0.01966	59.952	0.818
1850	4.478	0.01792	58.528	0.797
1600	4.960	0.01693	57.182	0.791
1350	5.705	0.01618	55.876	0.794
1100	6.891	0.01568	54.689	0.809
850	8.925	0.01543	53.462	0.831
600	12.814	0.01543	52.236	0.881
350	24.646	0.01717	50.771	0.988
159	50.492	0.01643	49.228	1.213
0		0.03908	42.540	2.039
0		0.21256	39.572 at 60°F	

*at 220°F and cell pressure
**at 60°F and 14.65 psia

Solution

[handwritten: R_{SD}, B_{oD}, Z, B_g @ 1100 psig]

All calculations shown will be at 2100 psig.
First, calculate solution gas-oil ratio.

[handwritten: $R_S D_b = \dfrac{\text{Tot.Vol gas (cm}}{\text{residual oil}}$]

$$R_{sD} = \frac{(0.21256 - 0.02265 - 0.01966)\text{scf}}{(39.572 \text{ cc residual oil})(6.29 \times 10^{-6} \text{ bbl/cc})}$$

$$R_{sD} = 684 \text{ scf/residual bbl}$$

Second, calculate relative oil volume.

[handwritten: $B_{od} = \dfrac{V_o}{V_{op}}$, oil in cell, oil in cell at 60°F]

$$B_{oD} = \frac{59.952 \text{ reservoir cc}}{39.572 \text{ residual cc}} = 1.515 \frac{\text{res bbl}}{\text{residual bbl}}$$

Third, calculate z-factor.

$$z = \frac{V_R p_R T_{sc}}{V_{sc} p_{sc} T_R} \qquad (6-1)$$

[handwritten: $RSd = \dfrac{\text{Total volume of gas removed} - \text{Volume removed from } P \text{ to } P_b}{\text{residual oil volume}}$]

$$z = \frac{(4.292 \text{ cc})(35.315 \times 10^{-6} \text{cu ft/cc})(2114.7 \text{ psia})(520°R)}{(0.01966 \text{ scf})(14.65 \text{ psia})(680°R)}$$

$$z = 0.851$$

Fourth, calculate formation volume factor of gas.

$$B_g = 0.0282 \frac{zT}{p} \frac{\text{cu ft}}{\text{scf}} \qquad (6-2)$$

$$B_g = \frac{(0.0282)(0.851)(680)}{(2114.7)} = 0.00771 \frac{\text{cu ft}}{\text{scf}}$$

Fifth, calculate relative total volume.

$$B_{tD} = B_{oD} + B_g(R_{sDb} - R_{sD}) . \qquad (10-1)$$

$$B_{tD} = \left(1.515 \frac{\text{res bbl}}{\text{residual bbl}}\right) + \left(\frac{0.00771 \text{ cu ft/scf}}{5.615 \text{ cu ft/bbl}}\right)\left(854 - 684 \frac{\text{scf}}{\text{residual bbl}}\right)$$

$$B_{tD} = 1.748 \frac{\text{res bbl}}{\text{residual bbl}}$$

See page 5 of Table 10–1.

Separator Tests

A sample of reservoir liquid is placed in the laboratory cell and brought to reservoir temperature and bubble-point pressure. Then the liquid is expelled from the cell through two stages of separation. See Figure 10–4. The vessel representing the stock tank is a stage of separation if it has lower pressure than the separator. Pressure in the cell is held constant at the bubble point by reducing cell volume as the liquid is expelled.

The temperatures of the laboratory separator and stock tank usually are set to represent average conditions in the field. The stock tank is always at atmospheric pressure. The pressure in the separator is selected by the operator.

The *formation volume factor of oil* is calculated as

$$B_{oSb} = \frac{\text{volume of liquid expelled from the cell}}{\text{volume of liquid arriving in the stock tank}} \qquad (10-2)$$

$$= \frac{\text{volume at } P_b \text{ and resv. temp}}{\text{volume at } 0 \text{ psig, } 60°F}$$

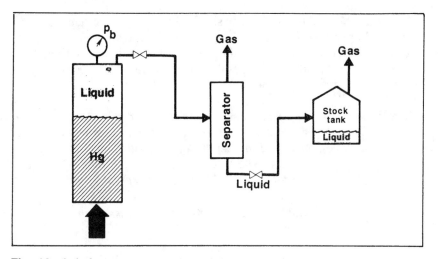

Fig. 10–4. Laboratory separator test.

The subscript S indicates that this is a result of a separator test, and the subscript b indicates bubble-point conditions in the reservoir. The volume of liquid expelled from the cell is measured at bubble-point conditions. The volume of stock-tank liquid is measured at standard conditions.

The *solution gas-oil ratio* is calculated as

$$R_{sSb} = \frac{\text{volume of separator gas} + \text{volume of stock-tank gas.}}{\text{volume of liquid in the stock tank}} \quad (10\text{–}3)$$

with all volumes adjusted to standard conditions. The subscripts S and b have the same meanings as discussed above.

The specific gravities of the separator gas and stock-tank gas are measured. Often the composition of the separator gas is determined.

Finally, a *separator volume factor* is calculated. It is the volume of separator liquid measured at separator conditions divided by the volume of stock-tank oil at standard conditions, SP bbl/STB.

The separator test usually is repeated for various values of separator pressure. The results of four separator tests are given on page 7 of Table 10–1.

Notice that the oil formation volume factor varies from 1.474 res bbl/STB to 1.495 res bbl/STB. Also, the solution gas-oil ratio varies from 768 scf/STB (676 + 92) for a separator pressure of 100 psig to 795 scf/STB for a separator pressure of 300 psig. This is disturbing in that it indicates that the values of these basic fluid properties are dependent on the method of surface separation. The differences are not great; however, these differences show that the fluid property calculations described later in this chapter should use the results of a separator test performed at expected field operating conditions.

✳ EXAMPLE 10–3: *Data from a separator test on a black oil are given below. Note that the volume of separator liquid was measured at separator pressure and temperature before it was released to the stock tank. Prepare a separator test table for the reservoir fluid study.*

Volume of oil at bubble-point pressure and reservoir temperature = 182.637 cc

Volume of separator liquid at 100 psig and 75°F = 131.588 cc

Volume stock-tank oil at 0 psig and 75°F = 124.773 cc

Volume of stock-tank oil at 0 psig and 60°F = 123.906 cc

Volume of gas removed from separator = 0.52706 scf

Volume of gas removed from stock tank = 0.07139 scf

Specific gravity of stock-tank oil = 0.8217

Specific gravity of separator gas = 0.786

Specific gravity of stock-tank gas = 1.363

Solution

First, calculate gas-oil ratio at separator and stock-tank conditions.

$$\text{Separator gas-oil ratio} = \frac{0.52706 \text{ scf}}{(131.588 \text{ cc})(6.29 \times 10^{-6} \text{ bbl/cc})}$$

$$= 637 \text{ scf/SP bbl}$$

$$\text{Stock-tank gas-oil ratio} = \frac{0.07139 \text{ scf}}{(124.773 \text{ cc})(6.29 \times 10^{-6} \text{ bbl/cc})}$$

$$= 91 \text{ scf/ST bbl}$$

Second, calculate gas-oil ratio based on stock-tank oil at standard conditions.

$$\text{Separator gas-oil ratio} = \frac{0.52706 \text{ scf}}{(123.906 \text{ cc})(6.29 \times 10^{-6} \text{ bbl/cc})}$$

$$= 676 \text{ scf/STB}$$

$$\text{Stock-tank gas-oil ratio} = \frac{0.07139 \text{ } \cancel{cc} \text{ } scf}{(123.906 \text{ cc})(6.29 \times 10^{-6} \text{ bbl/cc})}$$

$$= 92 \text{ scf/STB}$$

Third, calculate formation volume factor of oil.

$$B_{oSb} = \frac{\text{volume of liquid expelled from the cell}}{\text{volume of liquid arriving in stock tank}} \qquad (10\text{--}2)$$

$$B_{oSb} = \frac{182.637 \text{ res cc}}{123.906 \text{ ST cc}} = 1.474 \text{ } \frac{\text{res bbl}}{\text{STB}}$$

Fourth, calculate separator and stock-tank shrinkage factors.

$$\text{Separator shrinkage factor} = \frac{131.588 \text{ cc}}{123.906 \text{ cc}}$$

$$= 1.062 \text{ } \frac{\text{SP bbl}}{\text{STB}}$$

$$\text{Stock-tank shrinkage factor} = \frac{124.773 \text{ cc}}{123.906 \text{ cc}}$$

$$= 1.007 \text{ } \frac{\text{ST bbl}}{\text{STB}}$$

Fifth, convert specific gravity of stock-tank oil to °API.

$$°API = \frac{141.5}{\gamma_o} - 131.5 \qquad (8-2)$$

$$°API = \frac{141.5}{0.8217} - 131.5 = 40.7$$

Separator pressure, psig	Separator temperature, °F	Gas-oil ratio (1)	Gas-oil ratio (2)	Stock-tank gravity, °API	Formation Volume factor (3)	Separator Volume factor (4)	Specific gravity of flashed gas
100	75	637	676			1.062	0.786
to							
0	75	91	92	40.7	1.474	1.007	1.363

Oil Viscosity

Oil viscosity is measured in a rolling-ball viscosimeter or a capillary viscosimeter, either designed to simulate differential liberation. Measurements are made at several values of pressure in a stepwise process. The liquid used in each measurement is the liquid remaining after gas has been removed at that pressure. See page 6 of Table 10–1.

Gas Viscosity

Measurement of gas viscosity is very tedious. Obtaining accurate measurements on a routine basis is difficult. Thus, gas viscosity is estimated from correlations using the values of gas specific gravities measured in the differential liberation. Gas viscosity correlations as given in Chapter 6 or Appendix B are used.

Reservoir Fluid Properties from Reservoir Fluid Study

All the fluid properties required for a reservoir study using material balance equations can be calculated from the results of a reservoir fluid study.[3]

The underlying assumption is that at *pressures below the bubble point*, the process in the reservoir can be simulated by *differential vaporization*, and the process from the bottom of the well to the stock tank can be simulated by *the separator test*. Under this assumption, fluid properties

$P \geq P_b : \quad R_s = R_{ssb}$

$P < P_b : \quad R_s = R_{ssb} - (R_{sDb} - R_{sD}) \dfrac{B_{osb}}{B_{oDb}}$

$$P > P_b: \quad B_t = B_o$$
$$P < P_b: \quad B_t = B_{tD}\left(\frac{B_oSB}{B_oDb}\right)$$

at pressures below the bubble point can be calculated by combining the data from the differential vaporization and a separator test.

This assumption implies that gas and liquid in the reservoir separate as the gas is formed. One can argue that this is not strictly true. However, a laboratory process that more accurately represents the production process would be complicated and expensive and would require excessively large samples of reservoir liquids. Experience has shown that black oil properties calculated under this assumption are sufficiently accurate for reservoir engineering calculations.

At *pressures above the bubble point*, fluid properties are calculated by combining the data from the *flash vaporization* and a *separator test*.

There are a number of special symbols which are used only in referring to the results of a black oil reservoir fluid study. These symbols are defined in Table 10–3. The subscripts o and g refer to liquid and gas, as always. The subscripts F, D, and S refer to flash vaporization, differential vaporization, and separator test, respectively. The subscript b is added to indicate that the quantity is measured at the bubble point.

$$P > P_b \qquad B_o = \left(\frac{V_t}{V_b}\right)_F \cdot B_{osb}$$

$$P < P_b \qquad B_o = B_{oD} \cdot \frac{B_{osb}}{B_oDb}$$

TABLE 10–3

Nomenclature used in analysis of reservoir fluid studies

B_{oD} = relative oil volume by differential vaporization, page 5, column 3, Table 10–1

B_{oDb} = relative oil volume at bubble point by differential vaporization, page 5, column 3, Table 10–1

B_{oSb} = formation volume factor at bubble point from separator test (at selected separator pressure), page 7, column 6, Table 10–1

$(V_t/V_b)_F$ = relative total volume (oil and gas) by flash vaporization, page 4, column 2, Table 10–1

B_{tD} = relative total volume (oil and gas) by differential vaporization, page 5, column 4, Table 10–1

R_{sD} = gas remaining in solution by differential vaporization, page 5, column 2, Table 10–1

R_{sDb} = gas in solution at bubble point (and above) by differential vaporization, page 5, column 2, Table 10–1

R_{sSb} = sum of separator gas and stock-tank gas from separator test (at selected separator pressure), page 7, column 4, Table 10–1

$$R_{sSb} = \frac{Vol. \text{ of sep gas} + Vol\ ST\ gas\ SCF}{Vol.\ lie\ in\ ST\ (bbl)} \qquad \left[\frac{SCF}{STB}\right]$$

$$ST\ GOR = \frac{ST\ gas}{ST\ oil\ at\ higher\ T\ (not\ 60\ F)}$$

Selection of Separator Conditions

The first step in calculating fluid properties is selection of separator conditions. There may be circumstances for a particular field which dictate a specific separator pressure. If not, the separator pressure which produces the maximum amount of stock-tank liquid is selected. This pressure is known as *optimum separator pressure*. It is identified from the separator tests as the separator pressure which results in a minimum of total gas-oil ratio, a minimum in formation volume factor of oil (at bubble point), and a maximum in stock-tank oil gravity (°API). Most black oils have optimum separator pressures of 100 to 120 psig at normal temperatures.

The value of formation volume factor of oil at the selected separator pressure is B_{oSb} in the following calculations. The corresponding value of total gas-oil ratio is R_{sSb}. B_{oSb} will be used as the formation volume factor of oil at the bubble point, B_{ob}. R_{sSb} will be used as the solution gas-oil ratio at the bubble point, R_{sb}.

EXAMPLE 10–4: *Select optimum separator conditions for Good Oil Co. No. 4. Identify R_{sSb} and B_{oSb}.*

Solution

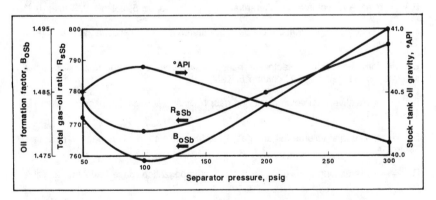

Fig. 10–5. Results of separator tests for Good Oil Co. No. 4 (part of solution to Example 10–4).

Optimum separator conditions = 100 psig at 75°F.

R_{sSb} = 676 scf/STB + 92 scf/STB = 768 scf/STB

B_{oSb} = 1.474 res bbl/STB

Formation Volume Factor of Oil B_o

At pressures *above* bubble-point pressure, oil formation volume factors are calculated from a combination of flash vaporization data and separator test data.

$$B_o = \left(\frac{V_t}{V_b}\right)_F B_{oSb} \text{ at } p \geq p_b \qquad (10-4)$$

The units involved in the calculations are

$$B_o = \left(\frac{\text{res bbl of oil at p}}{\text{res bbl of oil at } p_b}\right) \left(\frac{\text{res bbl of oil at } p_b}{\text{STB}}\right)$$

$$= \frac{\text{res bbl of oil at p}}{\text{STB}}. \qquad (10-5)$$

At pressures *below the bubble-point pressure*, oil formation volume factors are calculated from a combination of differential vaporization data and separator test data.

$$B_o = B_{oD} \frac{B_{oSb}}{B_{oDb}} \text{ at } p \leq p_b \qquad (10-6)$$

The units involved in the calculations are

$$B_o = \left(\frac{\text{res bbl of oil at p}}{\text{residual bbl by diff vap}}\right) \frac{\left(\dfrac{\text{res bbl of oil at } p_b}{\text{STB}}\right)}{\left(\dfrac{\text{res bbl of oil at } p_b}{\text{residual bbl by diff vap}}\right)}$$

$$= \frac{\text{res bbl at p}}{\text{STB}}. \qquad (10-7)$$

EXAMPLE 10–5: *Calculate formation volume factors of oil for Good Oil Co. No. 4. Use optimum separator conditions.*

Solution
First, calculate at pressures above bubble point. Only the calculation at 5000 psig will be shown.

$$B_o = \left(\frac{V_t}{V_b}\right)_F B_{oSb} \text{ at } p \geq p_b \qquad (10-4)$$

$$B_o = (0.9639) \left(1.474 \; \frac{\text{res bbl}}{\text{STB}}\right) = 1.421 \; \text{res bbl/STB}$$

Second, calculate at pressures below bubble point. Only the calculations at 2620 psig and 2100 psig will be shown.

$$B_o = B_{oD} \; \frac{B_{oSb}}{B_{oDb}} \quad \text{at} \quad p \leq p_b \tag{10-6}$$

At 2620 psig,

$$B_o = (1.600) \left(\frac{1.474}{1.600}\right) = 1.474 \; \frac{\text{res bbl}}{\text{STB}}$$

At 2100 psig,

$$B_o = (1.515) \left(\frac{1.474}{1.600}\right) = 1.396 \; \text{res bbl/STB}$$

Pressure, psig	Formation volume factor of oil, res bbl/STB
5000	1.421
4500	1.430
4000	1.440
3500	1.451
3000	1.464
2900	1.466
2800	1.469
2700	1.471
2620 = p_b	1.474
2350	1.432
2100	1.396
1850	1.363
1600	1.331
1350	1.301
1100	1.273
850	1.245
600	1.216
350	1.182
159	1.146

Solution Gas-Oil Ratio R_s

Solution gas-oil ratio at pressures *above* bubble-point pressure is a constant equal to the solution gas-oil ratio at the bubble point.

$$R_s = R_{sSb} \quad \text{at} \quad p \geq p_b \qquad (10\text{--}8)$$

Solution gas-oil ratios at pressures *below* bubble-point pressure are calculated from a combination of differential vaporization data and separator test data.

$$R_s = R_{sSb} - (R_{sDb} - R_{sD}) \frac{B_{oSb}}{B_{oDb}} \quad \text{at} \quad p < p_b \qquad (10\text{--}9)$$

Units are

$$R_s = \left(\frac{\text{scf}}{\text{STB}} \right) - \left(\frac{\text{scf}}{\text{residual bbl by diff vap}} \right) \frac{\left(\dfrac{\text{res bbl of oil at } p_b}{\text{STB}} \right)}{\left(\dfrac{\text{res bbl of oil at } p_b}{\text{residual bbl by diff vap}} \right)}$$

$$= \frac{\text{scf}}{\text{STB}} \cdot \qquad (10\text{--}10)$$

EXAMPLE 10–6: *Calculate solution gas-oil ratios for Good Oil Co. No. 4. Use optimum separator conditions.*

Solution
First, calculate at pressures above bubble-point pressure.

$$R_s = R_{sSb} \quad \text{at} \quad p \geq p_b \qquad (10\text{--}8)$$

$$R_s = 768 \text{ scf/STB}$$

Second, calculate at pressures below the bubble-point. Only the calculations at 2620 psig and 2100 psig will be shown.
At 2620 psig,

$$R_s = R_{sSb} \quad \text{at} \quad p \geq p_b \qquad (10\text{--}8)$$

$$R_s = 768 \frac{\text{scf}}{\text{STB}}$$

At 2100 psig,

$$R_s = R_{sSb} - (R_{sDb} - R_{sD}) \frac{B_{oSb}}{B_{oDb}} \quad \text{at} \quad p < p_b \qquad (10\text{--}9)$$

$$R_s = 768 - (854 - 684) \left(\frac{1.474}{1.600}\right) = 611 \text{ scf/STB}$$

Pressure, psig	Solution gas-oil ratio, scf/STB
5000	768
4500	768
4000	768
3500	768
3000	768
2900	768
2800	768
2700	768
2620 $= p_b$	768
2350	684
2100	611
1850	545
1600	482
1350	423
1100	364
850	307
600	250
350	187
159	126

Formation Volume Factor of Gas

Gas formation volume factors are calculated with z-factors measured with the gases removed from the cell at each pressure step during differential vaporization. Equation 6–2 is used. Usually B_g values as calculated are listed in the report.

$$B_g = 0.0282 \frac{zT}{p} \frac{\text{res cu ft}}{\text{scf}} \qquad (6\text{--}2)$$

Total Formation Volume Factor

Total formation volume factors may be calculated as

$$B_t = B_o + B_g(R_{sb} - R_s), \qquad (8\text{--}6)$$

using the fluid properties calculated from the reservoir fluid study.

If relative total volumes, B_{tD}, are reported as a part of the results of the differential vaporization, total formation factors can be calculated as

$$B_t = B_{tD} \frac{B_{oSb}}{B_{oDb}}. \qquad (10\text{--}11)$$

Units are

$$B_t = \left(\frac{\text{res bbl oil } + \text{ gas}}{\text{residual bbl by diff vap}} \right) \frac{\left(\dfrac{\text{res bbl of oil at } p_b}{\text{STB}} \right)}{\left(\dfrac{\text{res bbl of oil at } p_b}{\text{residual bbl by diff vap}} \right)}$$

$$= \frac{\text{res bbl oil } + \text{ gas}}{\text{STB}}. \qquad (10\text{--}12)$$

EXAMPLE 10–7: *Calculate total formation volume factors for Good Oil Co. No. 4. Use optimum separator conditions.*

Solution
First, $B_t = B_o$ at pressures above bubble point.
At 5000 psig,

$$B_t = 1.421 \text{ res bbl/STB. See Example } 10\text{--}5.$$

Second, calculate at pressures below bubble point. Only the calculation at 2100 psig will be shown.

$$B_t = B_{tD} \frac{B_{oSb}}{B_{oDb}}. \qquad (10\text{--}11)$$

$$B_t = (1.748) \left(\frac{1.474}{1.600} \right) = 1.610 \text{ res bbl/STB}$$

Pressure, psig		Total formation volume factor, res bbl/STB
5000		1.421
4500		1.430
4000		1.440
3500		1.451
3000		1.464
2900		1.466
2800		1.469
2700		1.471
2620	$= p_b$	1.474
2350		1.534
2100		1.610
1850		1.713
1600		1.857
1350		2.067
1100		2.389
850		2.920
600		3.919
350		6.426
159		13.536

Viscosities

Oil and gas viscosities as reported in the reservoir fluid study may be used directly. No calculations are required.

Coefficient of Isothermal Compressibility of Oil

Equation 8–11 may be used with the flash vaporization data to calculate oil compressibility at pressures *above* the bubble point.

$$c_o(p_2 - p_1) = - \ln \frac{V_2}{V_1} \quad \text{at} \quad p \geq p_b \qquad (8-11)$$

Relative volume from flash vaporization can be substituted for volume, resulting in

$$c_o = \frac{\ln \dfrac{(V_t/V_b)_{F2}}{(V_t/V_b)_{F1}}}{p_1 - p_2} . \qquad (10-13)$$

This results in a value of c_o which applies between pressures p_1 and p_2 at reservoir temperature.

Some applications require an average value of oil compressibility between a pressure and the bubble point. In these cases

$$c_o = \frac{\ln(V_t/V_b)_F}{p_b - p} , \qquad (10-14)$$

where c_o represents the fractional change of oil volume as pressure declines from p to the bubble point. Relative volume is determined at p.

EXAMPLE 10-8: *Calculate coefficients of isothermal compressibility at pressures above bubble point for Good Oil Co. No. 4.*

Solution.
Only the calculation between 5000 and 4500 psig will be shown.

$$c_o = \frac{\ln \dfrac{(V_t/V_b)_{F2}}{(V_t/V_b)_{F1}}}{p_1 - p_2} \qquad (10-13)$$

Between 5000 psig and 4500 psig,

$$c_o = \frac{\ln \dfrac{0.9639}{0.9703}}{(4500 - 5000)\ \text{psi}} = 13.24 \times 10^{-6}\ \text{psi}^{-1}$$

Coefficient of isothermal compressibility, $c_o \times 10^6$ psi^{-1}	Use for pressures between	
	psig	psig
13.24	5000	4500
13.97	4500	4000
15.29	4000	3500
16.79	3500	3000
17.11	3000	2900
18.08	2900	2800
19.05	2800	2700
21.27	2700	2620

At pressures *below* the bubble point, Equation 8-24 applies.

$$c_o = -\frac{1}{B_o}\left[\left(\frac{\partial B_o}{\partial p}\right)_T - B_g\left(\frac{\partial R_s}{\partial p}\right)_T\right] \quad \text{at } p < p_b \qquad (8-24)$$

The derivative of B_o with respect to p is the slope of a plot of B_o against p. The slope is measured at the pressure of interest. The derivative of R_s with respect to p is obtained by plotting R_s against p.

Equation 8–24 can be converted to

$$c_o = \frac{1}{B_o} \left(\frac{\partial R_s}{\partial p} \right)_T \left[B_g - \left(\frac{\partial B_o}{\partial R_s} \right)_T \right]. \qquad (10\text{–}15)$$

The derivative of B_o with respect to R_s is relatively easy to determine since the slope of a plot of B_o against R_s is virtually constant for most black oils.

Relative oil volume and solution gas-oil ratio from the differential vaporization can be used also to calculate c_o at pressures below the bubble point. The above equations become

$$c_o = - \frac{1}{B_{oD}} \left[\left(\frac{\partial B_{oD}}{\partial p} \right)_T - B_g \left(\frac{\partial R_{sD}}{\partial p} \right)_T \right] \qquad (10\text{–}16)$$

and $P < P_b$

$$c_o = \frac{1}{B_{oD}} \left(\frac{\partial R_{sD}}{\partial p} \right)_T \left[B_g - \left(\frac{\partial B_{oD}}{\partial R_{sD}} \right)_T \right]. \qquad (10\text{–}17)$$

EXAMPLE 10–9: *Calculate coefficients of isothermal compressibility at pressures below bubble point for Good Oil Co. No. 4.*

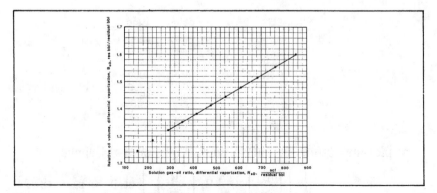

Fig. 10–6. B_{oD} vs. R_{sD} (part of solution to Example 10–9).

Solution

First, plot B_{oD} against R_{sD} and determine slope of the straight line.

$$\frac{\partial B_{oD}}{\partial R_{sD}} = 0.000498 \ \frac{\text{res bbl}}{\text{scf}} \quad \text{(See Figure 10–6)}$$

Second, plot R_{sD} against p and determine slopes of tangent lines at pressures of interest. Only the calculation at 2100 psig will be shown.

$$\frac{\partial R_{sD}}{\partial p} = 0.302 \ \frac{\text{scf}}{\text{residual bbl, psi}} \quad \text{(See Figure 10–7)}$$

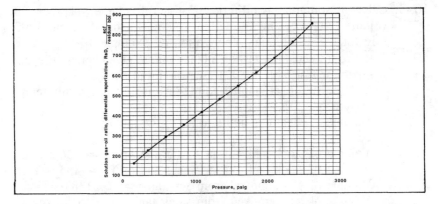

Fig. 10–7. R_{sD} vs. pressure (part of solution to Example 10–9).

Third, determine values of B_{oD} and B_g.

$$B_{oD} = 1.515 \ \frac{\text{res bbl}}{\text{residual bbl}} \quad \text{(See page 5 of Table 10–1)}$$

$$B_g = \frac{0.00771 \ \dfrac{\text{res cu ft}}{\text{scf}}}{5.615 \ \dfrac{\text{cu ft}}{\text{bbl}}} = 0.001373 \ \frac{\text{res bbl}}{\text{scf}}$$

Fourth, calculate c_o.

$$c_o = \frac{1}{B_{oD}} \left(\frac{\partial R_{sD}}{\partial p}\right)_T \left[B_g - \left(\frac{\partial B_{oD}}{\partial R_{sD}}\right)_T\right] \qquad (10\text{–}17)$$

$$c_o = \left(\cfrac{1}{1.515 \ \cfrac{res \ bbl}{residual \ bbl}}\right)\left(0.302 \ scf \ \cfrac{scf}{residual \ bbl, \ psi}\right)$$

$$\left(0.001373 \ \frac{res \ bbl}{scf} - 0.000498 \ \frac{res \ bbl}{scf}\right) = 174 \times 10^{-6} \ psi^{-1}$$

Exercises

10-1. Name the five laboratory procedures normally performed for a black oil reservoir fluid study.

10-2. Calculate the relative volume at 2900 psig for the flash vaporization of Example 10-1.

10-3. Calculate solution gas-oil ratio, relative oil volume, relative total volume, z-factor, and formation volume factor of gas at 1100 psig for the differential vaporization of Example 10-2.

10-4. Data from a separator test on a black oil are given below. Prepare a separator test table for a Reservoir Fluid Study.

Volume of oil at bubble-point pressure and reservoir temperature = 201.156 cc

Volume of separator liquid at 200 psig and 75°F = 150.833 cc

Volume of stock-tank oil at 0 psig and 75°F = 136.591 cc

Volume of stock-tank oil at 0 psig and 60°F = 135.641 cc

Volume of gas removed from separator = 0.51383 scf

Volume of gas removed from stock tank = 0.15186 scf

Specific gravity of stock-tank oil = 0.823

Specific gravity of separator gas = 0.732

Specific gravity of stock-tank gas = 1.329

10-5. You are preparing a reservoir study for Good Oil Co. Give values of the following properties as indicated by the Reservoir Fluid Study of Good Oil Co. No. 4.

Depth of the producing interval _____

Reservoir temperature _____

Bubble-point pressure at reservoir temperature _____

Viscosity of the reservoir liquid at the bubble point_____

Specific volume of the reservoir liquid at the bubble point _____

Solution gas-oil ratio at the bubble point using separator conditions of 50 psig and 75°F _____

Mole percent heptanes plus in the reservoir liquid_____

Specific gravity of the separator gas produced from a 300 psig, 75°F separator _____

Standard pressure and standard temperature _____

Mole fraction of methane in reservoir liquid _____

10–6. You are preparing a reservoir study for Good Oil Co. Give values of the following properties as indicated by the Reservoir Fluid Study of Good Oil Co. No. 4.

Formation volume factor of oil at bubble point using separator conditions of 200 psig and 75°F _____

Status of the well at the time of sampling _____

Mole fraction of nitrogen in separator gas for 100 psig, 75°F separator conditions _____

Gravity (°API) of the stock-tank oil from the 100 psig, 75°F two-stage separation _____

Gravity (°API) of the residual oil from the differential vaporization _____

Gross heating value of the gas from the 200 psig separator _____

Molecular weight of the heptanes plus fraction of the reservoir liquid _____

Coefficient of isothermal compressibility of the reservoir liquid at 220°F and 4500 psig _____

Separator temperature actually occurring in the field _____

Viscosity of the reservoir gas at 220°F and 850 psig _____

10–7. You are preparing a reservoir study for Good Oil Co. Give values of the following properties as indicated by the Reservoir Fluid Study of Good Oil Co. No. 4.

Thermal expansion at 5000 psig _____

Separator pressure actually used in field _____

Density of the oil in the reservoir at 1600 psig and 220°F _____

Weight fraction of carbon dioxide in reservoir liquid _____

Ratio of separator liquid volume to stock-tank liquid volume for a separator pressure of 300 psig _____

Oil-gas viscosity ratio at 600 psig and 220°F _____

Gross heating value of the separator gas produced from a 50 psig, 75°F separator _____

Specific gravity of heptanes plus in reservoir liquid sample _____

Formation volume factor of gas in reservoir at 220°F and 1350 psig _____

Specific gravity of the gas in the reservoir at 159 psig _____

10–8. Calculate formation volume factor of oil for Good Oil Co. No. 4 at 4500 psig. Use optimum separator conditions.

10–9. Calculate formation volume factor of oil for Good Oil Co. No. 4 at 1600 psig. Use optimum separator conditions.

10–10. Calculate solution gas-oil ratio for Good Oil Co. No. 4 at 4500 psig. Use optimum separator conditions.

10–11. Calculate solution gas-oil ratio for Good Oil Co. No. 4 at 1600 psig. Use optimum separator conditions.

10–12. Calculate total formation volume factor for Good Oil Co. No. 4 at 4500 psig. Use optimum separator conditions.

10–13. Calculate total formation volume factor for Good Oil Co. No. 4 at 1600 psig. Use optimum separator conditions.

10–14. Determine the viscosity of the oil in the reservoir of Good Oil Co. No. 4 at 4500 psig and 220°F.

10–15. Determine the viscosity of the oil in the reservoir of Good Oil Co. No. 4 at 1600 psig and 220°F.

10–16. Determine the viscosity of the gas in the reservoir of Good Oil Co. No. 4 at 1600 psig and 220°F.

10–17. Calculate a value of coefficient of isothermal compressibility for use at pressures between 4500 psig and 4000 psig at 220°F for Good Oil Co. No. 4.

10–18. Calculate a value of coefficient of isothermal compressibility for use at pressures between 4500 psig and bubble-point pressure at 220°F for Good Oil Co. No. 4.

10–19. Calculate a value of coefficient of isothermal compressibility for use at 1600 psig and 220°F for Good Oil Co. No. 4.

10–20. Determine the solution gas-oil ratio and formation volume factor of oil, both at bubble-point pressure, for Good Oil Co. No. 4. Field operation requires that the separator be operated at 200 psig and 75°F.

10–21. Field operation requires that the separator for Good Oil Co. No. 4 be operated at 200 psig and 75°F. Calculate formation volume factors of oil for pressures at and above the bubble point.

10–22. Field operation requires that the separator for Good Oil Co. No. 4 be operated at 200 psig and 75°F. Calculate formation volume factors of oil for pressures at and below the bubble point.

10–23. Plot the results of Exercises 10–21 and 10–22 as formation volume factor against pressure. Is the shape of the curve correct?

10–24. Field operation requires that the separator for Good Oil Co. No. 4 be operated at 200 psig and 75°F. Calculate solution gas-oil ratios for pressures at and above the bubble point.

10–25. Field operation requires that the separator for Good Oil Co. No. 4 be operated at 200 psig and 75°F. Calculate solution gas-oil ratios for pressures at and below the bubble point.

10–26. Plot the results of Exercises 10–24 and 10–25 as solution gas-oil ratio against pressure. Is the shape of the curve correct?

10–27. Plot the results of Exercises 10–22 and 10–25 as formation volume factor of oil against solution gas-oil ratio. What do you observe?

10–28. Plot the fluidity (reciprocal of viscosity) of the oil from Good Oil Co. No. 4 against pressure for pressures at and below the bubble point. What do you observe at pressures above about 1000 psig?

10–29. Plot the reciprocal of formation volume factor of gas from Good Oil Co. No. 4 against pressure. Estimate the formation volume factor of the gas at the bubble point.

10–30. Why is the relative volume of a flash liberation equal to 1.0000 at the bubble point? Why is the relative oil volume of a differential liberation not equal to 1.0000 at the bubble point?

References

1. Reudelhuber, F.O.: "Sampling Procedures for Oil Reservoir Fluids," *J. Pet. Tech.* (Dec. 1957) *9*, 15-18.

2. *API Recommended Practice For Sampling Petroleum Reservoir Fluids,* API (1966) *44*.

3. Moses, P.L. "Engineering Applications of Phase Behavior of Crude Oil and Condensate Systems," *J. Pet. Tech.* (July 1986) *38*, 715–723.

11

Properties of Black Oils—
Correlations

When a reservoir fluid study is unavailable, the engineer must rely on correlations to estimate values of the physical properties of interest. I have compared most of the published fluid property correlations with the results of hundreds of reservoir fluid studies. The best available correlations are given in this chapter.

This chapter begins with *bubble-point pressure* and *solution gas-oil ratio*, and then explains methods of estimating the *density* of reservoir liquids. The results of the density calculations are used to estimate *oil formation volume factors*. A technique for adjusting the results of the correlations to fit field derived bubble-point pressure is presented.

Next, procedures for estimating values of the *coefficient of isothermal compressibility* and *oil viscosity* are discussed. The chapter ends with methods of estimating hydrocarbon *liquid-gas interfacial tension*.

Bubble-Point Pressure

Figure 11–1 gives a correlation for bubble-point pressure.[1,3] The correlation is entered with solution gas-oil ratio derived from early production history. If the information is based on gas sales, the gas-oil ratio must be adjusted for stock-tank vent gas and for any gas lost or used in surface operations.

The specific gravity of the separator gas usually is measured as a part of the gas sales procedure. Separator gas specific gravity can be used as gas specific gravity in Figure 11–1.

The resulting value of bubble-point pressure will be accurate to within 15 percent.

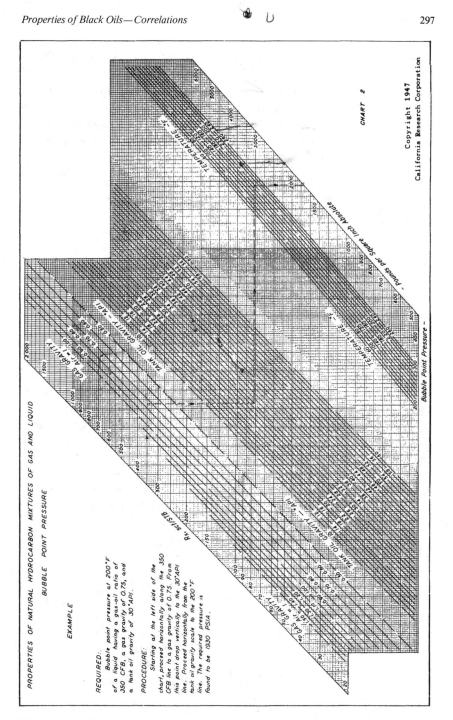

Fig. 11–1. Bubble-point pressure — solution gas-oil ratio correlation. (Copyright 1947 Chevron Oil Field Research Co., with permission.)

EXAMPLE 11–1: *Your recently completed well is producing 40.7 °API oil and 0.786 specific gravity gas at 768 scf/STB. Producing gas-oil ratio has remained constant, so you believe reservoir pressure is above the bubble-point pressure of the reservoir oil. Estimate bubble-point pressure given that reservoir temperature is 220°F.*

Solution

Enter Figure 11–1 with

$$R_{sb} = 768 \text{ scf/STB}$$
$$\gamma_g = 0.786$$
$$\gamma_o = 40.7°\text{API, and}$$
$$T = 220°\text{F}$$

and read

$$p_b = 2685 \text{ psia}$$

Solution Gas-Oil Ratio R_s

Figure 11–1 also can be used to estimate values of solution gas-oil ratios. Simply enter the right side of the graph with any pressure at or below bubble-point pressure. The result will be solution gas-oil ratio at that pressure. Remember that solution gas-oil ratio is constant above the bubble point.

EXAMPLE 11–2: *Estimate values of solution gas-oil ratio at various pressures below bubble-point pressure for the reservoir oil of Example 11–1.*

Solution

Enter Figure 11–1 with any desired pressure at or below p_b and with

$$T = 220°\text{F}$$
$$\gamma_o = 40.7°\text{API}$$
$$\gamma_g = 0.786$$
and read corresponding values of R_s

Pressure, psia	Solution gas-oil ratio, R_s, scf/STB
2685	768
2414	676
2165	594
1915	513
1665	434
1415	358
1165	284
915	214
665	147
415	86
165	31

Density of a Liquid

The petroleum engineer needs to be able to estimate the density of the reservoir liquid at reservoir conditions. Then the shrinkage in volume that a reservoir liquid undergoes while progressing from the reservoir to the stock tank can be estimated. There are several methods of calculating the volume occupied by a given mass of liquid at elevated pressures and temperatures. We will consider only one method: the method most applicable to the liquids encountered in the oil fields. This method is based on ideal-solution principles.

We will study *three* different applications of ideal solution theory to the calculation of density of a liquid. Each application will depend on the amount of information available. The *first* applies when the composition of the reservoir liquid is known. The *second* applies when solution gas-oil ratio, gas composition, and stock-tank oil gravity are known. The *third* is used when solution gas-oil ratio, gas specific gravity, and stock-tank oil gravity are known.

Calculation of Liquid Density Using Ideal-Solution Principles

An *ideal liquid solution* is a hypothetical mixture of liquids in which there is no special force of attraction between the components of the solution and for which no change in internal energy occurs on mixing. Under these circumstances no change in the character of the liquids is caused by mixing, merely a dilution of one liquid by the other.

When liquids are mixed to give an ideal solution, there is no heat effect and the properties are strictly additive. The volume of the ideal solution is the sum of the volumes of the liquid components. There is no shrinkage nor expansion when the liquids are mixed. Other physical properties of the solution (such as refractive index and viscosity) can be calculated directly by averaging the properties of the components which make up the ideal solution.

There are no ideal liquid solutions, just as there are no ideal gas mixtures. However, when liquids of similar chemical and physical characteristics are mixed, the behavior of the resulting solution is very much like the behavior of an ideal solution. Fortunately, most of the liquid mixtures encountered by petroleum engineers are mixtures of hydrocarbons with similar characteristics. Thus, ideal-solution principles can be applied to the calculation of the densities of these liquids.

The application of ideal-solution principles to the calculation of the density of a liquid is very easy. One simply calculates the mass and volume of each of the components of the mixture. Then these quantities are added to determine the mass and volume of the mixture. Density is simply mass divided by volume.

Calculations of this type usually are made for a temperature of 60°F and a pressure of 14.696 psia. The densities of typical pure hydrocarbons at these conditions may be obtained from Appendix A.

A standard pressure of 14.696 psia is used here because the data in Appendix A are based on that value. The value of standard pressure used in liquid volume calculations is not as important as for gas calculations. Liquid is not as compressible as gas, so the difference of a few tenths of a psi in standard pressure has a negligible effect on liquid density.

Later we will see how density at standard conditions can be used to calculate density at reservoir conditions.

EXAMPLE 11–3: *Use ideal-solution principles to calculate the density at 14.696 psia and 60°F of a hydrocarbon liquid of the following composition.*

Component	Mole fraction
n-Butane	0.270
n-Pentane	0.310
n-Hexane	0.420
	1.000

Solution

Determine the volume and mass of each component at 14.696 psia and 60°F. Divide the total volume into the total mass of the mixture to obtain density.

Component	Mole fraction, lb mol x_j	Molecular weight, lb/lb mol M_j	Mass, lb x_jM_j	Liquid density at 60°F and 14.696 psia, lb/cu ft ρ_{oj}	Liquid volume at 60°F and 14.696 psia, cu ft x_jM_j/ρ_{oj}
$n-C_4$	0.270	58.1	15.69	36.42	0.4307
$n-C_5$	0.310	72.2	22.38	39.36	0.5686
$n-C_6$	0.420	86.2	36.20	41.40	0.8745
	1.000		74.27 lb		1.8738 cu ft

$$\rho_o = \frac{74.27 \text{ lb}}{1.8738 \text{ cu ft}} = 39.64 \text{ lb/cu ft at } 60°F \text{ and } 14.696 \text{ psia}$$

Calculation of Reservoir Liquid Density at Saturation Pressure Using Ideal-Solution Principles

Now we will examine the methods of estimating the density of a reservoir liquid at reservoir conditions. First, we will consider liquids at their bubble points or liquids in contact with gas; in either case, we will call these saturated liquids. The *first* step in the calculation procedure is to determine the density of the liquid at *standard conditions*. The *next* step is to adjust this density to *reservoir conditions*.

There are *two problems* associated with this procedure. *First*, methane and ethane are not liquid at standard conditions, so a liquid density at standard conditions does not exist for either of these two components. *Second*, a hydrocarbon mixture which is liquid at reservoir conditions will partially vaporize at standard conditions.

The solution to the first problem is to use *apparent liquid densities* for methane and ethane.

These apparent liquid densities were derived through a study of mixtures containing methane and other heavier hydrocarbons and mixtures of ethane and other heavier hydrocarbons.[2]

The experimentally determined densities of these mixtures at numerous elevated pressures and temperatures were adjusted to atmospheric psia and 60°F using suitable compressibility and thermal expansion factors. Then the mass and volume contributed by the heavier components were subtracted. This left the contribution due to methane or ethane, i.e., their apparent liquid densities.

Thus, the apparent liquid densities are fictitious. However, they represent the contribution of methane and ethane to the density of the liquid mixture.

U

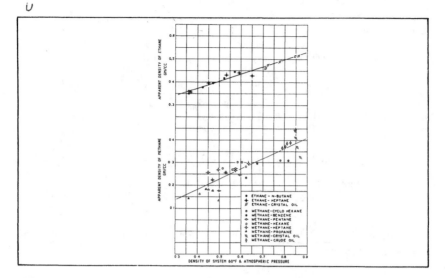

Fig. 11–2. Apparent liquid densities of methane and ethane. (Standing and Katz, *Trans.*, AIME, *146*, 159. Copyright 1942 SPE-AIME.)

Figure 11–2 gives the apparent liquid densities of methane and ethane at standard conditions. Note that the apparent liquid densities vary with the density of the total mixture.

The solution to the second problem is to create, by calculation, a *pseudoliquid* at standard conditions. The pseudoliquid has the same composition as the reservoir liquid even though a mixture of that composition is partially gas at standard conditions. The density of this pseudoliquid, called *pseudoliquid density*, is calculated with ideal solution procedures.

The density of the pseudoliquid is adjusted to reservoir pressure using the coefficient of isothermal compressibility and is adjusted to reservoir temperature using the coefficient of isobaric thermal expansion.

For convenience, the compressibility and thermal expansion factors have been put in graphical form.[3,4] These correlations are given as Figures 11–3 and 11–4. Fortunately, the coefficient of thermal expansion of hydrocarbon liquids is not affected greatly by pressure so that Figure 11–4 can be used regardless of the pressure involved.

Composition of the Saturated Liquid Known, Iterative Procedure

Use of the data of Figure 11–2 requires a trial and error solution in which a trial value of density at standard conditions must be selected so that the apparent densities of methane and ethane can be determined.

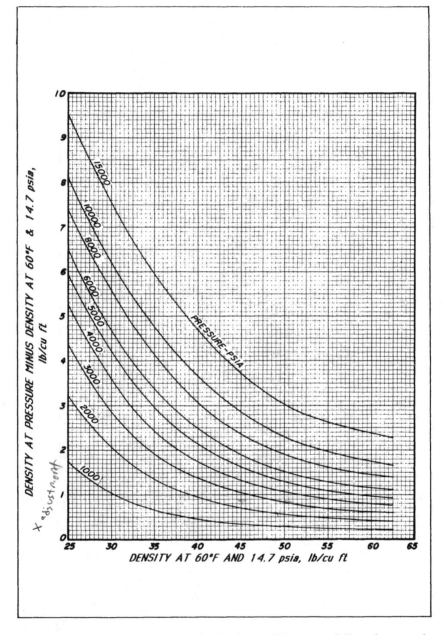

Fig. 11–3. Density adjustments for isothermal compressibility of reservoir liquids. (Standing, *Volumetric and Phase Behavior of Oil Field Hydrocarbon Systems,* SPE, Dallas, 1951. Copyright 1951 SPE-AIME.)

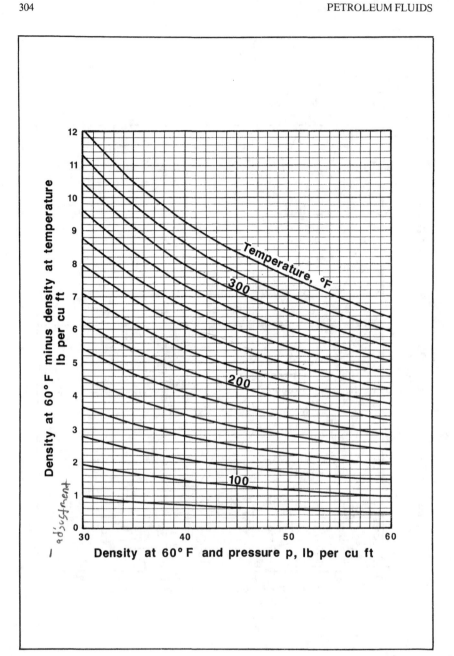

Fig. 11–4. Density adjustments for isobaric thermal expansion of reservoir liquids.

These apparent densities then are used to compute the density of the total mixture which must be checked against the trial value with which Figure 11–2 was entered.

Additional trial values of density must be selected and calculations repeated until the calculated value of density equals the trial value. This value is the pseudoliquid density of the mixture at standard conditions.

When nonhydrocarbons are present the usual procedure is to add the mole fraction of nitrogen to methane, the mole fraction of carbon dioxide to ethane, and the mole fraction of hydrogen sulfide to propane.

EXAMPLE 11–4: *Use the data given in Figure 11–2 to calculate the density of a reservoir liquid at its bubble point of 2635 psia at a reservoir temperature of 220°F. The composition of the well stream is as follows.*

Component	Composition, mole percent
Hydrogen sulfide	nil
Carbon dioxide	0.91
Nitrogen	0.16
Methane	36.47
Ethane	9.67
Propane	6.95
i-Butane	1.44
n-Butane	3.93
i-Pentane	1.44
n-Pentane	1.41
Hexanes	4.33
Heptanes plus	33.29
	100.00

Properties of heptanes plus
Specific gravity   0.8515
Molecular weight   218 lb/lb mole

Solution

First, calculate the pseudoliquid density of the mixture under the assumption that the mixture is all liquid at 60°F and 14.696 psia. This will be trial and error since pseudoliquid density must be known to enter Figure 11–2. Use a first trial value of 43.70 lb/cu ft which corresponds to 0.70 g/cc.

Apparent liquid density of methane = 0.32 g/cc = 19.98 lb/cu ft, Figure 11–2

Apparent liquid density of ethane = 0.47 g/cc = 29.34 lb/cu ft, Figure 11–2

Component	Mole fraction, lb mole x_j	Molecular weight, lb/lb mole M_j	Mass, lb $x_j M_j$	Liquid density, 60°F & 14.696 psia, lb/cu ft ρ_{oj}	Liquid volume, 60°F & 14.696 psia, cu ft $x_j M_j/\rho_{oj}$
C_1	0.3663	16.043	5.8766	11-2 19.98	0.2941
C_2	0.1058	30.070	3.1814	29.34	0.1084
C_3	0.0695	44.097	3.0647	31.62	0.0969
i-C_4	0.0144	58.123	0.8370	35.11	0.0238
n-C_4	0.0393	58.123	2.2842	36.42	0.0627
i-C_5	0.0144	72.150	1.0390	38.96	0.0267
n-C_5	0.0141	72.150	1.0173	39.36	0.0258
C_6	0.0433	86.177	3.7315	41.40	0.0901
C_{7+}	0.3329	218.000	72.5722	53.11	1.3665
	1.0000		93.6039 lb		2.0950 cu ft

The first calculated value of pseudoliquid density $=$

$$\frac{93.6039 \text{ lb}}{2.0950 \text{ cu ft}} = 44.68 \text{ lb/cu ft}$$

For a second trial value use 46.82 lb/cu ft, which corresponds to 0.75 g/cc.

1 g/cc = 62.43 lb/cu ft

Component	Mass, lb $x_j M_j$	Liquid density, 60°F & 14.696 psia ρ_{oj}	Liquid volume, 60°F & 14.696 psia $x_j M_j/\rho_{oj}$
C	5.8766	11-2 21.21	0.2771
C_2	3.1814	30.24	0.1052
C_{3+}	84.5459		1.6925
	93.6039 lb		2.0748 cu ft

The second calculated value of pseudoliquid density $=$

$$\frac{93.6039 \text{ lb}}{2.0748} = 45.11 \text{ lb/cu ft}$$

Construct a graph as in Figure 11–5 using trial values of pseudoliquid density and the resulting calculated values. The point at which a line through the calculated values crosses a line with slope of one is the correct value.

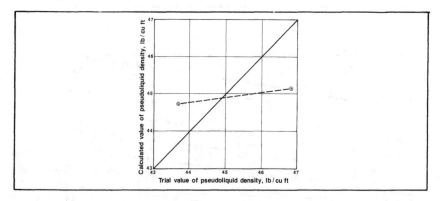

Fig. 11-5. Calculated values of pseudoliquid density vs. trial values of pseudoliquid density (part of solution to Example 11-4).

From Figure 11-5:

$$\rho_{po} = 44.85 \text{ lb/cu ft and } 60°F \text{ and } 14.696 \text{ psia}$$

Second, adjust pseudoliquid density at 60°F and 14.696 psia to 220°F and 2635 psia.

Compressibility adjustment = +0.91 lb/cu ft, at 2635 psia, Figure 11-3

$\rho_o = 44.85 + 0.91 = 45.76$ lb/cu ft at 60°F and 2635 psia

Thermal expansion adjustment = −4.79 lb/cu ft at 220°F, Figure 11-4

$\rho_o = 45.76 - 4.79 = 40.97$ lb/cu ft at 220°F and 2635 psia

or

$\rho_o = 40.97$ lb/cu ft × 5.615 cu ft/bbl = 230.0 lb/res bbl

Composition of the Saturated Liquid Known, Correlations

The iterative procedure given in Example 11-4 is time consuming; therefore, a calculating chart has been prepared using the method of Example 11-4.[4] This chart is given as Figure 11-6.

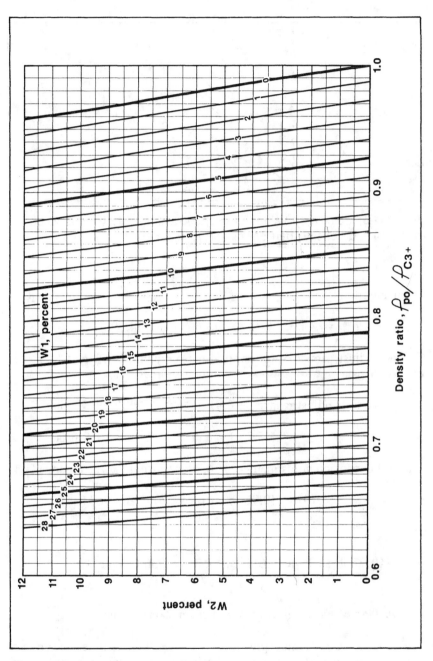

Fig. 11–6. Density ratios at standard conditions for pseudoliquids containing methane and ethane.

Use of this chart requires calculation of the *density of the propane-and-heavier fraction* of the mixture at standard conditions. If nonhydrocarbon components are present, hydrogen sulfide is included in the propane plus. Further, the *weight percent of methane in the mixture* and the *weight percent of ethane in the ethane-and-heavier fraction* must be determined.

The weight fraction of methane in the mixture is

$$W1 = \frac{w_{C1}}{w_{mix}},\qquad (11-1)$$

where w_{C1} is the mass of methane in lb methane/lb mole of mix (mole fraction times molecular weight) and w_{mix} is the mass of mixture in lb of mix/lb mole of mix.

Weight fraction of ethane in the ethane-and-heavier includes nonhydrocarbon components.

$$W2 = \frac{w_{C2} + w_{N2}}{w_{mix} - w_{C1} - w_{CO2}},\qquad (11-2)$$

where the terms on the right-hand side of the equation have meanings as given above. These two values are used to obtain the density ratio from the chart. The density ratio is multiplied by the density of the propane plus to calculate the density of the pseudoliquid.

This pseudoliquid density then is adjusted to density at reservoir temperature and pressure using Figures 11–3 and 11–4. Densities calculated by this method are as accurate as densities obtained experimentally for both black oils and volatile oils. If the liquid mixture contains hydrogen sulfide, an additional adjustment (given in Figure 11–7) is necessary.[4]

EXAMPLE 11–5: *Rework Example 11–4 using Figure 11–6.*

Solution

First, compute the density of the propane-and-heavier fraction, the weight fraction of ethane in the ethane-and-heavier fraction, the weight fraction of methane in the mixture, and then determine the density ratio.

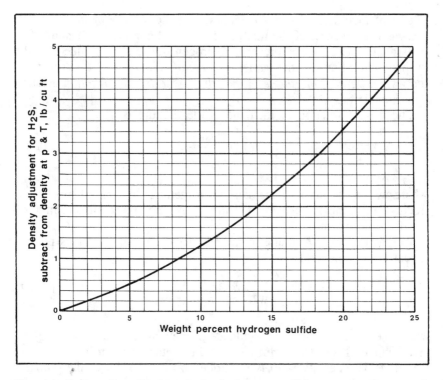

Fig. 11–7. Density adjustments for hydrogen sulfide content of reservoir liquids.

Component	Mole fraction, lb mole z_j	Molecular weight, lb/lb mole M_j	Mass, lb z_jM_j	Liquid density, 60°F & 14.696 psia, lb/cu ft ρ_{oj}	Liquid volume, 60°F & 14.696 psia, cu ft z_jM_j/ρ_{oj}
CO_2	0.0091	44.010	0.4005	use	
N_2	0.0016	28.013	0.0448	for	
C_1	0.3647	16.043	5.8509	$\rho_o'_2$	
C_2	0.0967	30.070	2.9078		
C_3	0.0695	44.097	3.0647	31.62	0.0969
i-C_4	0.0144	58.123	0.8370	35.11	0.0238
n-C_4	0.0393	58.123	2.2842	36.42	0.0627
i-C_5	0.0144	72.150	1.0390	38.96	0.0267
C_5	0.0141	72.150	1.0173	39.36	0.0258
C_6	0.0433	86.177	3.7315	41.40	0.0901
C_{7+}	0.3329	218.000	72.5722	53.11	1.3665
	1.0000		$M_a = 93.7499$	has -5132	1.6925

Standing and Katz method

density of propane plus = 84.5459 lb/1.6925 cu ft = 49.95 lb/cu ft

$$W2 = \frac{w_{C2} + w_{N2}}{w_{mix} - w_{C1} - w_{CO2}} \quad (11-2)$$

W2 = 2.9526 lb/87.4985 lb × 100 = 3.37 wt %

$$W1 = \frac{w_{C1}}{w_{mix}}, \quad (11-1)$$

W1 = 5.8509 lb/93.7499 lb × 100 = 6.24 wt %

ρ_{po}/ρ_{C3+} = 0.900, Figure 11–6

Second, determine pseudoliquid density at 14.696 psia and 60°F.

ρ_{po} = (49.95)(0.900) = 44.96 lb/cu ft $\quad \rho_{po} = \frac{\rho_{o}}{\rho_{C_3^+}} \cdot \rho_{C_3^+}$

Third, adjust to 2635 psia and 220°F as in Example 11–4.
pseudoliquid density = 44.96 lb/cu ft

pressure adjustment = $\frac{+0.91 \text{ lb/cu ft}}{45.87 \text{ lb/cu ft}}$ at p = 2635 psia, Figure 11–3

temperature adjustment = −4.78 lb/cu ft at T = 220°F, Figure 11–4

density of reservoir liquid = 41.09 lb/cu ft at 2635 psia and 220°F.

may need H2S adjustment (Fig 11-7)

Solution Gas-Oil Ratio, Gas Composition, and Stock-Tank Oil Gravity Known

Often, the only property of the surface liquid that is known is the specific gravity—although the composition of the gas associated with that liquid is known. In this situation, the procedure given in the previous section can be used to estimate the density of the liquid at reservoir conditions with comparable accuracy by simply considering the surface liquid as a single component.

EXAMPLE 11–6: *The solubility of the following natural gas in a 40.7°API stock-tank oil is 768 scf/STB at 2635 psia and 220°F. Calculate the density of the reservoir liquid at 2635 psia and 220°F*

Component	Composition, mole fraction
Hydrogen sulfide	nil
Carbon dioxide	0.0167
Nitrogen	0.0032
Methane	0.7108
Ethane	0.1552
Propane	0.0736
i-Butane	0.0092
n-Butane	0.0198
i-Pentane	0.0033
n-Pentane	0.0026
Hexanes	0.0027
Heptanes plus	0.0029
	1.0000

The usual assumption is that the molecular weight and specific gravity of heptanes plus in a separator gas are 103 lb/lb mole and 0.7.

Solution

First, compute the density of the propane-and-heavier fraction, the weight fraction of ethane in the ethane-and-heavier fraction, and the weight fraction of methane in the gas plus crude oil mixture. Remember that mole fraction equals volume fraction for a gas.

Component	Composition, volume fraction, y_j	Solubility, scf/STB $y_j R_s$	Mass, lb/STB $y_j R_s M_j / 380.7$	Liquid density, 60°F & 14.696 psia, lb/cu ft ρ_{oj}	Liquid volume, 60°F & 14.696 psia, cu ft/STB $\dfrac{y_j R_s M_j}{380.7 \rho_{oj}}$
CO_2	0.0167	12.8	1.48		
N_2	0.0032	2.5	0.18		
C_1	0.7108	545.9	23.00		
C_2	0.1552	119.2	9.41		
C_3	0.0736	56.5	6.55	31.62	0.207
$i\text{-}C_4$	0.0092	7.1	1.08	35.11	0.031
$n\text{-}C_4$	0.0198	15.2	2.32	36.42	0.064
$i\text{-}C_5$	0.0033	2.5	0.48	38.96	0.012
$n\text{-}C_5$	0.0026	2.0	0.38	39.36	0.010
C_6	0.0027	2.1	0.47	41.40	0.011
C_{7+}	0.0029	2.2	0.60	43.66	0.014
Stock-tank oil			*287.76		**5.615
		768.0	333.71 lb		5.964 cu ft

*Mass of 1 STB of 40.7°API stock-tank oil
**Volume of 1 barrel

Density of propane plus $= 299.64$ lb/5.964 cu ft $= 50.24$ lb/cu ft

$$W2 = \frac{w_{C2} + w_{N2}}{w_{mix} - w_{C1} - w_{CO2}} \qquad (11-2)$$

$$W2 = 9.59 \text{ lb}/309.23 \text{ lb} \times 100 = 3.1 \text{ wt } \%$$

$$W1 = \frac{w_{C1}}{w_{mix}} \qquad (11-1)$$

$$W1 = 23.00 \text{ lb}/333.71 \text{ lb} \times 100 = 6.9 \text{ wt } \%$$

Second, determine density ratio and calculate pseudoliquid density at 14.696 psia and 60°F.

$$\rho_{po}/\rho_{C3+} = 0.891, \text{ Figure } 11-6$$
$$\rho_{po} = (0.891)(50.24) = 44.76 \text{ lb/cu ft}$$

Third, adjust to 2635 psia and 220°F.

pseudoliquid density $\quad = 44.76$ lb/cu ft
pressure adjustment $\quad = +0.91$ at p $= 2635$ psia, Figure $11-3$

$\quad\quad\quad\quad\quad\quad\quad 45.67$ lb/cu ft
temperature adjustment $= -4.80$ at T $= 220°F$, Figure $11-4$

density of reservoir
liquid $\quad\quad\quad\quad\quad = 40.87$ lb/cu ft at 2635 psia and 220°F.

Solution Gas-Oil Ratio, Gas Specific Gravity, and Stock-Tank Gravity Known

The usual situation in black oil systems is that the specific gravities of the stock-tank liquid and the produced gas are known and the solubility of the gas in the liquid is known. When reservoir pressure is at or above the bubble point of the reservoir liquid, the gas solubility is simply equal to the producing gas-oil ratio.

However, if gas and liquid coexist in the reservoir, the produced gas-oil ratio includes both dissolved gas and free gas produced from the reservoir. In this situation gas solubility must be obtained by the method given earlier in this chapter.

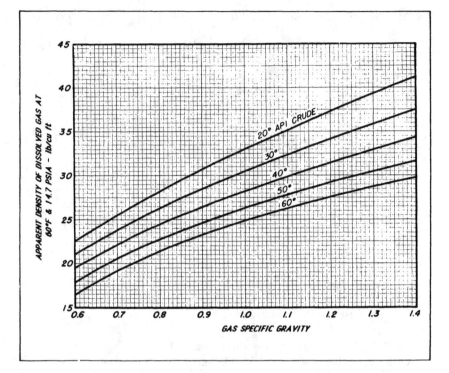

Fig. 11–8. Apparent liquid densities of natural gases. (Standing, *Volumetric and Phase Behavior of Oil Field Hydrocarbon Systems,* SPE, Dallas 1951. Copyright 1951 SPE-AIME.)

The concept of apparent liquid density has been extended to include natural gases. A correlation giving the apparent liquid density of gases which have come out of solution has been developed and is given as Figure 11–8.[5] Calculations of liquid densities at reservoir conditions using this correlation are almost as accurate as the methods using the composition of the mixture and normally will give results to within 5 percent of experimental determination.

EXAMPLE 11–7: *The producing gas-oil ratio of a well is 768 scf/STB, and the specific gravities of the gas and stock-tank oil are 0.786 and 40.7°API, respectively. The liquid in the reservoir is at its bubble point at reservoir conditions of 2635 psia and 220°F. Calculate the density of this liquid at reservoir conditions.*

Solution

First, calculate the mass of produced gas and stock-tank oil on the basis of one barrel of stock-tank oil.

Apparent molecular weight of the gas

$$M_a = 29\gamma_g = (29)(0.786) = 22.8 \text{ lb/lb mole} \qquad (3-38)$$

Mass of gas

$$\left(768 \frac{\text{scf}}{\text{STB}}\right)\left(\frac{\text{lb mole}}{380.7 \text{ scf}}\right)\left(22.8 \frac{\text{lb}}{\text{lb mole}}\right) = 46.0 \frac{\text{lb gas}}{\text{STB}}$$

Specific gravity of stock-tank oil

$$\gamma_{STO} = \frac{141.5}{131.5 + {}^\circ API} = \frac{141.5}{131.5 + 40.7} = 0.822 \qquad (8-2)$$

Mass of stock-tank oil

$$\left(5.615 \frac{\text{cu ft}}{\text{STB}}\right)\left(0.822 \frac{\text{lb oil/cu ft oil}}{\text{lb water/cu ft water}}\right)\left(62.37 \frac{\text{lb water}}{\text{cu ft water}}\right)$$

$$= 287.9 \frac{\text{lb oil}}{\text{STB}}$$

Second, calculate the pseudoliquid density of 60°F and 14.65 psia.

Component	Mass, lb/STB	Liquid density, 60°F & 14.65 psia lb/cu ft	Liquid volume, 60°F & 14.65 psia cu ft/STB
Surface gas	46.0	*24.2	1.901
Stock-tank oil	287.9		5.615
	333.9 lb/STB		7.516 cu ft/STB

*Figure 11–8.

$$\rho_{po} = \frac{333.9 \text{ lb/STB}}{7.516 \text{ cu ft/STB}} = 44.43 \text{ lb/cu ft}$$

Third, adjust pseudoliquid density to density at reservoir conditions.

Compressibility adjustment

$\rho_o = 44.43 + 0.93 = 45.36$ lb/cu ft at 60°F and 2635 psia, Figure 11−3

Thermal expansion correction

$\rho_o = 45.36 - 4.83 = 40.53$ lb/cu ft at 220°F and 2635 psia, Figure 11−4

or

$\rho_o = 40.53$ lb/cu ft $\times$ 5.615 cu ft/bbl $= 227.6$ lb/res bbl

The procedure of Example 11−7 is represented by

$$\rho_{po} = \frac{R_{sb}\gamma_g + 4600\ \gamma_{STO}}{73.71 + R_{sb}\gamma_g/\rho_a} \qquad (11-3)$$

where ρ_a is the apparent liquid density of the gas from Figure 11−8.

Calculation of Reservoir Liquid Density at Pressures Above the Bubble Point

The calculation of liquid density at pressures above the bubble point is a two-step procedure. First, the density at the bubble point must be computed using one of the methods previously described. Then this density must be adjusted to take into account the compression due to the increase in pressure from bubble-point pressure to the pressure of interest.

This second step is accomplished using the coefficient of isothermal compressibility. Methods of estimating values of this compressibility coefficient will be discussed later in this chapter. However, we now will show the use of compressibility in computing density changes corresponding to pressure changes.

Equation 8−19 follows from the definition of the coefficient of isothermal compressibility of a liquid above the bubble point.

$$\rho_o = \rho_{ob}\ \text{EXP}\ [c_o\ (p - p_b)] \qquad (8-19)$$

Equation 8−19 may be used to compute the density of a liquid above its bubble point. The value of c_o should be determined at the average pressure between the bubble-point pressure and the pressure of interest.

EXAMPLE 11–8: *The producing gas-oil ratio of the well given in Example 11–7 is 768 scf/STB, and the gravities of the gas and stock-tank oil are 0.786 and 40.7°API, respectively. Reservoir conditions are 5000 psig and 220°F, but the bubble point of the reservoir liquid is 2635 psia at 220°F. Calculate the density of this liquid at reservoir conditions given that the coefficient of isothermal compressibility of the reservoir liquid is 15.4 × 10⁻⁶ psia⁻¹.*

Solution

First, calculate the density of the reservoir liquid at bubble-point pressure of 2635 psia at 220°F.

$\rho_{ob} = 40.53$ lb/cu ft, see Example 11–7.

Second, compute ρ_o at 5015 psia and 220°F.

$$\rho_o = \rho_{ob} \, \text{EXP} \, [c_o \, (p - p_b)] \tag{8–19}$$

$$\rho_o = (40.53 \text{ lb/cu ft}) \, \text{EXP}[(15.4 \times 10^{-6})(5015 - 2635)] =$$

$$42.04 \text{ lb/cu ft}$$

Formation Volume Factor of Oil B_o

The results of the reservoir liquid density calculations can be used to calculate oil formation volume factors.

Remember that *three* situations were covered in the discussion of liquid density:

reservoir liquid composition known;

solution gas-oil ratio, gas composition, and stock-tank oil gravity known;

and solution gas-oil ratio, gas specific gravity, and stock-tank oil gravity known.

When the composition of the reservoir liquid is known, oil formation volume factor can be calculated very accurately using the procedure given in Chapter 13. The other two situations will be discussed here.

Estimation of Formation Volume Factor of Oil at Saturation Pressure Using Ideal-Solution Principles

The oil formation factor can be calculated using the results of ideal-solution calculations of the liquid density at reservoir conditions.

$$B_o = \frac{\text{mass of 1.0 STB} + \text{mass of gas evolved from 1.0 STB}}{\text{mass of 1.0 reservoir barrel}} \qquad (11-4)$$

results from the definition of oil formation volume factor.

EXAMPLE 11–9: *Calculate the formation volume factor of the oil described in Example 11–7 at its bubble-point pressure of 2635 psia at 220°F.*

Solution

Select the necessary quantities from the solution of Example 11–7.

$$B_o = \frac{287.9 \text{ lb oil/STB} + 46.0 \text{ lb gas/STB}}{227.6 \text{ lb/res bbl}} \qquad (11-4)$$

$$B_o = 1.467 \text{ res bbl/STB}$$

Equation 11–4 can be rewritten as

$$B_o = \frac{\rho_{STO} + 0.01357 R_s \gamma_g}{\rho_{oR}}, \qquad (11-5)$$

where ρ_{STO} is the density of the stock-tank oil and ρ_{oR} is the density of the oil at reservoir conditions, both in lb/cu ft, and R_s has the units of scf/STB. B_o has units of res bbl/STB.

EXAMPLE 11–10: *Calculate the formation volume factor of the oil described in Example 11–7 at its bubble-point pressure of 2635 psia at 220°F.*

Solution

Select the necessary quantities from the solution of Example 11–7.

$$B_o = \frac{\rho_{STO} + 0.01357 R_s \gamma_g}{\rho_{oR}} \qquad (11-5)$$

$$B_o = \frac{51.25 + (0.01357)(768)(0.786)}{40.53} = 1.467 \text{ res bbl/STB}$$

Since these calculations are applicable only for a liquid at a pressure equal to or below its bubble-point pressure, this method is useful only for saturated liquids.

This method of calculating the formation volume factor has a probable error of about 5 percent.

Estimation of Formation Volume Factor of Oil at Saturation Pressures by Correlation

Figure 11–9 may be used to obtain an accurate estimate of formation volume factor of an oil at its bubble point if the producing gas-oil ratio, gas specific gravity, stock-tank oil gravity, and reservoir temperature are known.[1,3] Reservoir pressure must be equal to the bubble-point pressure of the oil because the value of gas-oil ratio used to enter the chart must represent the solubility of the gas at the bubble point. If reservoir pressure is below the bubble point, some of the produced gas may come from free gas in the reservoir, and the use of producing gas-oil ratio in this correlation will give incorrect results.

Formation volume factors computed with this correlation should be within about 5 percent of the experimentally determined values.

If the solution gas-oil ratio is known at some pressure below the bubble point, it can be used to enter Figure 11–9, and an accurate estimate of the oil formation volume factor at that reservoir pressure can be determined. Solution gas-oil ratio can be obtained from Figure 11–1. However, the accuracy of the final result is a combination of the 5 percent attributed to Figure 11–8 and the 15 percent attributed to Figure 11–1.

EXAMPLE 11–11: *Estimate values of oil formation volume factor at various pressures below bubble-point pressure for the reservoir oil of Example 11–1.*

Solution

Enter Figure 11–9 with any desired solution gas-oil ratio and with

$$\gamma_g = 0.786$$
$$\gamma_o = 40.7°API$$
$$T = 220°F$$

and read corresponding values of B_o. We will use the values of R_s determined in Example 11–2.

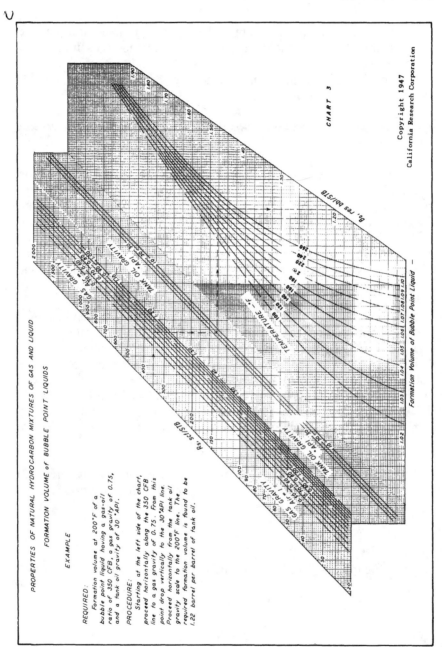

Fig. 11-9. Formation volume factors of saturated black oils. (Copyright 1947 Chevron Oil Field Research Co., with permission.)

Pressure, psia	Solution* gas-oil ratio, scf/STB	Formation volume factor, res bbl/STB
2685	768	1.469
2415	676	1.418
2165	594	1.372
1915	513	1.329
1665	434	1.287
1415	358	1.248
1165	284	1.211
915	214	1.176
665	147	1.144
415	86	1.116
165	31	1.091

*From Exercise 11–2

Estimation of Formation Volume Factor of Oil at Pressures Above the Bubble-Point Pressure

The oil formation volume factor at pressures above the bubble-point pressure is less than the formation volume factor at bubble-point pressure because of the contraction of the oil as reservoir pressure is increased. This compression due to increased reservoir pressure is the only factor which affects formation volume factor at pressures above the bubble point.

The normal procedure for estimating formation volume factor at pressures above the bubble point is first to estimate the factor at bubble-point pressure and reservoir temperature using one of the methods just described. Then, adjust the factor to higher pressure through the use of the coefficient of isothermal compressibility. The equation used for this adjustment follows directly from the definition of the compressibility coefficient at pressures above the bubble point.

$$c_o = -\frac{1}{B_o}\left(\frac{\partial B_o}{\partial p}\right)_T \qquad (8\text{–}9)$$

Integration from bubble-point pressure to pressure, p, higher than the bubble point, results in

$$B_o = B_{ob}\ \text{EXP}\ [c_o(p_b - p)] \qquad (11\text{–}6)$$

or

$$\ln B_o = \ln B_{ob} + c_o(p_b - p). \qquad (11\text{–}7)$$

EXAMPLE 11–12: *Estimate the formation volume factor of the oil in Example 11–10 at a reservoir pressure of 5,000 psig and reservoir temperature of 220°F. Use a value of 15.4 × 10⁻⁶ psi⁻¹ for the coefficient of isothermal compressibility of the oil between 5015 psia and the bubble point.*

Solution

$$B_o = B_{ob} \; EXP \; [c_o(p_b - p)] \qquad (11-6)$$

$$B_o = (1.467)EXP[(15.4 \times 10^{-6})(2635 - 5015)] = 1.414 \text{ res bbl/STB}$$

Adjustment of Formation Volume Factor of Oil and Solution Gas-Oil Ratio for Field Derived Bubble-Point Pressure

The accuracy of the results of the use of correlations to estimate oil formation volume factor and solution gas-oil ratio can be improved if an accurate value of bubble-point pressure is available. The method described in Chapter 9 can be used to get a reasonably accurate value of bubble-point pressure if reservoir pressure has been measured regularly during the life of the field.

Tables of oil formation volume factor and solution gas-oil ratio tabulated against pressure are adjusted by changing the values of pressure. A *delta pressure* is calculated as the difference between field derived bubble-point pressure and bubble-point pressure from correlation.

$$\Delta \text{pressure} = p_{b \text{ corr}} - p_{b \text{ field}} \qquad (11-8)$$

This delta pressure is then subtracted from every value of pressure in the tabulation.

EXAMPLE 11–13: *Examination of reservoir pressure measurements shows that the bubble-point pressure of the reservoir oil of Example 11–1 is 2635 psia and 220°F. The table below gives results of Examples 11–1, 11–2, and 11–11. Adjust the table to agree with the field derived bubble-point pressure.*

Pressure, psia	Formation volume factor, res bbl/ STB	Solution gas-oil ratio, scf/STB
2685 = p_b	1.469	768
2415	1.418	676
2165	1.372	594
1915	1.329	513
1665	1.287	434
1415	1.248	358
1165	1.211	284
915	1.176	214
665	1.144	147
415	1.116	86
165	1.091	31

Solution

$$\Delta \text{pressure} = 2685 - 2635 = 50 \text{ psia} \qquad (11-8)$$

Subtract Δ pressure from each pressure in the table.

Pressure, psia	Formation volume factor, res bbl/STB	Solution gas-oil ratio, scf/STB
2635 = p_b	1.469	768
2365	1.418	676
2115	1.372	594
1865	1.329	513
1615	1.287	434
1365	1.248	358
1115	1.211	284
865	1.176	214
615	1.144	147
365	1.116	86
115	1.091	31

Total Formation Volume Factor

Total formation volume factor includes the volume of the liquid, represented by B_o, plus the volume of the gas which has evolved from the liquid in the reservoir, represented by the term $B_g(R_{sb} - R_s)$.

$$B_t = B_o + B_g(R_{sb} - R_s) \qquad (8-6)$$

or Fig 11-10

The total formation volume factor may be estimated by combining estimates of B_o, R_{sb}, and R_s from this chapter with an estimate of B_g from Chapter 6. The accuracy of the estimate of B_t will be only as good as the accuracy of the estimates of these quantities.

A value of specific gravity of the reservoir gas is required to calculate B_g. The specific gravity of the separator gas may be used as an estimate of the specific gravity of the reservoir gas at reservoir pressures above about 1000 psia.

EXAMPLE 11–14: *Calculate total formation volume factor for the reservoir oil of Example 11–1 at 2115 psia and 220°F. Use the results of Example 11–13.*

Solution

First, calculate a value of gas formation volume factor at 2115 psia and 220°F.

$$z = 0.866, \text{ method of Example } 3–11 \text{ at } \gamma_g = 0.786$$

$$B_g = 0.00502 \, \frac{zT}{p} \, \frac{\text{res bbl}}{\text{scf}} \qquad (6–3)$$

$$B_g = \frac{(0.00502)(0.866)(220+460)}{2115} = 0.00140 \frac{\text{res bbl}}{\text{scf}}$$

Second, calculate total formation volume factor.

$$B_t = B_o + B_g(R_{sb} - R_s) \qquad (8–6)$$

$$B_t = 1.372\frac{\text{res bbl}}{\text{STB}} + 0.00140 \frac{\text{res bbl}}{\text{scf}} (768-594)\frac{\text{scf}}{\text{STB}}$$

$$= 1.616 \text{ res bbl/STB}$$

Figure 11–10 gives a chart which can be used to estimate B_t directly with probable error of about 5 percent.[1,3] The use of this chart eliminates the need for estimating the individual terms in Equation 8–6.

EXAMPLE 11–15: *Repeat Example 11–14 using Figure 11–10.*

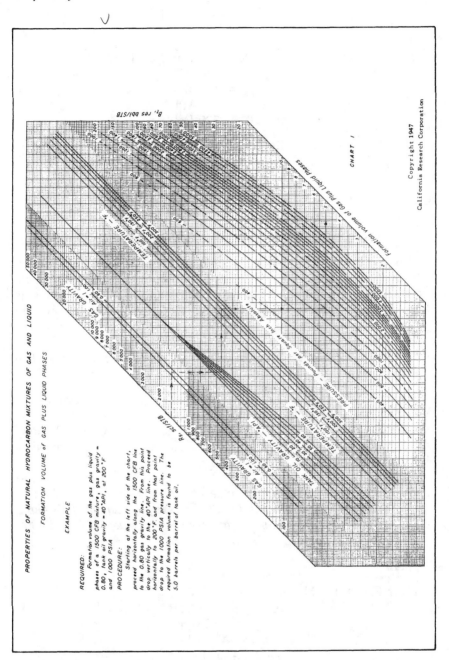

Fig. 11–10. Total formation volume factors of black oils. (Copyright 1947 Chevron Oil Field Research Co., with permission.)

Solution

Enter Figure 16–10 with data from Example 11–1,

$$R_s = 768 \text{ scf/STB}$$
$$\gamma_g = 0.786$$
$$\gamma_o = 40.7°\text{API}$$
$$T = 220°\text{F}$$
$$p = 2115 \text{ psia}$$

and read

$$B_t = 1.65 \text{ res bbl/STB}$$

The Coefficient of Isothermal Compressibility of Oil

The coefficient of isothermal compressibility of a liquid is defined in Chapter 8. Equations 8–7 apply to a liquid at pressures above its bubble point. Equation 8–24 applies to a reservoir liquid at pressures below its bubble point. Figure 8–7 shows the effect a decline in reservoir pressure has on oil compressibility.

Coefficient of Isothermal Compressibility of Oil at Pressures Above Bubble-Point Pressure

Figure 11–11 can be used to estimate values of oil compressibility at pressures *above* the bubble point.[6] This is the best available correlation considering both accuracy and ease of use. The results are generally low, by as much as 50 percent at high pressures. Accuracy is improved as bubble-point pressure is approached.

EXAMPLE 11–16: *Estimate a value of the coefficient of isothermal compressibility of the reservoir liquid of Example 11–1 for use at pressures between 5000 and 4000 psig.*

Solution

Enter Figure 11–11 with

$$R_s = 768 \text{ scf/STB}$$
$$\gamma_g = 0.786$$
$$\gamma_o = 40.7°\text{API}$$
$$T = 220°\text{F}$$

$$p = \frac{5000 + 4000}{2} + 14.7 = 4515 \text{ psia}$$

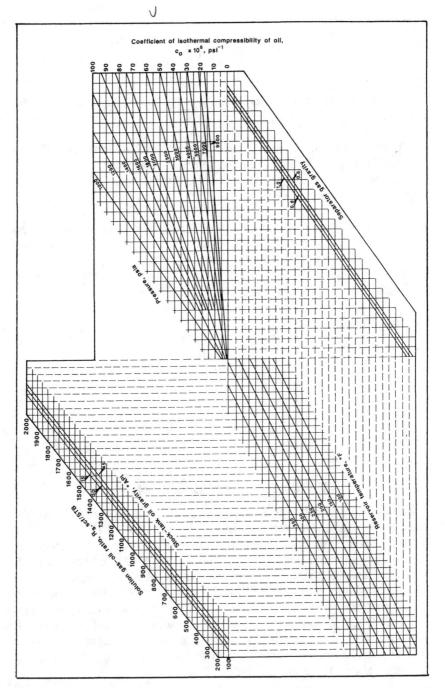

Fig. 11–11. Coefficients of isothermal compressibility of undersaturated black oils.

$P > P_b$

and read

$$c_o = 12.8 \times 10^{-6} \text{ psi}^{-1}$$

Coefficient of Isothermal Compressibility at Pressures Below Bubble-point Pressure

Figure 11–12 can be used to estimate values of oil compressibility at pressures below the bubble point.[7] The results are accurate to within 10 percent at pressures above 500 psia. Below 500 psia the accuracy is within 20 percent.

EXAMPLE 11–17: *Continue Example 11–1. Estimate a value of the coefficient of isothermal compressibility at 2100 psig.*

Solution

Enter Figure 11–12 with
$$R_{sb} = 768 \text{ scf/STB}$$
$$T = 220°F$$
$$\gamma_o = 40.7°API$$
$$p = 2115 \text{ psia}$$

and read

$$c_o = 180 \times 10^{-6} \text{ psi}^{-1}$$

The Coefficient of Viscosity of Oil

The *viscosity* of a liquid is directly related to the type and size of the compounds which make up the liquid. The variation of liquid viscosity with molecular structure is not well understood; however, the viscosities of liquids which are members of a homologous series of compounds will vary in a regular manner, as do most other physical properties. The viscosities of liquids composed of large complex molecules will be much higher than the viscosities of liquids composed of smaller molecules.

Figure 8–8 shows how changes in reservoir pressure affect oil viscosity. The figure also shows that oil viscosities at pressures below the bubble-point must be correlated differently than oil viscosities at pressures above the bubble point.

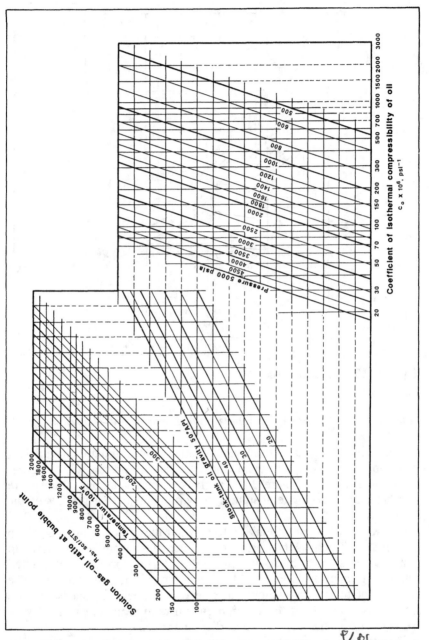

Fig. 11–12. Coefficients of isothermal compressibility of saturated black oils. (McCain et al., SPE *Form. Eval.*, *3*, 1988. Copyright 1988 SPE-AIME.)

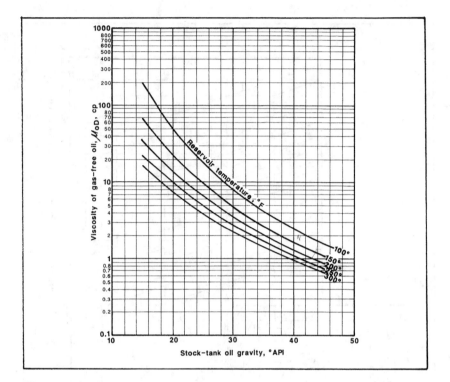

Fig. 11–13. Dead oil viscosities.

Estimation of Oil Viscosity at Bubble-Point Pressure and Below

If experimental data of viscosities of a reservoir liquid are unavailable, an estimate can be made using the correlations presented in Figures 11–13 and 11–14. Figure 11–13 gives viscosity at one atmosphere pressure and reservoir temperature as a function of the gravity of the stock-tank oil.[8]

Figure 11–14 is used to take into account the effect of the gas which goes into solution as the pressure is increased from atmospheric to saturation pressure at reservoir temperature.[9] This correlation is entered with the viscosity from Figure 11–13 and the solution gas-oil ratio. The viscosity of the reservoir liquid at reservoir temperature and saturation pressure is obtained. If the solution gas-oil ratio is unknown, it may be estimated using methods given earlier in this chapter.

The resulting values of oil viscosity are not particularly accurate. The results should be of the correct order of magnitude.

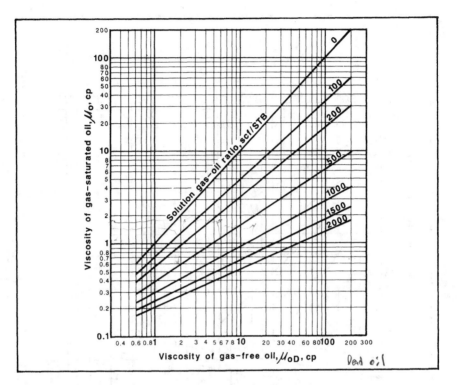

Fig. 11-14. Viscosities of saturated black oils.

EXAMPLE 11-18: *Estimate a value of oil viscosity for the reservoir oil of Example 11-1 at the bubble-point.*

Solution use Fig 11-13 then 11-14

First, estimate dead oil viscosity.

Enter Figure 11-13 with

$T = 220°F$ and $\gamma_o = 40.7°API$

and read

$$\mu_{oD} = 1.15 \text{ cp}$$

Second, estimate oil viscosity at the bubble point.

Enter Figure 11-14 with

$\mu_{oD} = 1.15 \text{ c}_p$ and $R_s = 768 \text{ scf/STB}$

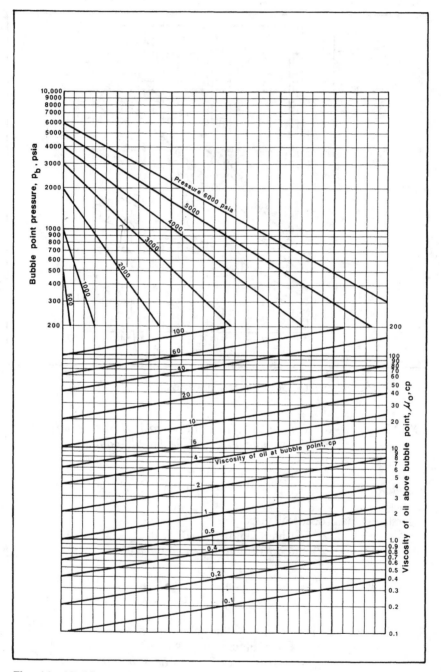

Fig. 11–15. Viscosities of undersaturated black oils.

and read

$$\mu_{ob} = 0.36 \text{ cp at } 220°F \text{ and } 2635 \text{ psia.}$$

Estimation of Oil Viscosity at Pressures Above Bubble-Point Pressure

Often the pressure in a reservoir is higher than the bubble-point pressure of the reservoir liquid. Figure 11–15 may be used to determine the viscosity of the liquid for pressures above the bubble-point pressure.[6] Figure 11–15 is entered with bubble-point pressure and with the viscosity at reservoir temperature and bubble-point pressure from Figure 11–14. The isobars are for any pressure greater than bubble point. The viscosity of the liquid above the bubble point is obtained. This viscosity will be greater than the viscosity at the bubble point due to the compression of the liquid caused by the increase in pressure.

The results obtained from Figure 11–15 should be considered to be only of the correct order of magnitude.

EXAMPLE 11–19: *Estimate a value of oil viscosity for the reservoir oil of Example 11–1 at 5000 psig.*

Solution

Enter Figure 11–15 with

$$p_b = 2635 \text{ psia}$$

$$p = 5015 \text{ psia}$$

$$\mu_{ob} = 0.36 \text{ cp, from Example 11–18,}$$

and read

$$\mu_o = 0.46 \text{ cp at } 5015 \text{ psia and } 220°F.$$

Interfacial Tension

The imbalance of molecular forces at the interface between two phases is known as interfacial tension. This is explained in Chapter 8.

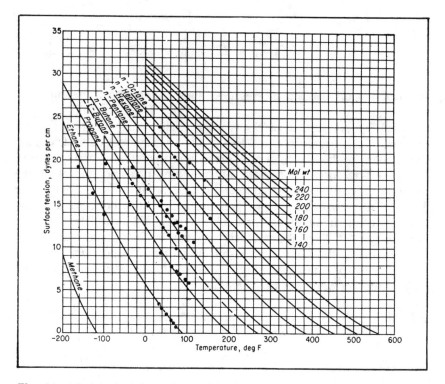

Fig. 11–16. Interfacial tensions of hydrocarbons. (Adapted from Katz, et al., *J. Pet. Tech.*, Sept. 1943.)

Liquid-Gas Interfacial Tension, Pure Substances

Interfacial tension (surface tension) only exists when two phases are present. This occurs along the vapor pressure line for pure substances. The interfacial tension for pure substances decreases as temperature increases and becomes zero at the critical point. See Figure 11–16.[10] Interfacial tension for pure substances may be calculated as[11,12]

$$\sigma = \left(P \frac{\rho_L - \rho_g}{M} \right)^4 \qquad (11-9)$$

where σ is interfacial tension in dynes/cm and the densities are in g/cc. P is called the parachor. The parachors for pure substances are considered to be constant. Parachors for some hydrocarbons of interest are given in Table 11–1.[13] Figure 11–17 can be used to estimate parachors for other pure hydrocarbons.

Table 11–1.
Parachors for Computing
Interfacial Tension[13]

Constituent	Parachor
Methane..............	77.0
Ethane................	108.0
Propane.............	150.3
i-Butane..............	181.5
n-Butane..............	189.9
i-Pentane.............	225.0
n-Pentane.............	231.5
n-Hexane..............	271.0
n-Heptane.............	312.5
n-Octane..............	351.5
Hydrogen........... 34 (approx.)
Nitrogen............. 41 (approx.)
Carbon dioxide...	78

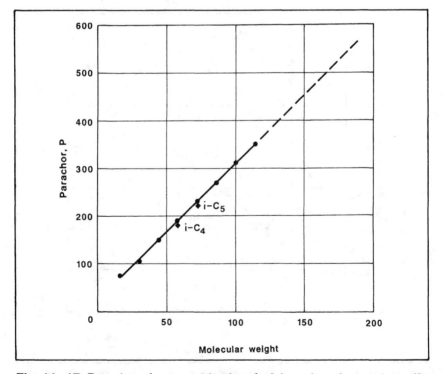

Fig. 11–17. Parachors for computing interfacial tension of normal paraffin hydrocarbons. (Data from Katz et al., *Handbook of Natural Gas Engineering,* McGraw-Hill, 1959.)

EXAMPLE 11–20: *Propane gas and liquid are in equilibrium at 100°F and 190 psia. Estimate the interfacial tension between the liquid and the gas.*

Solution

$\rho_L = 0.476$ g/cc, $\rho_g = 0.026$ g/cc, Figure 2–12
P = 150.3, Table 11–1

$$\sigma = \left(P \frac{\rho_L - \rho_g}{M} \right)^4 \tag{11–9}$$

$$\sigma = \left(150.3 \frac{0.476 - 0.026}{44.1} \right)^4 = 5.5 \text{ dyne/cm}$$

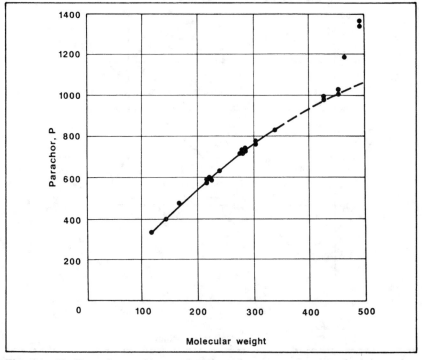

Fig. 11–18. Parachors of heavy fractions for computing interfacial tension of reservoir liquids. (Data from Firoozabadi et al., SPE *Res. Eng., 3,* 1988, 265.)

Liquid-Gas Interfacial Tension, Hydrocarbon Mixtures

Interfacial tension of hydrocarbon liquid and gas at equilibrium may be estimated with[14]

$$\sigma = \left[\sum_j P_j \left(x_j \frac{\rho_L}{M_L} - y_j \frac{\rho_g}{M_g} \right) \right]^4, \qquad (11-10)$$

where ρ_L is the density in g/cc and M_L is the apparent molecular weight of the equilibrium liquid and ρ_g and M_g have similar meaning for the equilibrium gas. Values of parachor for the components of a hydrocarbon mixture are obtained from Table 11-1 or Figure 11-17. Parachor values for the heptanes plus fraction may be obtained from the correlation line of Figure 11-18.[15]

EXAMPLE 11-21: *The compositions of reservoir liquid and gas at 2315 psia and 190°F for a volatile oil are given below. Calculate the liquid-gas interfacial tension.*

Component	Liquid composition, mole fraction	Gas composition, mole fraction
Carbon dioxide	0.0159	0.0259
Nitrogen	0.0000	0.0022
Methane	0.3428	0.8050
Ethane	0.0752	0.0910
Propane	0.0564	0.0402
i-Butane	0.0097	0.0059
n-Butane	0.0249	0.0126
i-Pentane	0.0110	0.0039
n-Pentane	0.0141	0.0044
Hexanes	0.0197	0.0040
Heptanes plus	0.4303	0.0049
	1.0000	1.0000

Properties of heptanes plus of liquid
Specific gravity   0.868
Molecular weight   217 lb/lb mole

Solution

First, calculate densities and molecular weights of the liquid and gas.

ρ_L = 0.719 g/cc, methods of this chapter

ρ_g = 0.137 g/cc, methods of Chapter 3

M_L = 110.1 g/g mole, Equation 3–35

M_g = 21.1 g/g mole, Equation 3–35

Second, calculate interfacial tension.

Component	x_j	y_j	P_j	Equation 11–10
CO_2	0.0159	0.0259	78.0	−0.0050
N_2	0.0000	0.0022	41.0	−0.0006
C_1	0.3428	0.8050	77.0	−0.2301
C_2	0.0752	0.0910	108.0	−0.0108
C_3	0.0564	0.0402	150.3	0.0161
i-C_4	0.0097	0.0059	181.5	0.0046
n-C_4	0.0249	0.0126	189.9	0.0154
i-C_5	0.0110	0.0039	225.0	0.0105
n-C_5	0.0141	0.0044	231.5	0.0147
C_6	0.0197	0.0040	271.0	0.0278
C_{7+}	0.4303	0.0049	*586.6	1.6297
	1.0000	1.0000		1.4723

*from Figure 11–18

$$\sigma = (1.4723)^4 = 4.7 \text{ dynes/cm} \tag{11-10}$$

The results of calculations as illustrated in Example 11–21 should be within 10 percent of laboratory measured values. Better accuracy will be obtained if the heptanes plus fraction is divided into several fractions and the parachor for each fraction obtained from Figure 11–18 according to the molecular weight of the fraction. This is the manner in which the data of Figure 11–18 were obtained. The deviation of the heavier fractions from the line on Figure 11–18 is attributed to the collection of asphaltenes in the heaviest fraction of the liquid.[15] The correlation line obviously does not give good values of parachors for this fraction.

Exercises

11–1. Your recently completed well is producing 40.3°API oil and 0.756 specific gravity gas at 1000 scf/STB. Producing gas-oil ratio has remained constant, so you believe reservoir pressure is above the bubble-point pressure of the reservoir oil. Estimate bubble-point pressure given that reservoir temperature is 205°F.

11–2. Your recently completed well is producing 39.0°API oil and 0.950 specific gravity gas at 589 scf/STB. Producing gas-oil ratio has remained constant so you believe reservoir pressure is above the bubble-point pressure of the reservoir oil. Estimate bubble-point pressure given that reservoir temperature is 250°F.

11–3. Estimate values of solution gas-oil ratio for pressures below bubble-point pressure for the reservoir oil of Exercise 11–1. Start at a pressure of 3150 psig and decrease pressure in 500 psi increments to a final pressure of 150 psig.

11–4. Estimate values of solution gas-oil ratio for various pressures below bubble-point pressure for the reservoir oil of Exercise 11–2. Start at a pressure of 1900 psig and reduce pressure in 150 psi increments to a final pressure of 150 psig.

11–5. Use ideal-solution principles to calculate the density at 14.696 psia and 60°F of a hydrocarbon liquid of the following composition.

Component	Composition, mole fraction
Propane	0.025
i-Butane	0.325
n-Pentane	0.650
	1.000

11–6. Use ideal-solution principles to calculate the density at 14.696 psia and 60°F of a hydrocarbon liquid of the following composition.

Component	Composition, mole fraction
n-Pentane	0.225
Hexanes	0.425
Heptanes plus	0.350
	1.000

Properties of heptanes plus
Specific gravity   0.8525
Molecular weight   219 lb/lb mole

11–7. Use the data given in Figure 11–2 to calculate the density of a reservoir liquid at its bubble point of 3385 psia at a reservoir temperature of 205°F. The composition of the well stream is as follows.

Component	Composition, mole fraction
Methane	0.4642
Ethane	0.0637
Propane	0.0561
i-Butane	0.0119
n-Butane	0.0383
i-Pentane	0.0168
n-Pentane	0.0195
Hexanes	0.0285
Heptanes plus	0.3010
	1.0000

Properties of heptanes plus
Specific gravity   0.8500
Molecular weight   202 lb/lb mole

11–8. Repeat Exercise 11–7. Use Figure 11–6. Compare your answer with laboratory measurement of 39.05 lb/cu ft.

11–9. Use the correlation of Figure 11–6 to calculate the density of a reservoir liquid at its bubble point of 1763 psia at a reservoir temperature of 250°F. The composition of the well stream is as follows.

Component	Composition, mole fraction
Hydrogen sulfide	0.0879
Carbon dioxide	0.0270
Nitrogen	0.0009
Methane	0.2112
Ethane	0.0763
Propane	0.0703
i-Butane	0.0147
n-Butane	0.0428
i-Pentane	0.0171
n-Pentane	0.0237
Hexanes	0.0248
Heptanes plus	0.4033
	1.0000

Properties of heptanes plus
Specific gravity   0.8500
Molecular weight   215 lb/lb mole

Compare your answer with laboratory measurement of 42.12 lb/cu ft.

11–10. The solubility of the following gas in a 40.3°API stock-tank oil is 1000 scf/STB at 3385 psia and 205°F. Calculate the density of the reservoir liquid at 3385 psia and 205°F.

Component	Composition, mole fraction
Methane	0.7833
Ethane	0.0965
Propane	0.0663
i-Butane	0.0098
n-Butane	0.0270
i-Pentane	0.0063
n-Pentane	0.0064
Hexanes	0.0037
Heptanes plus	0.0007
	1.0000

11–11. The solubility of the following gas in a 39.0°API stock-tank oil is 589 scf/STB at 1763 psia and 250°F. Calculate the density of the reservoir liquid at 1763 psia and 250°F.

Component	Composition, mole fraction
Hydrogen sulfide	0.1521
Carbon dioxide	0.0592
Nitrogen	0.0022
Methane	0.5039
Ethane	0.1431
Propane	0.0857
i-Butane	0.0109
n-Butane	0.0261
i-Pentane	0.0053
n-Pentane	0.0081
Hexanes	0.0025
Heptanes plus	0.0009
	1.0000

11–12. The producing gas-oil ratio of a well is 1000 scf/STB, and the specific gravities of the gas and stock-tank oil are 0.756 and 40.3°API, respectively. The liquid in the reservoir is at its bubble point at reservoir conditions of 3385 psia and 205°F. Calculate the density of this liquid at reservoir conditions.

11–13. The producing gas-oil ratio of a well is 589 scf/STB, and the specific gravities of the gas and stock-tank oil are 0.950 and 39.0°API, respectively. The liquid in the reservoir is at its bubble point at reservoir conditions of 1763 psia and 250°F. The reservoir liquid has 3 weight percent hydrogen sulfide. Calculate the density of this liquid at reservoir conditions.

11–14. Calculate the density of the reservoir liquid of Exercise 11–13 at the initial reservoir pressure of 4800 psia given that the coefficient of isothermal compressibility between bubble-point pressure and 4800 psia is 14.5×10^{-6} psi^{-1}.

11–15. Calculate the formation volume factor of the oil described in Exercise 11–10 at its bubble point of 3385 psia at 205°F.

11–16. Calculate the formation volume factor of the oil described in Exercise 11–11 at its bubble point of 1763 psia and 250°F.

11–17. Calculate the formation volume factor of the oil described in Exercise 11–12 at its bubble point of 3385 psia and 205°F.

11–18. Calculate the formation volume factor of the oil described in Exercise 11–13 at its bubble point of 1763 psia and 250°F. Use the procedure of Example 11–9.

11–19. Repeat Exercise 11–18. Use Equation 11–5.

11–20. Calculate the density of a reservoir liquid at its bubble point of 2361 psia and 225°F given R_{sb} = 973 scf/STB, B_{ob} = 1.591 res bbl/STB, γ_g = 1.010, γ_{STO} = 37.0°API.

11–21. Estimate values of oil formation volume factor at pressures below bubble-point pressure for the reservoir oil of Exercise 11–1. Start at a pressure of 3150 psig and decrease pressure by 500 psi increments to a final pressure of 150 psig.

11–22. Estimate values of oil formation volume factor at pressures below the bubble-point pressure for the reservoir oil of Exercise 11–2. Start at a pressure of 1900 psig and reduce pressure in 150 psi increments to a final pressure of 150 psig.

11–23. A black oil has a bubble-point pressure of 4000 psia at 225°F. The oil formation volume factor at the bubble point is 1.519 res bbl/STB. Estimate the oil formation volume factor at initial reservoir pressure of 6250 psia. Use a value of 13 microsips for the coefficient of isothermal compressibility.

11–24. An oil has B_{ob} = 1.255 res bbl/STB, R_{sb} = 410 scf/STB for p_b = 2265 psig at 214°F. Produced gas specific gravity is 0.811 and stock-tank oil gravity is 23.7°API. Estimate B_o at p_i = 4070 psig. Use c_o = 9 × 10^{-6} psi^{-1}.

11–25. Correlations have been used to estimate the following properties of a black oil at reservoir temperature of 214°F.

Pressure, psia	Formation volume factor, B_o res bbl/STB	Solution gas-oil ratio, R_s scf/STB
2502 = p_b	1.263	420
2415	1.255	393
2165	1.231	345
1915	1.208	298
1665	1.187	252
1415	1.166	208
1165	1.146	165
915	1.128	124
665	1.111	85
415	1.095	50
165	1.082	18

Field data indicate the bubble-point pressure to be 2265 psia. Adjust the table above to the field bubble point.

11–26. Estimate a value of the coefficient of isothermal compressibility for use at 4000 psia for the black oil of Exercise 11–1.

11–27. Estimate a value of the coefficient of isothermal compressibility for use at 3500 psia for the black oil of Exercise 11–2.

11–28. Continue Exercise 11–26. Estimate a value of the coefficient of isothermal compressibility at 2000 psia.

11-29. Continue Exercise 11-27. Estimate a value of the coefficient of isothermal compressibility at 1000 psia.

11-30. Estimate a value of viscosity for the black oil of Exercise 11-24 at the bubble point.

11-31. Estimate a value of viscosity for the black oil of Exercise 11-24 at 3000 psia.

11-32. A black oil has a bubble-point pressure of 1250 psia at reservoir temperature of 142°F. Producing gas-oil ratio is 370 scf/STB of 0.872 specific gravity gas and 40.2°API stock-tank oil. Estimate a value of viscosity at initial reservoir pressure of 2880 psia.

11-33. Estimate the interfacial tension of n-octane at 300°F.

11-34. The compositions of the equilibrium gas and liquid of a black oil at reservoir conditions of 1500 psia and 150°F are given below. Calculate the liquid-gas interfacial tension.

Component	Liquid composition, mole fraction	Gas composition, mole fraction
Methane	0.3105	0.8742
Ethane	0.0638	0.0634
Propane	0.0646	0.0335
i-Butane	0.0146	0.0049
n-Butane	0.0478	0.0129
i-Pentane	0.0218	0.0035
n-Pentane	0.0255	0.0034
Hexanes	0.0382	0.0026
Heptanes plus	0.4132	0.0016
	1.0000	1.0000

Properties of heptanes plus of the liquid
Specific gravity 0.850
Molecular weight 202 lb/lb mole

11–35. The compositions of the equilibrium gas and liquid of a black oil at reservoir conditions of 1200 psia and 160°F are given below. Calculate the liquid-gas interfacial tension.

Component	Liquid composition, mole fraction	Gas composition, mole fraction
Carbon dioxide	0.0075	0.0147
Nitrogen	0.0011	0.0034
Methane	0.2389	0.8025
Ethane	0.0937	0.1072
Propane	0.0769	0.0436
i-Butane	0.0169	0.0058
n-Butane	0.0469	0.0127
i-Pentane	0.0177	0.0028
n-Pentane	0.0175	0.0023
Hexanes	0.0547	0.0036
Heptanes plus	0.4282	0.0014
	1.0000	1.0000

Properties of the heptanes plus of the liquid	
Specific gravity	0.8515
Molecular weight	218 lb/lb mole

References

1. Standing, M.B.: "A Pressure-Volume-Temperature Correlation for Mixtures of California Oils and Gases," *Drilling and Production Practice*, API (1947) 275–287.

2. Standing, M.B. and Katz, D.L.: "Density of Crude Oils Saturated with Natural Gas," *Trans.*, AIME (1942) *146*, 159–165.

3. Standing, M.B.: *Volumetric and Phase Behavior of Oil Field Hydrocarbon Systems*, Reinhold Publishing Corp., New York City (1952).

4. Witte, T.W., Jr.: "The Development of a Correlation for Determining Oil Density in High Temperature Reservoirs," M.S. Thesis (Dec. 1987), Texas A&M University.

5. Katz, D.L.: "Prediction of the Shrinkage of Crude Oils," *Drilling and Production Practice*, API (1942) 137–147.

6. Vazquez, M. and Beggs, H.D.: "Correlations for Fluid Physical Property Prediction," *J. Pet. Tech.* (June 1980) *32*, 968–970.

7. McCain, W.D., Jr., Rollins, J.B., and Villena, A.J.: "The Coefficient of Isothermal Compressibility of Black Oils at Pressures Below the Bubble Point," *SPE Form. Eval.* (Sept. 1988) *3*, No. 3, 659–662.

8. Ng, J.T.H. and Egbogah, E.O.: "An Improved Temperature-Viscosity Correlation for Crude Oil Systems," paper 83–34–32 presented at Petroleum Society of CIM 34th Annual Technical Meeting, Banff, May 10–13, 1983.

9. Beggs, H.D. and Robinson, J.R.: "Estimating the Viscosity of Crude Oil Systems," *J. Pet. Tech.* (Sept. 1975) *27*, 1140–1141.

10. Katz, D.L., and Saltman, W.: "Surface Tension of Hydrocarbons," *Ind. Eng. Chem.* (Jan. 1939) *31*, 91–94.

11. Macleod, D.B.: "On a Relation Between Surface Tension and, Density," *Trans.*, Faraday Soc. (1923) *19*, 38–42.

12. Fowler, R.H.: "A Tentative Statistical Theory of Macleod's Equation for Surface Tension and the Parachor," *Proc.*, Royal Soc. of London, Series A (1937) 229–246.

13. Katz, D.L., et al.: *Handbook of Natural Gas Engineering*, McGraw-Hill Book Co. Inc., New York City (1959).

14. Weinaug, C.F. and Katz, D.L.: "Surface Tension of Methane-Propane Mixtures," *Ind. Eng. Chem.* (1943) *35*, 239–242.

15. Firoozabadi, A., Katz, D.L., Soroosh, H., and Sajjadian, V.A.: "Surface Tension of Reservoir Crude-Oil/Gas Systems Recognizing the Asphalt in the Heavy Fraction," *SPE Res. Eng.* (Feb. 1988) *3*, No. 1, 265–272.

Gas-Liquid Equilibria

12

The area bounded by the bubble point and dew point curves on the phase diagram of a multicomponent mixture defines the conditions for gas and liquid to exist in equilibrium. This was discussed in Chapter 2. The quantities and compositions of the two phases vary at different points within the limits of this phase envelope.

In this chapter we will consider methods of calculating the behavior of hydrocarbon mixtures in this two-phase region. Three types of calculations will be examined:

calculation of conditions for which a mixture exhibits a dew point,

calculation of conditions for which a mixture exhibits a bubble point, and

calculation of quantities and compositions of the gas and liquid at conditions within the two-phase region.

These calculation methods enable us to predict reservoir fluid behavior and to determine conditions for processing reservoir fluids at the surface.

We will begin, as we have in other chapters, with a discussion of the behavior of a hypothetical fluid known as an ideal solution. Then a study of the factors which cause real solutions to deviate from the behavior of ideal solutions will guide us in the development of methods to predict the behavior of real solutions.

Ideal Solutions

An *ideal liquid solution* is a solution for which

mutual solubility results when the components are mixed,
no chemical interaction occurs upon mixing,
the molecular diameters of the components are the same,

and

the intermolecular forces of attraction and repulsion
are the same between unlike as between like molecules.

These properties of ideal solutions lead to two practical results. First, there is no heating effect when the components of an ideal solution are mixed. Second, the volume of the ideal solution equals the sum of the volumes the components would occupy as pure liquids at the same temperature and pressure.

As in the case of ideal gases, ideal liquid solutions do not exist. Actually, the only solutions which approach ideal solution behavior are gas mixtures at low pressures. Liquid mixtures of components of the same homologous series approach ideal-solution behavior only at low pressures. However, studies of the phase behavior of ideal solutions help us understand the behavior of real solutions.

Raoult's Equation

Raoult's equation states that the partial pressure of a component in the gas is equal to the mole fraction of that component in the liquid multiplied by the vapor pressure of the pure component. Raoult's equation is valid only if both the gas and liquid mixtures are ideal solutions. The mathematical statement is

$$p_j = x_j p_{vj} , \qquad P_j = n_j \frac{RT}{V} = y_j P \qquad (12\text{--}1)$$

where p_j represents the partial pressure of component j in the gas in equilibrium with a liquid of composition x_j. The quantity p_{vj} represents the vapor pressure that pure component j exerts at the temperature of interest.

Dalton's Equation

We saw that *Dalton's equation* can be used to calculate the partial pressure exerted by a component of an ideal gas mixture:

$$p_j = y_j p \ . \tag{3-30}$$

Compositions and Quantities of the Equilibrium Gas and Liquid Phases of an Ideal Solution

Raoult's and Dalton's equations both represent the partial pressure of a component in a gas mixture. In the case of Raoult's equation, the gas must be in equilibrium with a liquid. These equations may be combined by eliminating partial pressure. The resulting equation relates the compositions of the gas and liquid phases in equilibrium to the pressure and temperature at which the gas-liquid equilibrium exists.

$$y_j p = x_j p_{vj}$$

or

$$\frac{y_j}{x_j} = \frac{p_{vj}}{p} \ . \tag{12-2}$$

Equation 12–2 relates the ratio of the mole fraction of a component in the gas to its mole fraction in the liquid. To determine values of y_j and x_j, this equation must be combined with another equation relating these two quantities. Such an equation can be developed through consideration of a material balance on the jth component of the mixture.

Since n represents the total number of moles in the mixture,

n_L represents the total number of moles in the liquid,

n_g represents the total number of moles in the gas;

and

z_j represents the mole fraction of the jth component in the total mixture including both liquid and gas phases,

x_j represents the mole fraction of the jth component in the liquid,

y_j represents the mole fraction of the jth component in the gas;

then
$$z_j = \frac{m_j/M_j}{n_{total}}$$

(handwritten: $P_{vj} - Fig\ 2-7$ Pg. 56)

$z_j n$ represents the moles of the jth component in the total mixture,

$x_j n_L$ represents the moles of the jth component in the liquid, and

$y_j n_g$ represents the moles of the jth component in the gas.

Thus, a material balance on the jth component results in

$$z_j n = x_j n_L + y_j n_g . \qquad (12-3)$$

Combination of Equations 12-2 and 12-3 to eliminate y_j results in

$$z_j n = x_j n_L + x_j \frac{p_{vj}}{p} n_g \qquad (12-4)$$

or

(handwritten left margin: $y_j = \dfrac{z_j}{1 + \bar{n}_L\left(\frac{p}{p_{vj}} - 1\right)}$)

(handwritten: $x_j = z_j / \left(1 + \bar{n}_g\left(\frac{p_{vj}}{p} - 1\right)\right)$)

$$x_j = \frac{z_j n}{n_L + \frac{p_{vj}}{p} n_g} . \qquad (12-5)$$

Since by definition $\sum_j x_j = 1$, Equation 12-6 follows.

$$\sum_j x_j = \sum_j \frac{z_j n}{n_L + \frac{p_{vj}}{p} n_g} = 1 . \qquad (12-6)$$

Equations 12-2 and 12-3 also can be combined to eliminate x_j, with the result

$$\sum_j y_j = \sum_j \frac{z_j n}{n_g + \frac{p}{p_{vj}} n_L} = 1 . \qquad (12-7)$$

Either Equation 12-6 or 12-7 is utilized to calculate the compositions of the gas and liquid phases of a mixture at equilibrium. In either case, a trial-and-error solution is required. The calculation is simplified if one mole of total mixture is taken as a basis so that $\bar{n}_L = n_L/n$, $\bar{n}_g = n_g/n$, and $\bar{n}_L + \bar{n}_g = 1$. This results in the reduction of Equations 12-6 and 12-7 to

(handwritten: $M_L = \sum M y_j \times \bar{n}_L \times n_{total}$)

$$\sum_j x_j = \sum_j \frac{z_j}{1 + \bar{n}_g \left(\dfrac{p_{vj}}{p} - 1 \right)} = 1 \qquad (12\text{--}8)$$

and

$$\sum_j y_j = \sum_j \frac{z_j}{1 + \bar{n}_L \left(\dfrac{p}{p_{vj}} - 1 \right)} = 1. \qquad (12\text{--}9)$$

If Equation 12–8 is selected for the computation, a trial value of $\bar{n}_g$ between zero and one is chosen, vapor pressures of the individual components at the given temperature are obtained, and the summation is carried out. If the sum is equal to 1.0, each term in the sum is equal to x_j, and the total mass of gas is equal to the product of the selected value of $\bar{n}_g$ multiplied by the total moles in the mixture. If the summation does not equal 1.0, a new trial value of $\bar{n}_g$ must be selected and the computation repeated until the summation equals 1.0.

If Equation 12–9 is used for the calculation, trial values of $\bar{n}_L$ must be selected. When a value of $\bar{n}_L$ is found that causes the summation to equal 1.0, the resulting terms in the sum represent the values of y_j, and $\bar{n}_L$ represents the number of moles in the liquid per mole of total mixture.

EXAMPLE 12–1: *Calculate the compositions and quantities of the gas and liquid when 1.0 lb mole of the following mixture is brought to equilibrium at 150°F and 200 psia. Assume ideal-solution behavior.*

Component	Composition, mole fraction
Propane	0.610
n-Butane	0.280
n-Pentane	0.110
	1.000

Solution

By trial and error, determine a value of $\bar{n}_L$ which will satisfy the equation

$$\sum_j \frac{z_j}{1 + \bar{n}_L \left(\dfrac{p}{p_{vj}} - 1 \right)} = 1. \qquad (12\text{--}9)$$

In order to save space, only the calculation with the final trial value of $\bar{n}_L$ = 0.487 will be shown.

Component	Composition of mixture, mole fraction z_j	Vapor pressure at 150°F, p_{vj}	Composition of gas, mole fraction $y_j = z_j/1 + \bar{n}_L(p/p_{vj}-1)$	Composition of liquid, mole fraction $x_j = y_j(p/p_{vj})$
C_3	0.610	350	0.771	0.441
$n\text{-}C_4$	0.280	105	0.194	0.370
$n\text{-}C_5$	0.110	37	0.035	0.189
			$\Sigma y_j = 1.000$	1.000

The summation equals one for $\bar{n}_L$ = 0.487; thus there are 0.487 moles of liquid and 0.513 moles of vapor for each mole of total mixture. And the compositions of the equilibrium gas and liquid are given in columns 4 and 5 of the above table.

Calculation of the Bubble-Point Pressure of an Ideal Liquid Solution

The bubble point is the point at which the first bubble of gas is formed. For all practical purposes, the quantity of gas is negligible. Thus we can take n_g to be equal to zero and n_L to equal the total moles of mixture. Substitution of $n_g = 0$, $n_L = n$, and $p = p_b$ into Equation 12–7 results in

$$\sum_j \frac{z_j}{p_b/p_{vj}} = 1 \qquad (12\text{--}10)$$

or

$$p_b = \sum_j z_j p_{vj} . \qquad (12\text{--}11)$$

Therefore, the bubble-point pressure of an ideal liquid solution at a given temperature is simply the summation of the products of mole fraction times vapor pressure for each component.

EXAMPLE 12–2: *Calculate the bubble-point pressure at a temperature of 150°F for the mixture given in Example 12–1. Assume ideal-solution behavior.*

Solution

Solve the equation

$$p_b = \sum_j z_j p_{vj} . \qquad (12\text{--}11)$$

Component	Composition, mole fraction z_j	Vapor pressure at 150°F, p_{vj}	$z_j p_{vj}$
C_3	0.610	350	213.5
n-C_4	0.280	105	29.4
n-C_5	0.110	37	4.1
	1.000	from fig 2-7	p_b = 247 psia

49,56

Calculation of the Dew-Point Pressure of an Ideal Gas Solution

At the dew point, the quantity of liquid essentially is negligible, so for $p = p_d$ we can substitute $n_L = 0$ and $n_g = n$ into Equation 12–6. This results in

$$\sum_j \frac{z_j}{p_{vj}/p_d} = 1 \qquad (12\text{--}12)$$

or

$$p_d = \frac{1}{\sum_j (z_j/p_{vj})} . \qquad (12\text{--}13)$$

The dew point of an ideal gas mixture at a given temperature is simply the reciprocal of the summation of the mole fraction divided by vapor pressure for each component.

EXAMPLE 12–3: *Calculate the dew-point pressure at 150°F of the mixture given in Example 12–1. Assume ideal-solution behavior.*

Solution

Solve the equation

$$p_d = \frac{1}{\sum_j (z_j/p_{vj})} . \qquad (12\text{--}13)$$

Component	Composition, mole fraction z_j	Vapor pressure at 150°F, p_{vj}	z_j/p_{vj}
C_3	0.610	350	0.00174
n-C_4	0.280	105	0.00267
n-C_5	0.110	37	0.00297
	1.000	$\Sigma_j(z_j/p_{vj})$ =	0.00738

$$p_d = \frac{1}{0.00738} = 136 \text{ psia} \qquad (12\text{–}13)$$

Nonideal Solutions

The use of Equations 12–6 or 12–7 to predict gas-liquid equilibria is restricted severely for several reasons. First, Dalton's equation is based on the assumption that the gas behaves as an ideal solution of ideal gases. For practical purposes the ideal-gas assumption limits the equations to pressures below about 100 psia and moderate temperatures.

Second, Raoult's equation is based on the assumption that the liquid behaves as an ideal solution. Ideal-solution behavior is approached only if the components of the liquid mixture are very similar chemically and physically.

Third, a pure compound does not have a vapor pressure at temperatures above its critical temperature. Thus, these equations are limited to temperatures less than the critical temperature of the most volatile component in the mixture. For example, if methane is a component of the mixture, Equations 12–6 and 12–7 cannot be applied at temperatures above −116°F.

Various theoretical methods to overcome these problems have been tried. But the only reasonably accurate way to predict gas-liquid equilibria is through correlations based on experimental observations of gas-liquid equilibria behavior. These correlations usually involve the *equilibrium ratio*, K, which is defined as

$$K_j = \frac{y_j}{x_j}, \qquad \text{read off Figs A3, A5, A7} \qquad (12\text{–}14)$$

where y_j and x_j are experimentally determined values of the compositions of the gas and liquid that exist at equilibrium at a given pressure and temperature. Equilibrium ratios are sometimes called *vaporization equilibrium constants, equilibrium vapor-liquid distribution ratios, distribution coefficients,* or, simply, *K-factors* or *K-values*.

Also, the term *equilibrium constant* often is used to describe K-factor. This term is misleading since K-factors definitely are not constant. K-factors change with pressure, temperature, and type of mixture. Thus we have elected to use the terms *equilibrium ratio* or *K-factor*.

We will discuss equilibrium-ratio correlations in some detail later. First, however, we will look at their use in gas-liquid equilibria calculations.

Compositions and Quantities of the Equilibrium Gas and Liquid Phases of a Real Solution

Examination of Equation 12–2 indicates that we can modify Equations 12–6 and 12–7 to account for real-solution behavior simply by replacing the pressure ratios with experimentally determined equilibrium ratios. Thus,

$$\sum_j x_j = \sum_j \frac{z_j n}{n_L + n_g K_j} = 1 \qquad (12-15)$$

and

$$\sum_j y_j = \sum_j \frac{z_j n}{n_g + \dfrac{n_L}{K_j}} = 1 . \qquad (12-16)$$

at bubble point $n_g = 0$
$n_L = n$

We saw earlier that the calculation is simplified if it is based on one mole of total mixture. The following modifications of Equations 12–8 and 12–9 usually are used.

$$\sum_j x_j = \sum_j \frac{z_j}{1 + \bar{n}_g (K_j - 1)} = 1 \qquad (12-17)$$

$$\sum_j y_j = \sum_j \frac{z_j}{1 + \bar{n}_L \left(\dfrac{1}{K_j} - 1 \right)} = 1 \qquad (12-18)$$

The choice between Equations 12–17 and 12–18 is strictly arbitrary; either works with equal efficiency. And either equation requires a trial-and-error solution. Normally, successive trial values of either $\bar{n}_g$ or $\bar{n}_L$ are selected until the summation equals 1.0. The value of $\bar{n}_g$ or $\bar{n}_L$ which causes the summation to equal 1.0 is the correct value of $\bar{n}_g$ or $\bar{n}_L$. Then, the terms in the summation represent the composition of either the liquid or the gas, depending upon the equation used. The accuracy of the results depends strictly on the accuracy of the values of equilibrium ratios used.

Determine values of K_j
use first trial value of $\bar{n}_L$

if $\Sigma < 1 \rightarrow$ increase $\bar{n}_L$
$> 1 \rightarrow$ decrease $\bar{n}_L$

want $\Sigma \geq 1.00$

✱ EXAMPLE 12-4: *Repeat Example 12-1. Do not assume ideal solution behavior. Assume that the equilibrium ratio charts given in Appendix A can be used for this mixture.*

Solution

By trial and error, determine a value of $\bar{n}_L$ which will satisfy the equation

$$\sum_j \frac{z_j}{1 + \bar{n}_L \left(\dfrac{1}{K_j} - 1 \right)} = 1 \qquad (12\text{--}18)$$

Determine values of K_j at 150°F and 200 psia from Appendix A. Use a first trial value of 0.5 for $\bar{n}_L$. This will result in a summation of 0.992. Then try a value of 0.6 for $\bar{n}_L$. This will result in a summation of 1.010. Further trials result in $\bar{n}_L = 0.547$. Calculations are given in the table below.

Component	Composition of mixture, mole fraction z_j	Equilibrium ratio K_j	$z_j/1 + \bar{n}_L(1/K_j-1)$ for $\bar{n}_L = 0.5$	$z_j/1 + \bar{n}_L(1/K_j-1)$ for $\bar{n}_L = 0.6$	$z_j/1 + \bar{n}_L(1/K_j-1)$ for $\bar{n}_L = 0.547$
C_3	0.610	1.55	0.742	0.775	0.757
$n\text{-}C_4$	0.280	0.592	0.208	0.198	0.203
$n\text{-}C_5$	0.110	0.236	0.042	0.037	0.040
	1.000		0.992	1.010	$\Sigma_j y_j = 1.000$

The summation equals one for a value of $\bar{n}_L = 0.547$. Thus, there are 0.547 moles of liquid for every mole of total mixture, and the compositions of the liquid and gas phases are

Component	Composition of gas, mole fraction $y_j = z_j/1 + \bar{n}_L(1/K_j-1)$	Composition of liquid, mole fraction $x_j = y_j/K_j$
C_3	0.757	0.488
$n\text{-}C_4$	0.203	0.344
$n\text{-}C_5$	0.040	0.168
	1.000	1.000

Note that the assumption of ideal-solution behavior in Example 12–1 resulted in an error of over ten percent in $\tilde{n}_L$.

Examination of the way Equations 12–17 and 12–18 are affected by incorrect trial values of $\tilde{n}_g$ or $\tilde{n}_L$ will provide guidance for rapid convergence to the correct values. Figure 12–1 shows the results of calculations with Equation 12–17. The correct value of $\tilde{n}_g$ is the value which makes the summation equal 1.0. Regardless of the pressure or temperature selected, a trivial solution is obtained at $\tilde{n}_g$ equals zero.

The mixture will be all liquid if the selected temperature is less than the bubble-point temperature or if the selected pressure is greater than the bubble-point pressure. The upper dashed line on Figure 12–1 shows that at these conditions no solution exists. This situation occurs when $\sum\limits_j z_j K_j$ is less than 1.0.

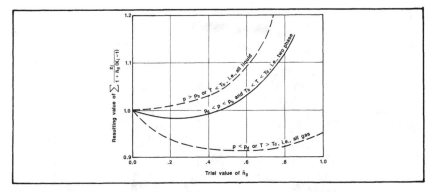

Fig. 12–1. Reaction of Equation 12–17 to incorrect trial values of $\tilde{n}_g$ and calculations at pressures or temperatures outside of two-phase region.

The mixture will be all gas if the selected temperature is greater than the dew-point temperature or the selected pressure is less than the dew-point pressure. The lower dashed line on Figure 12–1 shows that no solution is obtained for these conditions. This situation occurs when $\sum\limits_j z_j/K_j$ is less than 1.0.

The upper dashed line occurs if $\sum\limits_j z_j K_j$ is less than 1.0, and the lower dashed line occurs if $\sum\limits_j z_j/K_j$ is less than 1.0. Therefore, the solid line, which represents conditions in the two-phase region, can occur only if both quantities are greater than 1.0. A simple check to ensure that conditions are in the two-phase region is

$$\sum_j z_j K_j > 1.0 \qquad (12\text{–}19)$$

and

$$\sum_j z_j/K_j > 1.0 . \qquad (12\text{--}20)$$

This check should be routinely made before any gas-liquid equilibrium calculation.

EXAMPLE 12–5: *Perform the check which should have been performed prior to making the calculations in Example 12–4.*

Solution

$$\sum_j z_j K_j > 1.0 \qquad (12\text{--}19)$$

$$\sum_j z_j/K_j > 1.0 \qquad (12\text{--}20)$$

Component	Composition of mixture, mole fraction z_j	Equilibrium ratio K_j	$z_j K_j$	z_j/K_j
C_3	0.610	1.55	0.946	0.394
$n\text{-}C_4$	0.280	0.592	0.166	0.473
$n\text{-}C_5$	0.110	0.236	0.026	0.466
	1.000		1.138	1.333

Both summations are greater than 1.0, so the mixture is two-phase at 150°F and 200 psia.

The solid line on Figure 12–1 represents the results if the pressure and temperature are in the two-phase region. The correct value of $\bar{n}_g$ is obtained at the point where the solid curve crosses the $\sum_j x_j = 1.0$ line. If the summation results in a value less than 1.0 for a particular trial value of $\bar{n}_g$, increase the next trial value of $\bar{n}_g$. Conversely, if the summation is greater than 1.0, decrease the next trial value of $\bar{n}_g$. Fortunately, the curve in the vicinity of the correct value is relatively straight so that linear interpolation can be used once calculations have been made for two trial values. This should result in rapid convergence on the true value.

Figure 12–2 shows the results of Equation 12–18. The meanings of the dashed curves are similar to the meanings of the dashed curves in Figure 12–1. The solid curve represents the solution to Example 12–4 in which a correct value of 0.547 moles is obtained for $\tilde{n}_L$.

A slightly different formulation of Equations 12–17 and 12–18 and a technique for determining trial values of $\tilde{n}_g$ or $\tilde{n}_L$ are given elsewhere.[1]

Calculation of the Bubble-Point Pressure of a Real Liquid

Again we will consider that the quantity of gas at the bubble point is negligible. Thus we can substitute $n_g = 0$ and $n_L = n$ into Equation 12–16 to obtain an equation which can be used to calculate the bubble-point pressure at a given temperature or the bubble-point temperature at a given pressure.

$$\text{iterate until} \quad \sum_j z_j K_j = 1 \tag{12–21}$$

Pressure does not appear implicitly in Equation 12–21. Pressure is represented in the equilibrium ratio. Thus bubble-point pressure cannot be calculated directly as in the case of ideal solutions.

The bubble-point pressure at a given temperature may be determined by selection of a trial value of pressure, from which values of equilibrium ratios are obtained. Then the summation of Equation 12–21 is computed. If the sum is less than 1.0, the calculation is repeated at a lower pressure. If the sum is greater than 1.0, a higher trial value pressure is chosen.

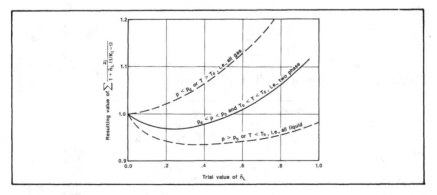

Fig. 12–2. Reaction of Equation 12–18 to incorrect trial values of $\tilde{n}_L$ and calculations at pressures or temperatures outside of two-phase region.

An approximate value of bubble-point pressure for use as the initial trial value can be calculated using Equation 12–11. When the correct bubble-point pressure has been established by this trial-and-error process, the composition of the infinitesimal amount of gas at the bubble point is given by the terms in the summation.

EXAMPLE 12–6: *Estimate the bubble-point pressure of the mixture given in Example 12–1 at a temperature of 150°F. Assume that the equilibrium ratio charts given in Appendix A can be used for this mixture.*

Solution

By trial and error, determine a pressure which will satisfy the equation

$$\sum_j z_j K_j = 1 \qquad\qquad (12\text{–}21)$$

First, the initial trial value of bubble-point pressure is 247 psia since this is the bubble-point pressure calculated in Example 12–2 under the assumption of ideal-solution behavior. Determine K_j values at 150°F and 247 psia from Appendix A.

Component	Composition of mixture, mole fraction z_j	Equilibrium ratio of 150°F and 247 psia, K_j	$z_j K_j$
C_3	0.610	1.32	0.805
$n\text{-}C_4$	0.280	0.507	0.142
$n\text{-}C_5$	0.110	0.204	0.022
	1.000		0.969

$\sum > 1$

repeat at higher p

Second, try a new trial value of bubble-point pressure of 220 psia. Determine K_j values at 150°F and 220 psia.

Component	Composition of mixture, mole fraction z_j	Equilibrium ratio at 150°F and 220 psia, K_j	$z_j K_j$
C_3	0.610	1.44	0.878
$n\text{-}C_4$	0.280	0.553	0.155
$n\text{-}C_5$	0.110	0.218	0.024
	1.000		1.057

$z_j \quad p_{v_j} \quad z_j p_{v_j}$

fig 2-7
pg 56

$\sum = \underline{\quad}$

use $\sum z_j p_{v_j}$ as first guess of pressure

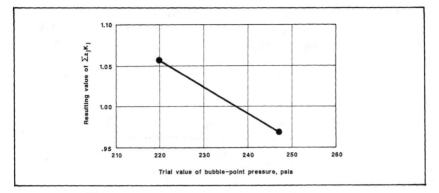

Fig. 12-3. $\Sigma z_j K_j$ vs. trial value of bubble-point pressure (part of solution to Example 12-6).

Third, plot trial values of pressure against the resulting summation as in Figure 12-3, and interpolate to determine a third trial value of bubble-point pressure of 237 psia.

Component	Composition of mixture, mole fraction z_j	Equilibrium ratio at 150°F and 237 psia, K_j	$z_j K_j$
C_3	0.610	1.36	0.830
n-C_4	0.280	0.523	0.146
n-C_5	0.110	0.208	0.023
	1.000		0.999

$$\Sigma z_j K_j = 1.0 \text{ so } p_b = 237 \text{ psia}$$

Calculation of the Dew-Point Pressure of a Real Gas

At the dew point, the mixture is entirely in the gas phase except for an infinitesimal amount of liquid. Thus $n_L = 0$ and $n_g = n$ so that Equation 12-15 becomes 　iterate until

$$\sum_j \frac{z_j}{K_j} = 1 \qquad (12\text{-}22)$$

The dew-point pressure is the pressure at a given temperature for which Equation 12-22 is satisfied. Again the calculation involves a trial-and-error process. A fairly good initial trial value of dew-point pressure may be obtained through the use of Equation 12-13.

EXAMPLE 12–7: *Estimate the dew-point pressure of the mixture given in Example 12–1 at a temperature of 150°F. Assume that the equilibrium ratio charts given in Appendix A can be used for this mixture.*

Solution $P_{\text{initial guess}} \simeq 1/\sum z_j/P_{V_j}$

By trial and error, determine a pressure which will satisfy the equation

$$\sum_j \frac{z_j}{K_j} = 1 \tag{12–22}$$

In order to save space, only the last calculation with a final trial value of $p_d = 141$ psia will be shown. Determine K_j values at 141 psia and 150°F from Appendix A.

Component	Composition of mixture, mole fraction z_j	Equilibrium ratio at 150°F and 141 psia, K_j	z_j/K_j
C_3	0.610	2.11	0.289
n-C_4	0.280	0.785	0.357
n-C_5	0.110	0.311	0.354
	1.000		1.000

$\sum_j (z_j/K_j) = 1$, thus $p_d = 141$ psia

Flash Vaporization

Calculations of gas-liquid equilibria using either Equation 12–17 or Equation 12–18 often are called *flash vaporization calculations*. Usually the term is reduced to *flash calculations*.

This seems to imply that the pressure on a liquid is reduced suddenly, causing a flash of vapor to form. This implication is incorrect. The path that the changes in pressure and temperature follow in arriving at a particular condition are immaterial. Only the existing conditions affect the gas-liquid behavior at an instant in time. Of course, the equations are valid only if the gas and liquid are in equilibrium.

A flash vaporization process is described in Chapter 10 as a part of the discussion of black-oil laboratory procedures. The results of this process can be calculated by performing calculations like Example 12–4 for a series of pressures. Overall composition and temperature are held constant, and successively lower pressures are used.

The results of the flash calculations are in terms of pound-moles. The molecular weights and densities of the coexisting gas and liquid also must be calculated in order to reproduce page 4 of Table 10–1. The compositions of the equilibrium gas and liquid are used to calculate these molecular weights and densities.

Differential Vaporization

Differential vaporization is a process in which the gas is removed from contact with the liquid as the gas is formed. A constant-temperature differential vaporization is illustrated in Figure 12–4.

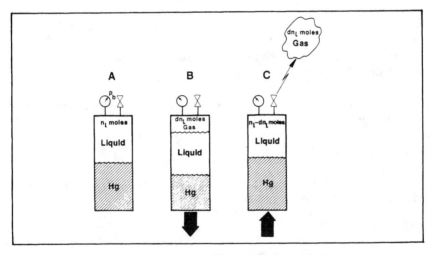

Fig. 12–4. Constant temperature differential vaporization.

Figure 12–4A shows n_L moles of a liquid mixture contained in a constant-temperature cell. Pressure is set at the bubble-point pressure of the liquid. When an increment of mercury is removed from the cell (as shown in Figure 12–4B), the pressure is reduced slightly, and dn_L moles of liquid are vaporized. Mercury is then reinjected into the cell, and pressure is maintained constant by allowing the gas to escape through a valve.

This leaves the cell full of a liquid which has a bubble-point pressure equal to the pressure in the cell. Only n_L - dn_L moles of liquid remain since dn_L moles have been removed as gas. Repeating the process results in the differential vaporization of the liquid.

Each successive increment of gas which is formed and removed causes a change in the composition of the liquid. These composition changes may be calculated through consideration of a material balance on the jth component of the mixture.

n_L represents the total number of moles of liquid in the cell at the beginning of a differential vaporization

and

dn_L represents the number of moles of gas formed during the differential vaporization,

then

$n_L - dn_L$ represents the number of moles of liquid remaining at the end of the increment.

x_j represents the mole fraction of the jth component in the liquid at the beginning of the differential vaporization

and

dx_j represents the change in the mole fraction of the jth component caused by the loss of an increment of vapor,

Thus,

$x_j - dx_j$ represents the mole fraction of the jth component in the liquid following the differential vaporization.

Also,

$x_j n_L$ represents the moles of the jth component in the liquid at the beginning of a differential vaporization,

$(x_j - dx_j)(n_L - dn_L)$ represents the moles of the jth component in the liquid at the end of the increment,

and

y$_j$dn$_L$ represents the moles of the jth component in the gas removed.

We can equate the number of moles of the jth component in the gas to the number of moles of the jth component lost by the liquid to give

$$y_j dn_L = x_j n_L - (x_j - dx_j)(n_L - dn_L). \qquad (12-23)$$

This equation may be expanded to

$$y_j dn_L = x_j n_L - x_j n_L + n_L dx_j + x_j dn_L - dx_j dn_L. \quad (12-24)$$

Since dx$_j$ is small, the term dx$_j$dn$_L$ may be neglected. Also, K$_j$x$_j$ can be substituted for y$_j$ to give

$$K_j x_j dn_L = n_L dx_j + x_j dn_L \qquad (12-25)$$

or

$$\frac{dn_L}{n_L} = \frac{1}{K_j - 1}\left(\frac{dx_j}{x_j}\right). \qquad (12-26)$$

If we assume that the values of K$_j$ remain constant for a reasonable change in pressure, Equation 12–26 can be integrated as follows:

$$\int_{n_{Li}}^{n_{Lf}} \frac{dn_L}{n_L} = \frac{1}{K_j - 1} \int_{x_{ji}}^{x_{jf}} \frac{dx_j}{x_j} \qquad (12-27)$$

$$\ln \frac{n_{Lf}}{n_{Li}} = \frac{1}{K_j - 1} \ln \frac{x_{jf}}{x_{ji}}, \qquad (12-28)$$

where the subscripts i and f indicate initial and final conditions of the differential vaporization. Equation 12–28 can also be written as

$$\frac{x_{jf}}{x_{ji}} = \left(\frac{n_{Lf}}{n_{Li}}\right)^{K_j - 1}. \qquad (12-29)$$

Since either Equation 12–28 or 12–29 involves two unknowns, x_{jf} and n_{Lf}, another equation is required to complete the calculation. Equation 12–30 is suitable.

$$\sum_j x_{jf} = 1.0 \tag{12–30}$$

Differential vaporization calculations of two types interest petroleum engineers. In both cases differential vaporization is at constant temperature, and the composition of the initial liquid is known. In the first case, initial conditions and final pressure are given and the number of moles to be vaporized is calculated. In the second case, the initial conditions and the number of moles to be vaporized are given and the final pressure is calculated. Either case requires a trial-and-error solution.

Differential Vaporization—Calculation Procedure, Final Pressure Known

When the final pressure of a differential vaporization is known, the final and initial pressures are averaged to give a value of pressure for which equilibrium ratios are obtained. Then a first trial value of n_{Lf} is selected, and either Equation 12–28 or Equation 12–29 is used to calculate values of x_{jf}.

The calculated liquid compositions are summed as indicated in Equation 12–30. If the summation is equal to 1.0, the trial value of n_{Lf} is the correct quantity of moles of liquid remaining after the differential vaporization has been carried out to the final pressure. If the summation of the mole fractions of the liquid components does not equal 1.0,

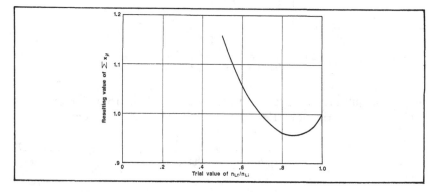

Fig. 12–5. Reaction of Equations 12–29 and 12–30 to incorrect trial values of n_{Lf}/n_{Li}.

another trial value of n_{Lf} must be selected and the procedure repeated until summation x_{jf} equals 1.0.

Notice that a trial value of n_{Lf}/n_{Li} equal to 1.0 results in $\sum_j x_{jf} = 1.0$, regardless of the values of K_j. This is a trivial solution. Also as the value of n_{Lf}/n_{Li} approaches zero, $\sum_j x_{jf}$ becomes very large. Thus the curve relating trial values of n_{Lf}/n_{Li} to the resulting values of $\sum_j x_{jf}$ takes the shape shown in Figure 12–5. A study of the shape of this curve should assist in the selection of trial values of n_{Lf}/n_{Li}.

EXAMPLE 12–8: *The following hydrocarbon mixture has a bubble point of 600 psia at 80°F. Determine the fraction of the gas that will be vaporized when this liquid is differentially vaporized to 400 psia at 80°F. Use K-values from Appendix A.*

Component	Composition, mole fraction
Methane	0.160
Propane	0.380
n-Pentane	0.460
	1.000

Solution

Determine by trial and error, a value of n_{Lf}/n_{Li} that satisfies equations

$$\frac{x_{jf}}{x_{ji}} = \left(\frac{n_{Lf}}{n_{Li}} \right)^{K_j - 1} \qquad (12–29)$$

and

$$\sum_j x_{jf} = 1.0. \qquad (12–30)$$

Start with a trial value of $n_{Lf}/n_{Li} = 0.90$. Use the shape of Figure 12–5 as a guide for successive trial values.

Component	Composition, mole fraction x_{ji}	Equilibrium ratio at 80°F and 500 psia K_j	First trial x_{jf} from Eq. 12-29 for $n_{Lf}/n_{Li}=0.90$	Second trial x_{jf} from Eq. 12-29 for $n_{Lf}/n_{Li}=0.95$	Third trial x_{jf} from Eq. 12-29 for $n_{Lf}/n_{Li}=0.93$
C_1	0.160	6.17	0.0928	0.1227	0.1099
C_3	0.380	0.388	0.4053	0.3921	0.3973
n-C_5	0.460	0.0520	0.5083	0.4829	0.4928
	1.000		1.0064	0.9977	$\sum_j x_{jf} = 1.0000$

Since $\sum_j x_{jf} = 1.0$, $n_{Lf}/n_{Li} = 0.93$ is the correct value. Thus 7 mole percent of the liquid has vaporized, and the composition of the remaining liquid is given in the last column of the table above.

Differential Vaporization—Calculation Procedure, Number of Moles to Be Vaporized Known

In this case, a trial value of final pressure must be selected and averaged with the initial pressure to give a pressure value for determining values of equilibrium ratios. Either Equation 12–28 or 12–29 can be used to calculate the composition of the liquid remaining after the differential vaporization, x_{jf}.

Summation of x_{jf} must equal 1.0; if it does not, a new trial value of final pressure must be selected and the calculation repeated.

The results of this trial-and-error calculation do not have a characteristic curve such as the one given previously in Figure 12–5. However, the relationship between $\sum_j x_{jf}$ and p_f is essentially linear so that linear interpolation or extrapolation can be used with good effect to obtain succeeding trial values of p_f.

EXAMPLE 12–9: *Determine the final pressure required to differentially vaporize 10 mole percent of the mixture given in Example 12–8 at 80°F. Use K-values from Appendix A.*

$$\frac{n_{Lf}}{n_{Li}} = 0.90$$

Solution

Determine by trial and error a value of p_f which will give values of K_j that satisfy equations

$$\frac{x_{jf}}{x_{ji}} = \left(\frac{n_{Lf}}{n_{Li}} \right)^{K_j - 1} \tag{12–29}$$

and

$$\sum_j x_{jf} = 1.0. \tag{12–30}$$

Start with a trial value of $p_f = 300$ psia, i.e., $p_{avg} = 450$ psia.

Component	Composition, mole fraction x_{ji}	first trial		second trial	
		Equilibrium ratio at 80°F & 450° psia K_j	x_{jf} from Eq. 12–29 for $p_f = 300$ psia & $n_{Lf}/n_{Li} = 0.90$	Equilibrium ratio at 80°F & 455 psia K_j	x_{jf} from Eq. 12–29 for $p_f = 310$ psia & $n_{Lf}/n_{Li} = 0.90$
C_1	0.160	6.81	0.0868	6.74	0.0874
C_3	0.380	0.413	0.4042	0.410	0.4044
n-C_5	0.460	0.0536	0.5082	0.053	0.5082
	1.000		0.9992	$\sum_j x_{jf} =$	1.0000

Since $\sum_j x_{jf} = 1.0$, $p_f = 310$ psia is the pressure for which 10 mole percent of the original mixture has differentially vaporized, and the composition of the remaining liquid is given in the last column of the table above.

The key assumption in the derivation of the equations used to predict the results of differential vaporization is given just above Equation 12–27; K_j remains constant for reasonable changes in pressure. Equilibrium ratios do not remain constant as pressure changes. In fact, the change in equilibrium ratio can be quite large, as shown in Figures 14–11 and 14–2.

The error in differential vaporization calculations caused by variable K_j can be minimized by making the calculations illustrated by Examples 12–8 and 12–9 in stepwise manner. Select relatively small changes in pressure, obtain values of K_j at the average pressure, then calculate the resulting n_{Lf} and x_{jf} by trial and error. These values of n_{Lf} and x_{jf} then are used as n_{Li} and x_{ji} for the next calculation over another small change in pressure.

When the calculated value of either the final pressure or the final quantity of liquid reaches the desired value, the calculation is completed.

A *differential vaporization* process is described in Chapter 10 as a part of the discussion of black-oil laboratory procedures. The laboratory process is really a series of flash vaporizations in which the gas is removed at the end of each flash. Of course, a series of many flash vaporizations with very small quantities of gas removed at each step approximates a differential vaporization.

However, in order to calculate the results of the differential vaporization of the black-oil laboratory analysis, one should calculate a series of flash vaporizations with the gas removed at the end of each step.

Exercises

12–1. Use Raoult's equation and $\sum_j p_j = \sum_j x_j p_{vj} = p$ to derive an equation with which the composition of the liquid phase of an ideal two-component mixture may be calculated directly, i.e., without trial and error. How can the composition of the gas phase be calculated for this mixture?

12–2. Use the equations developed in Exercise 12–1 along with the substitution of K_j for p_{vj}/p to derive an equation with which the composition of the gas phase of a nonideal two-component mixture may be calculated directly, i.e., without trial and error. How can the composition of the liquid phase be calculated for this mixture?

12–3. Derive Equation 12–7. Show all steps.

12–4. Develop an equation for computation of the ratio n_L/n for a nonideal two-component mixture given that the compositions of the total mixture, the equilibrium gas, and the equilibrium liquid are known.

12–5. Develop an equation for computation of the ratio n_g/n for a nonideal two-component mixture given that the compositions of the total mixture, the equilibrium phase, and the equilibrium liquid are known.

12–6. Three pounds of 2,2-dimethylbutane and 2 lb of 2,2,4-trimethyl-pentane are mixed in a sealed container. The temperature and pressure are adjusted to 5 psia and 100°F. Calculate the compositions and weights of the gas and liquid at equilibrium. Assume that the mixture acts like an ideal solution.

12–7. Calculate the bubble-point pressure and the dew-point pressure of the mixture in Exercise 12–6 at 100°F.

12–8. Calculate the bubble-point temperature and the dew-point temperature of the mixture in Exercise 12–6 at 5 psia.

12–9. Six pound moles of ethane, three lb moles of propane and one lb mole of n-butane are mixed in a closed container and the temperature is adjusted to 75°F. What is the bubble-point pressure? What is the dew-point pressure? Calculate the compositions of the gas and liquid at equilibrium at 75°F and 300 psia. You may assume that the mixture behaves like an ideal solution.

12–10. Determine the compositions of the gas and liquid phases of the mixture in Exercise 12–9 at 75°F and 5000 psia. What do your results indicate about the phase of this system?

12–11. Determine the compositions of the gas and liquid phases of the mixture in Exercise 12–9 at 75°F and 100 psia. What do your results indicate about the phase of this system?

12–12. One standard cubic foot of n-pentane and two scf of n-butane are mixed in a sealed container and the temperature is adjusted to 60°F. What is the bubble-point pressure? What is the dew-point pressure? Compute the compositions and volumes of the gas and liquid at 60°F and 14.7 psia. Use the equilibrium ratios of Appendix A.

12–13. Recalculate Exercise 12–9. Do not assume that the mixture acts like an ideal solution. Use equilibrium ratios from Appendix A.

12–14. A mixture with the total composition given below is at equilibrium at 160°F and 250 psia. Calculate the compositions of the gas and liquid. Use equilibrium ratios from Appendix A.

Component	Composition, mole fraction
Propane	0.650
n-Butane	0.300
n-Pentane	0.050
	1.000

12–15. Calculate the bubble-point pressure of the following mixture at 140°F. Use equilibrium ratios from Appendix A.

Component	Composition, mole fraction
Propane	0.083
n-Butane	0.376
n-Pentane	0.101
n-Hexane	0.257
n-Heptane	0.183
	1.000

12–16. Calculate the dew-point pressure of the mixture in Exercise 12–15 at 140°F.

12–17. A nonideal solution has the following composition.

Component	Mass, lb
Propane	25.0
n-Butane	15.0
n-Pentane	20.0
n-Hexane	10.0

Make the following computations at 160°F:
a.   What is the bubble-point pressure?
b.   What is the composition of the vapor at the bubble point?
c.   What is the dew-point pressure?
d.   What is the composition of the liquid at the dew point?
e.   Calculate the composition of the liquid and the gas at 150 psia.
f.   Calculate the masses of the two phases at 150 psia.

12–18. Calculate the solubility of ethane in n-hexane in pound-moles of ethane per pound-mole of n-hexane at 75°F. and 300 psia. Assume that the mixture behaves as ideal solution.

12–19. Calculate the solubility of methane in a liquid of the following composition at 100°F and 100 psia. Give your answer in lb moles of methane per lb mole of liquid. Use equilibrium ratios from Appendix A.

Component	Composition, mole fraction
n-Butane	0.100
n-Pentane	0.300
n-Hexane	0.600
	1.000

12–20. The compositions of a gas and a liquid are given below. Determine the solubility of the gas in the liquid at 80°F and 200 psia. Give your answer in scf/STB. Use equilibrium ratios from Appendix A.

Component	Composition, mole fraction	
	Gas	Oil
Methane	0.87	
Ethane	0.08	
Propane	0.04	0.06
Butanes	0.01	0.17
Pentanes		0.32
Hexanes plus		0.45
	1.00	1.00

Use properties of n-butane and n-pentane for butanes and pentanes and of octane for hexane plus.

12–21. One pound mole of a solution with composition given below at its bubble point of 200 psia and 160°F is differentially vaporized to 100 psia and 160°F. Calculate the amount of gas vaporized. The composition is as follows:

Component	Composition, mole fraction
Ethane	0.173
n-Butane	0.352
n-Hexane	0.475
	1.000

12–22. To what pressure must the mixture in Exercise 12–21 be reduced to differentially vaporize 25 mole percent?

Reference

1. Rachford, H.H., Jr. and Rice, J.D.: "Procedure for Use of Electronic Digital Computers in Calculating Flash Vaporization Hydrocarbon Equilibrium," *J. Pet. Tech.* (Oct. 1952) *4*, 19–20.

13

Surface Separation

A common example of flash vaporization is the separation of gas and liquid in surface equipment in an oil or gas field.[1] The fluid from the wellhead is brought to equilibrium in a separator at separator temperature and pressure. This fluid is called separator feed.

The pressure of the separator is controlled with a pressure-regulating device through which the produced gas flows. Normally, separator temperature is determined by the temperature of the feed, the atmospheric temperature, and cooling due to vaporization and expansion of part of the feed stream. Separator temperature can be controlled by heating or refrigeration.

Sometimes several separators are operated in series at successively lower pressures to obtain the maximum amount of liquid. Figure 13–1 shows examples of two-stage and three-stage separation processes. If the pressure of the last separator is greater than atmospheric pressure, the stock tank acts as a stage of separation.

A series of flash separations in which the gas is removed at each stage simulates a differential vaporization process.

Separator calculations are performed to determine the optimum operating pressure for processing a particular hydrocarbon mixture. Normally for a black oil, the *composition of the produced gases*, the *gravity of the stock-tank oil*, the *producing gas-oil ratio*, and the *formation volume factor of the oil* are calculated. Other physical properties can be calculated from the compositions of the gases or liquids.

Surface Separator Calculations for Black Oils

Usually a black oil is produced through a two-stage separator system. A two-stage separator calculation requires a six-step procedure. This procedure is discussed as follows.

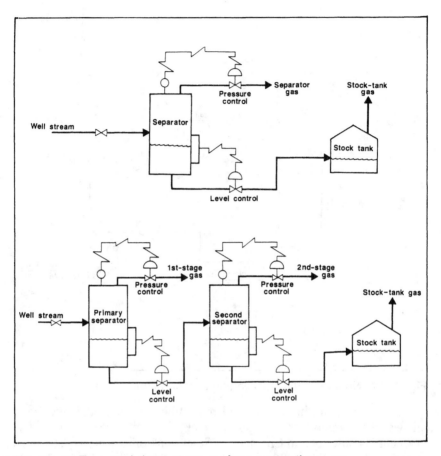

Fig. 13–1. Two- and three-stage surface separations.

Step 1: Calculate the quantities and compositions of the gas and liquid leaving the separator. Use the composition of the feed stream to the separator, separator temperature, and separator pressure.

Chapter 12 gives the equations for nonideal gas-liquid equilibria calculations. The quantities of gas and liquid leaving the separator on a basis of lb moles per lb mole of separator feed will be designated as $\bar{n}_{g1}$ and $\bar{n}_{L1}$. See Figure 13–2.

Step 2: Calculate the quantities and compositions of the gas and liquid in the stock tank. Use the composition of the liquid leaving the separator calculated in Step 1, stock-tank temperature, and atmospheric pressure.

Use nonideal gas-liquid equilibria calculations. The quantities of gas and liquid formed in the stock tank on a basis of lb moles per lb mole of liquid leaving the separator will be designated as $\bar{n}_{g2}$ and $\bar{n}_{L2}$.

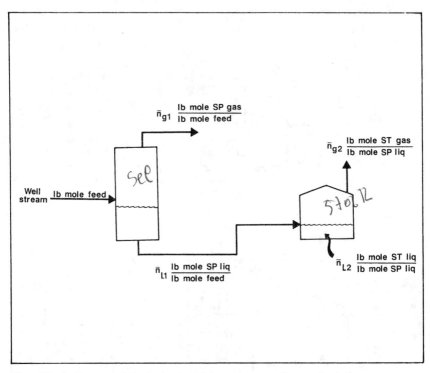

Fig. 13–2. Nomenclature for two-stage separation calculations.

Step 3: Calculate the density and molecular weight of the stock-tank oil. Use the composition of the stock-tank oil calculated in Step 2. The procedure is described in Chapter 11. Stock-tank oil density is calculated at standard conditions. Stock-tank oil gravity can be computed from stock-tank oil density at standard conditions.

Step 4: Calculate the total gas-oil ratio using quantities calculated in Steps 1, 2, and 3.

Normally, gas-oil ratio is calculated as standard cubic feet of gas per barrel of stock-tank oil. The ratio

$$\frac{\bar{n}_{g1} \dfrac{\text{lb mole SP gas}}{\text{lb mole SP feed}}}{\left(\bar{n}_{L1} \dfrac{\text{lb mole SP oil}}{\text{lb mole SP feed}}\right)\left(\bar{n}_{L2} \dfrac{\text{lb mole STO}}{\text{lb mole SP oil}}\right)} \qquad (13-1)$$

gives the lb moles of separator gas per lb mole of stock-tank oil.

This ratio can be converted to standard cubic feet of separator gas per barrel of stock-tank oil.

$$R_{SP} = \left(\frac{\bar{n}_{gl}}{\bar{n}_{L1}\bar{n}_{L2}} \frac{\text{lb mole SP gas}}{\text{lb mole STO}} \right) \left(\frac{380.7 \text{ scf SP gas}}{\text{lb mole SP gas}} \right)$$

$$\left(\frac{5.615 \, \rho_{STO}}{M_{STO}} \frac{\text{lb mole STO}}{\text{STB}} \right) \qquad (13-2)$$

Thus the separator gas-oil ratio is

$$R_{SP} = \frac{2138 \, \bar{n}_{gl} \, \rho_{STO}}{\bar{n}_{L1} \, \bar{n}_{L2} \, M_{STO}} . \qquad (13-2)$$

The ratio

$$\frac{\bar{n}_{g2} \dfrac{\text{lb mole ST gas}}{\text{lb mole SP oil}}}{\bar{n}_{L2} \dfrac{\text{lb mole STO}}{\text{lb mole SP oil}}} \qquad (13-3)$$

gives the lb moles of stock-tank gas per lb mole of stock-tank oil. This may be converted to standard cubic feet of stock-tank gas per barrel of stock-tank oil, as in Equation 13–2, to give

$$R_{ST} = \frac{2138 \bar{n}_{g2} \rho_{STO}}{\bar{n}_{L2} M_{STO}} \frac{\text{scf}}{\text{STB}} . \qquad (13-4)$$

Total producing gas-oil ratio is the sum of the gas from the separator and the stock tank,

$$R = R_{SP} + R_{ST} . \qquad (13-5)$$

If the separator feed stream is from an oil reservoir with pressure equal to or greater than bubble-point pressure, the total producing gas-oil ratio is the solution gas-oil ratio at the bubble point, R_{sb}.

The calculation may be terminated at this point; however, if the formation volume factor of the oil is required, Steps 5 and 6 must be completed.

Step 5: Calculate the density of the reservoir liquid at reservoir conditions using the feed stream composition from Step 1, reservoir temperature, and bubble-point pressure. The procedure is described in Chapter 11.

This calculation is correct only if the feed to the separator was a liquid at its bubble point at reservoir conditions. In the course of this calculation the molecular weight of the reservoir liquid will be determined.

Step 6: Calculate the oil formation volume factor using quantities calculated in Steps 1, 2, 3, and 5.

The equation will be developed below. The formation volume factor may be calculated as

$$B_{ob} = \frac{\text{res bbl/lb mole res oil}}{\text{STB/lb mole res oil}}. \qquad (13\text{-}6)$$

The numerator of Equation 13–6 may be developed as follows.

$$\frac{\text{res bbl}}{\text{lb mole res oil}} = \frac{M_{oR} \dfrac{\text{lb res oil}}{\text{lb mole res oil}}}{\left(\rho_{oR} \dfrac{\text{lb res oil}}{\text{cu ft res oil}}\right)\left(5.615 \dfrac{\text{cu ft res oil}}{\text{res bbl}}\right)} \qquad (13\text{-}7)$$

The denominator of Equation 13–6 may be developed as follows.

$$\frac{\text{STB}}{\text{lb mole res oil}} =$$

$$\frac{\left(M_{STO} \dfrac{\text{lb STO}}{\text{lb mole STO}}\right)\left(\bar{n}_{L1} \dfrac{\text{lb mole SP oil}}{\text{lb mole res oil}}\right)\left(\bar{n}_{L2} \dfrac{\text{lb mole STO}}{\text{lb mole SP oil}}\right)}{\left(\rho_{STO} \dfrac{\text{lb STO}}{\text{cu ft STO}}\right)\left(5.615 \dfrac{\text{cu ft STO}}{\text{STB}}\right)} \qquad (13\text{-}8)$$

Thus the oil formation volume factor may be calculated as

$$B_{ob} = \frac{M_{oR}\rho_{STO}}{M_{STO}\rho_{oR}\bar{n}_{L1}\bar{n}_{L2}} \quad \frac{\text{res bbl}}{\text{STB}}. \qquad (13\text{-}9)$$

$= B_o$ (same formula)

This is the formation volume factor of the oil at its bubble-point pressure in the reservoir. The density of reservoir oil, ρ_{oR}, was calculated at its bubble-point pressure.

EXAMPLE 13–1: *Calculate the producing gas-oil ratio, stock-tank oil gravity, and oil formation volume factor which will result from a two-stage separation of the hydrocarbon mixture below. Use separator conditions of 75°F and 100 psig and a stock-tank temperature of 75°F. The mixture is a liquid at its bubble point at reservoir conditions of 2620 psig and 220°F. Use K-factors from Appendix A. Use decane K-factors for heptanes plus.*

Component	Composition reservoir oil, mole fraction
Carbon dioxide	0.0091
Nitrogen	0.0016
Methane	0.3647
Ethane	0.0967
Propane	0.0695
i-Butane	0.0144
n-Butane	0.0393
i-Pentane	0.0144
n-Pentane	0.0141
Hexanes	0.0433
Heptanes plus	0.3329
	1.0000

Properties of heptanes plus	
Specific gravity	0.8515
Molecular weight	218 lb/lb mole

Solution

Follow the stepwise procedure given above.

Step 1: Calculate the compositions and quantities of separator gas and liquid using Equation 12–17. Trial and error on $\bar{n}_g$ is required.

$$\sum_j x_j = \sum_j \frac{z_j}{1 + \bar{n}_g (K_j - 1)} = 1.0 \qquad (12\text{–}17)$$

Only the final trial value of $\bar{n}_g = 0.4919$ will be shown.

Component	Composition, separator feed z_j	K-factor at 114.7 psia & 75°F K_j	Composition, separator liquid $z_j/(1+\bar{n}_g(K_j-1))$ x_j	Composition, separator gas $K_j x_j$ y_j
CO_2	0.0091	9.87	0.0017	0.0167
N_2	0.0016	64.03	0.0000	0.0032
C_1	0.3647	23.45	0.0303	0.7102
C_2	0.0967	4.15	0.0379	0.1574
C_3	0.0695	1.17	0.0641	0.0751
i-C_4	0.0144	0.448	0.0198	0.0089
n-C_4	0.0393	0.331	0.0585	0.0194
i-C_5	0.0144	0.135	0.0251	0.0034
n-C_5	0.0141	0.106	0.0252	0.0027
C_6	0.0433	0.0329	0.0826	0.0027
C_7^+	0.3329	0.00047	0.6548	0.0003
	1.0000		1.0000	1.0000

$$\bar{n}_{g1} = 0.4919 \frac{\text{lb mole SP gas}}{\text{lb mole SP feed}}$$

$$\bar{n}_{L1} = 1 - 0.4919 = 0.5081 \frac{\text{lb mole SP liq}}{\text{lb mole SP feed}}$$

Step 2: Calculate the compositions and quantities of stock-tank gas and liquid using Equation 12–17. Use the composition of the separator liquid for z_j. Trial and error on $\bar{n}_g$ is required.

$$\sum_j x_j = \sum_j \frac{z_j}{1 + \bar{n}_g (K_j - 1)} = 1.0 \qquad (12-17)$$

Only the final trial value of $\bar{n}_g = 0.1234$ will be shown.

Component	Composition, stock-tank feed z_j	K-factor at 14.7 psia & 75°F K_j	Composition, stock-tank liquid $z_j/(1+\bar{n}_g(K_j-1))$ x_j	Composition, stock-tank gas $K_j x_j$ y_j
CO_2	0.0017	74.3	0.0002	0.0126
C_1	0.0303	181.1	0.0013	0.2363
C_2	0.0379	30.5	0.0081	0.2493
C_3	0.0641	8.46	0.0334	0.2822
i-C_4	0.0198	3.05	0.0158	0.0481
n-C_4	0.0585	2.22	0.0509	0.1130
i-C_5	0.0251	0.852	0.0255	0.0218
n-C_5	0.0252	0.681	0.0262	0.0179
C_6	0.0826	0.187	0.0918	0.0172
C_7+	0.6548	0.0021	0.7468	0.0016
	1.0000		1.0000	1.0000

$$\bar{n}_{g2} = \frac{0.1234 \text{ lb mole ST gas}}{\text{lb mole SP liq}}$$

$$\bar{n}_{L2} = 1 - 0.1234 = 0.8766 \frac{\text{lb mole ST liq}}{\text{lb mole SP liq}}$$

Step 3: Calculate the density and molecular weight of the stock-tank oil at standard conditions.

Component	Composition, mole fraction x_j	Molecular weight M_j	Weight x_jM_j	Liquid density @60°F and 14.7 psia ρ_{oj}	Liquid volume @ 60°F and 14.7 psia x_jM_j/ρ_{oj}
CO_2	0.0002	44.010	0.0074		
C_1	0.0013	16.043	0.0209		
C_2	0.0081	30.070	0.2455		
C_3	0.0334	44.097	1.4716	31.62	0.0465
i-C_4	0.0158	58.123	0.9169	35.12	0.0261
n-C_4	0.0509	58.123	2.9580	36.42	0.0812
i-C_5	0.0255	72.150	1.8415	38.96	0.0473
n-C_5	0.0262	72.150	1.8904	39.36	0.0480
C_6	0.0918	86.177	7.9111	41.40	0.1911
C_7+	0.7468	218.000	162.8024	53.11	3.0656
	1.0000	M_{STO} = 180.0657			3.5058

Density of propanes plus $= 179.7919/3.5058 = 51.28$ lb/cu ft

$$W2 = 0.2455/180.0374 \times 100 = 0.14 \text{ wt } \%$$

$$W1 = 0.0209/180.0657 \times 100 = 0.01 \text{ wt } \%$$

$$\rho_{po}/\rho_{C3+} = 0.9996, \text{ Figure } 11-6$$

Pseudoliquid density $= (51.28)(0.9996) = 51.26$ lb/cu ft

No temperature and pressure adjustment necessary at standard conditions.

$$\text{Specific gravity of stock-tank oil} = \frac{51.26}{62.37} = 0.822 \quad (8-1)$$

$$\text{Gravity of stock-tank oil} = \frac{141.5}{0.822} - 131.5 = 40.6°\text{API} \quad (8-2)$$

Step 4: Calculate gas-oil ratio using Equations 13–2, 13–4, and 13–5.

$$R_{SP} = \frac{2138 \, \bar{n}_{g1} \, \rho_{STO}}{\bar{n}_{L1} \, \bar{n}_{L2} \, M_{STO}} \quad (13-2)$$

$$R_{SP} = \frac{(2138)(0.4919)(51.26)}{(0.5081)(0.8766)(180.1)} = 672 \text{ scf/STB}$$

$$R_{ST} = \frac{2138\bar{n}_{g2}\rho_{STO}}{\bar{n}_{L2}M_{STO}} \quad (13-4)$$

$$R_{ST} = \frac{(2138)(0.1234)(51.26)}{(0.8766)(180.1)} = 86 \text{ scf/STB}$$

$$R = R_{SP} + R_{ST}. \qquad (13\text{-}5)$$

$$R = 672 \text{ scf/STB} + 86 \text{ scf/STB} = 758 \text{ scf/STB}$$

Step 5: Calculate the density and molecular weight of the reservoir liquid at reservoir conditions.
See Example 11–5

$$\rho_{oR} = 41.09 \text{ lb/cu ft, } M_{oR} = 93.75 \text{ lb/lb mole}$$

Step 6: Calculate formation volume factor using Equation 13–9.

$$B_{ob} = \frac{M_{oR}\rho_{STO}}{M_{STO}\rho_{oR}\bar{n}_{L1}\bar{n}_{L2}} \qquad (13\text{-}9)$$

$$B_{ob} = \frac{(93.75)(51.26)}{(180.1)(41.09)(0.5081)(0.8766)} = 1.458 \text{ res bbl/STB}$$

Step 7: The compositions of the gases and liquids can be used to calculate various properties. Example 6–12 is a calculation of the heating value of the separator gas. Example 7–9 is a calculation of liquids content of the separator gas. Specific gravities of the gases can be calculated.

Component	Molecular weight, M_j	Composition, mole fraction, separator gas y_{SPj}	$y_{SPj}M_j$	Composition, mole fraction, stock-tank gas y_{STj}	$y_{STj}M_j$
CO_2	44.010	0.0167	0.735	0.0126	0.555
N_2	28.013	0.0032	0.093	0.0000	0.000
C_1	16.043	0.7102	11.394	0.2363	3.791
C_2	30.070	0.1574	4.733	0.2493	7.496
C_3	44.097	0.0751	3.312	0.2822	12.444
i-C_4	58.123	0.0089	0.517	0.0481	2.796
n-C_4	58.123	0.0194	1.128	0.1130	6.568
i-C_5	72.150	0.0034	0.245	0.0218	1.573
n-C_5	72.150	0.0027	0.195	0.0179	1.291
C_6	86.177	0.0027	0.233	0.0172	1.482
C_{7+}	218.000	0.0003	0.065	0.0016	0.349
		1.0000	22.650	1.0000	38.345

$$\gamma_{SP} = 22.65/29 = 0.781 \tag{3-38}$$

$$\gamma_{ST} = 38.34/29 = 1.322 \tag{3-38}$$

The table below gives a comparison of these calculated values with the results of a laboratory separation of this hydrocarbon mixture. The agreement is remarkably good.

	Experimental	Calculated	Deviation, %
Gas-oil ratio, scf/STB	768	758	1.3
Oil formation volume factor, res bbl/STB	1.474	1.458	1.1
Stock-tank oil gravity, °API	40.7	40.6	0.2
Separator gas specific gravity	0.786	0.781	0.6
Separator gas gross heating value, BTU/scf	1321	1321	0.0
Separator gas liquids content, GPM	7.516	7.488	0.4
Stock-tank gas specific gravity	1.363	1.322	3.0

The calculations of Example 13–1 could be repeated for different separator pressures to determine the separator pressure which produces the largest amount of stock-tank liquid. Results of a laboratory separation of the hydrocarbon mixture of Example 13–1 are given in Figure 10–5. Optimum separator pressure is about 100 psig. This corresponds to minima in gas-to-oil ratio and formation volume factor and maxima in quantity of stock-tank oil and stock-tank oil gravity.

Further examination of gas-oil ratios is informative. Figure 13–3 shows changes in separator gas-oil ratio, stock-tank gas-oil ratio, and total gas-oil ratio as separator pressure is changed.

Notice that, as separator pressure is increased, the separator gas-oil ratio decreases. This forces more light molecules into the liquid going to the stock tank. These molecules leave the liquid when the pressure is reduced to atmospheric, causing an increase in stock-tank gas-oil ratio. The sum of the separator gas and the stock-tank gas goes through a minimum as Figure 13–3 shows.

Separator Calculations for Other Reservoir Fluid Types

Volatile oils and retrograde gases are produced routinely through at least three stages of separation. Thus, the calculation procedure requires three gas-liquid equilibria calculations. The equations for producing gas-oil ratios are

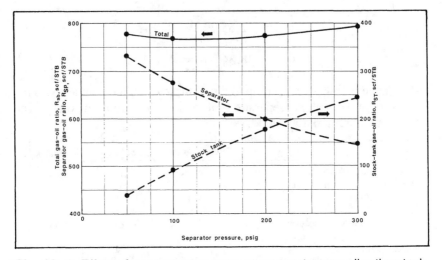

Fig. 13-3. Effect of separator pressure on separator gas-oil ratio, stock-tank gas-oil ratio, and total gas-oil ratio for a two-stage separation of the black oil of Example 13-1. (Data from Core Laboratories, Inc., Good Oil Co. No. 4)

$$R_{SP1} = \frac{2138 \bar{n}_{g1} \rho_{STO}}{\bar{n}_{L1} \bar{n}_{L2} \bar{n}_{L3} M_{STO}}, \; \frac{scf}{STB}, \quad (13\text{--}10)$$

$$R_{SP2} = \frac{2138 \bar{n}_{g2} \rho_{STO}}{\bar{n}_{L2} \bar{n}_{L3} M_{STO}}, \; \frac{scf}{STB}, \quad (13\text{--}11)$$

$$R_{ST} = \frac{2138 \bar{n}_{g3} \rho_{STO}}{\bar{n}_{L3} M_{STO}}, \; \frac{scf}{STB}, \quad (13\text{--}12)$$

and

$$R = R_{SP1} + R_{SP2} + R_{ST}, \; \frac{scf}{STB}. \quad (13\text{--}13)$$

The total producing gas-oil ratio is equal to solution gas-oil ratio at the bubble point, R_{sb}, if the feed to the primary separator comes from an oil reservoir with pressure equal to or greater than bubble point.

The formation volume factor at the bubble point is

$$B_{ob} = \frac{M_{oR} \rho_{STO}}{M_{STO} \rho_{oR} \bar{n}_{L1} \bar{n}_{L2} \bar{n}_{L3}}, \; \frac{res \, bbl}{STB}, \quad (13\text{--}14)$$

for a three-stage separator system.

If the reservoir is a gas reservoir, Equation 13–14 does not apply. In this situation, liquid recovery in STB/MMscf often is calculated. When liquid recovery based on surface gas recovery is desired, the equation is

$$\frac{M_{STO}\bar{n}_{L1}\bar{n}_{L2}\bar{n}_{L3}}{0.002138\rho_{STO}\ (\bar{n}_{g1}\ +\ \bar{n}_{g2}\bar{n}_{L1}\ +\ \bar{n}_{g3}\bar{n}_{L1}\bar{n}_{L2})}\ \frac{STB}{MMscf\ surface\ gas}\ . \qquad (13-15)$$

When liquid recovery based on reservoir gas production is desired, the equation is

$$\frac{M_{STO}\bar{n}_{L1}\bar{n}_{L2}\bar{n}_{L3}}{0.002138\rho_{STO}}\ \frac{STB}{MMscf\ reservoir\ gas}\ . \qquad (13-16)$$

Wet gas formation volume factor as defined in Chapter 7 can be calculated by converting reservoir gas in scf to reservoir gas at reservoir conditions, res bbl.

$$B_{wg}\ =\ \frac{10.732\ \rho_{STO}\ zT_R}{M_{STO}\bar{n}_{L1}\bar{n}_{L2}\bar{n}_{L3}p_R}\ \frac{res\ bbl\ gas}{STB}\ , \qquad (13-17)$$

where z-factor is determined at reservoir pressure and temperature, p_R is reservoir pressure in psia, T_R is reservoir temperature in °R, and ρ_{STO} is stock-tank liquid density in lb/cu ft. The calculated value of B_{wg} applies only at p_R.

K-Factors for Separator Calculations

Later we will see that *K-factors* are correlated with a property termed *convergence pressure*. Convergence pressure will be defined in Chapter 14. The set of K-factor correlations in Appendix A is for use with petroleum mixtures having convergence pressures close to 5000 psia.

Typical *black oils* have convergence pressures near 10,000 psia. Typical *retrograde gases* and *wet gases* have convergence pressures of about 5000 psia. The convergence pressures of *volatile oils* are between 5000 and 10,000 psia.

A K-factor correlation based on a convergence pressure of 5000 psia can be used in separator calculations for all of these reservoir systems. Surface separators for black oils typically operate at low pressures. Convergence pressure has little effect on K-factors at low pressure.

Appendix A has no chart for K-factors of carbon dioxide. At low concentrations of carbon dioxide use

$$K_{CO2}\ =\ \sqrt{K_{C1}K_{C2}}\ . \qquad (13-18)$$

Virtually all of the heptanes plus will remain in the liquid at surface conditions. Thus, the choice of K-factor for heptanes plus is not critical. K-factors for decane give good results.

Exercises

13–1. Gas-liquid equilibria calculations have been made for a two-stage separator system as shown in Figure 13–2,

$$\text{separator } \bar{n}_g \; = \; 0.595$$

$$\text{stock-tank } \bar{n}_g \; = \; 0.079$$

The feed rate to the separator is 2500 lb moles per day. Calculate the quantity of separator gas, stock-tank gas and stock-tank liquid, all in lb moles per day.

13–2. Gas-liquid equilibria calculations have been made for a three-stage separator system,

$$\text{primary separator } \bar{n}_g \; = \; 0.123$$

$$\text{second separator } \bar{n}_g \; = \; 0.542$$

$$\text{stock-tank } \bar{n}_g \; = \; 0.014$$

The feed rate to the primary separator is 1500 lb moles per day. Calculate the quantity of primary separator gas, second separator gas, stock-tank gas and stock-tank liquid, all in lb moles per day.

13–3. Calculate the quantity of primary separator liquid and the quantity of second separator liquid, both in lb moles per day, for the three-stage separator of Exercise 13–2.

13–4. Repeat Example 13–1. Use separator conditions of 50 psig at 75°F. Compare your answers with the laboratory results given on pages 7 and 8 of Table 10–1.

13–5. Repeat Example 13–1. Use separator conditions of 200 psig at 75°F. Compare your answers with the laboratory results given on pages 7 and 10 of Table 10–1.

13–6. Repeat Example 13–1. Use separator conditions of 300 psig at 75°F. Compare your answers with the laboratory results given on pages 7 and 11 of Table 10–1.

13-7. The gas-liquid equilibria calculations for a two-stage separation of a black oil have been completed. Results are given below.

Component	Composition, separator, mole fraction @ 150 psig & 120°F				Composition, stock-tank, mole fraction @ 0 psig & 80°F			
	feed	K-factor	liquid	gas	feed	K-factor	liquid	gas
H_2S	0.0879	3.20	0.0475	0.1521	0.0475	25.8	0.0087	0.2243
CO_2	0.0270	8.87	0.0067	0.0592	0.0067	77.5	0.0005	0.0351
N_2	0.0009	42.42	0.0000	0.0022	0.0000	–	0.0000	0.0003
C_1	0.2112	18.81	0.0268	0.5039	0.0268	184.8	0.0008	0.1453
C_2	0.0763	4.18	0.0342	0.1431	0.0342	32.5	0.0051	0.1668
C_3	0.0703	1.41	0.0606	0.0857	0.0606	9.06	0.0247	0.2241
i-C_4	0.0147	0.639	0.0171	0.0109	0.0171	3.32	0.0121	0.0400
n-C_4	0.0428	0.489	0.0533	0.0261	0.0533	2.41	0.0425	0.1026
i-C_5	0.0171	0.215	0.0245	0.0053	0.0245	0.941	0.0248	0.0233
n-C_5	0.0237	0.241	0.0335	0.0081	0.0335	0.759	0.0351	0.0266
C_6	0.0248	0.0637	0.0389	0.0025	0.0389	0.211	0.0453	0.0096
C_{7+}	0.4033	0.00140	0.6569	0.0009	0.6569	0.00251	0.8004	0.0020
	1.0000		1.0000	1.0000	1.0000		1.0000	1.0000

$$\bar{n}_g = 0.3866 \qquad \bar{n}_g = 0.1798$$

The heptanes plus of the well stream has molecular weight of 215 lb/lb mole and specific gravity of 0.850. The bubble-point pressure of the reservoir liquid is 1748 psig at 250°F.
Calculate

a. Separator gas-oil ratio, scf/STB
b. Stock-tank gas-oil ratio, scf/STB
c. Total gas-oil ratio, R_{sb}, scf/STB
d. Oil formation volume factor, B_{ob}, res bbl/STB
e. Specific gravity of separator gas
f. Specific gravity of stock-tank gas
g. Gravity of stock-tank oil, °API

Compare your answers with laboratory results of

$R_{sb} = 587$ scf/STB, $B_{ob} = 1.425$ res bbl/STB, $\gamma_{gSP} = 0.936$, $\gamma_{gST} = 1.388$, $\gamma_{STO} = 39.1$°API, $\rho_{oR} = 42.12$ lb/cu ft.

13–8. Perform a two-stage separator calculation for the following black oil.

Component	Composition, mole fraction
Methane	0.2648
Ethane	0.0951
Propane	0.0961
i-Butane	0.0173
n-Butane	0.0501
i-Pentane	0.0188
n-Pentane	0.0281
Hexanes	0.0378
Heptanes plus	0.3919
	1.0000

Properties of heptanes plus	
Gravity	35.2°API
Molecular weight	208 lb/lb mole

Use K-factors of decane for the heptanes plus fraction. The black oil is produced from a reservoir operating at its bubble point of 1951 psig at 250°F. Separator conditions are 105 psig and 100°F; stock tank is at 100°F.

Calculate

a. Composition of separator gas, mole fraction
b. Specific gravity of separator gas
c. Formation volume factor of oil, B_{ob}, res bbl/STB
d. Separator gas-oil ratio, scf/STB
e. Stock-tank gas-oil ratio, scf/STB
f. Total gas-oil ratio, R_{sb}, scf/STB
g. Gravity of the stock-tank oil, °API

Compare your answer with laboratory results of B_{ob} = 1.478 res bbl/STB, R_{sb} = 618 scf/STB, ρ_{oR} = 40.31 lb/cu ft, γ_{STO} = 40.7°API, and γ_{gSP} = 0.849.

13–9. Perform a two-stage separator calculation for the following black oil.

Component	Composition, mole fraction
Methane	0.3827
Ethane	0.0846
Propane	0.0742
i-Butane	0.0105
n-Butane	0.0381
i-Pentane	0.0138
n-Pentane	0.0209
Hexanes	0.0261
Heptanes plus	0.3491
	1.0000
Properties of heptanes plus	
Gravity	31.5°API
Molecular weight	238 lb/lb mole

Use K-factors of decane for the heptanes plus fraction. The black oil is produced from a reservoir operating at its bubble point of 2405 psig at 129°F. Separator conditions are 100 psig and 71°F; stock tank is at 68°F.

Calculate

a. Composition of separator gas, mole fraction
b. Specific gravity of separator gas
c. Formation volume factor of oil, B_{ob}, res bbl/STB
d. Separator gas-oil ratio, scf/STB
e. Stock-tank gas-oil ratio, scf/STB
f. Total gas-oil ratio, R_{sb}, scf/STB
g. Gravity of the stock-tank oil, °API

Compare your answers with laboratory results of $B_{ob} = 1.356$ res bbl/STB, $R_{sb} = 701$ scf/STB, $\rho_{oR} = 44.78$ lb/cu ft, $\gamma_{STO} = 37.0°API$, and $\gamma_{gSP} = 0.743$.

13–10. Perform a three-stage separator calculation for the following black oil.

Component	Composition, mole percent
Hydrogen sulfide	9.14
Carbon dioxide	8.91
Nitrogen	0.25
Methane	25.15
Ethane	8.17
Propane	5.74
i-Butane	1.14
n-Butane	4.03
i-Pentane	1.34
n-Pentane	1.94
Hexanes	2.40
Heptanes plus	31.79
	100.00

Properties of heptanes plus	
Specific gravity	0.8525
Molecular weight	219 lb/lb mole

Use K-factor of decane for heptanes plus. The black oil is produced from a reservoir at its bubble point of 2361 psia at 225°F.

Primary separator conditions: 250 psig and 160°F
Second separator conditions: 50 psig and 148°F
Stock-tank conditions: 140°F

Calculate
a. Composition of primary separator gas, mole fraction
b. Specific gravity of primary separator gas
c. Composition of second separator gas, mole fraction
d. Specific gravity of second separator gas
e. Composition of stock-tank gas, mole fraction
f. Specific gravity of stock-tank gas
g. Primary separator gas-oil ratio, scf/STB
h. Second separator gas-oil ratio, scf/STB
i. Stock-tank gas-oil ratio, scf/STB
j. Total gas-oil ratio, R_{sb}, scf/STB
k. Formation volume factor of oil, B_{ob}, res bbl/STB
l. Gravity of stock-tank oil, °API

Compare your answers with laboratory results of
B_{ob} = 1.604 res bbl/STB, R_{sb} = 986 scf/STB,
ρ_{oR} = 41.31 lb/cu ft, γ_{STO} = 38.0°API,
γ_{gSP1} = 0.986, γ_{gSP2} = 1.257, and
γ_{gST} = 1.557.

13-11. Show the derivation of Equation 13-10.
13-12. Show the derivation of Equation 13-11.
13-13. Show the derivation of Equation 13-12.
13-14. Show the derivation of Equation 13-14.
13-15. Show the derivation of Equation 13-15.
13-16. Show the derivation of Equation 13-16.
13-17. The gas-liquid equilibria calculations for a three-stage separation of a volatile oil have been completed. Results are given below.

Component	Composition, mole fraction primary separator @ 450 psia and 70°F			Composition, mole fraction second separator @ 55 psia and 75°F			Composition, mole fraction stock tank @ 14.7 psia and 75°F		
	feed	liquid	gas	feed	liquid	gas	feed	liquid	gas
C_1	0.6092	0.1192	0.8373	0.1192	0.0094	0.4535	0.0094	0.0005	0.0844
C_2	0.1023	0.0907	0.1077	0.0907	0.0325	0.2679	0.0325	0.0078	0.2392
C_3	0.0569	0.1005	0.0366	0.1005	0.0754	0.1770	0.0754	0.0420	0.3551
i-C_4	0.0161	0.0376	0.0061	0.0376	0.0389	0.0337	0.0389	0.0319	0.0973
n-C_4	0.0257	0.0637	0.0080	0.0637	0.0703	0.0438	0.0703	0.0622	0.1381
i-C_5	0.0132	0.0375	0.0019	0.0375	0.0461	0.0113	0.0461	0.0468	0.0399
n-C_5	0.0116	0.0336	0.0014	0.0336	0.0419	0.0082	0.0419	0.0434	0.0295
C_6	0.0186	0.0566	0.0009	0.0566	0.0738	0.0043	0.0738	0.0808	0.0151
C_{7+}	0.1464	0.4606	0.0001	0.4606	0.6117	0.0003	0.6117	0.6846	0.0014
	1.0000	1.0000	1.0000	1.0000	1.0000	1.0000	1.0000	1.0000	1.0000

$\bar{n}_g$ = 0.6823 $\bar{n}_g$ = 0.2472 $\bar{n}_g$ = 0.1066

The heptanes plus fraction of the primary separator feed has a molecular weight of 177 lb/lb mole and a specific gravity of 0.809. Bubble-point pressure of the reservoir liquid is 4475 psia at 250°F.

Calculate

a. Primary separator gas-oil ratio, scf/STB
b. Second separator gas-oil ratio, scf/STB
c. Stock-tank gas-oil ratio, scf/STB
d. Total gas-oil ratio, R_{sb}, scf/STB

e. Oil formation volume factor, B_{ob}, res bbl/STB
f. Gravity of stock-tank oil, °API

Compare your answers with the laboratory measured specific volume of reservoir liquid at bubble point, 0.0353 cu ft/lb.

13–18. The gas-liquid equilibrium calculations for a two-stage separation of a retrograde gas have been completed. Results are given below.

Component	Composition, mole fraction primary separator @ 300 psia and 90°F			Composition, mole fraction stock tank @ 14.7 psia and 70°F		
	feed	liquid	gas	feed	liquid	gas
C_1	0.4850	0.0649	0.6542	0.0649	0.0008	0.1457
C_2	0.1724	0.1036	0.2001	0.1036	0.0078	0.2243
C_3	0.1207	0.1660	0.1025	0.1660	0.0410	0.3234
i-C_4	0.0170	0.0351	0.0097	0.0351	0.0196	0.0546
n-C_4	0.0502	0.1149	0.0241	0.1149	0.0786	0.1607
i-C_5	0.0118	0.0335	0.0031	0.0335	0.0373	0.0288
n-C_5	0.0209	0.0615	0.0045	0.0615	0.0743	0.0454
C_6	0.0186	0.0609	0.0016	0.0609	0.0966	0.0160
C_{7+}	0.1034	0.3596	0.0002	0.3596	0.6440	0.0011
	1.0000	1.0000	1.0000	1.0000	1.0000	1.0000

$$\bar{n}_g = 0.7128 \qquad\qquad \bar{n}_g = 0.4425$$

The heptanes plus of the well stream has a molecular weight of 171 lb/lb mole and a specific gravity of 0.801. The dew-point pressure of the reservoir gas is 2973 psig at reservoir temperature of 184°F.

Calculate
a. Separator gas-oil ratio, scf/STB
b. Stock-tank gas-oil ratio, scf/STB
c. Total gas-oil ratio, scf/STB
d. Specific gravity of reservoir gas
e. Specific gravity of separator gas
f. Specific gravity of stock-tank gas
g. Gross heating value of separator gas, BTU/scf
h. "Plant products" of separator gas, GPM
i. Gravity of stock-tank oil, °API
j. Liquid recovery, STB/MMscf res gas

    k. Liquid recovery, STB/MMscf surface gas

    l. Wet gas formation volume factor at the dew point, B_{wg}, bbl res gas/STB

Compare your answers with laboratory data of z = 0.708 at dew point and liquid recovery = 217.12 STB/MMscf of reservoir gas.

13–19. The gas-liquid equilibrium calculations for a three-stage separation of a retrograde gas have been completed. Results are given below.

Component	Composition, mole fraction primary separator @ 350 psia and 90°F			Composition, mole fraction second separator @ 55 psia and 85°F			Composition, mole fraction stock tank @ 14.7 psia and 70°F		
	feed	liquid	gas	feed	liquid	gas	feed	liquid	gas
C_1	0.4850	0.0764	0.6680	0.0764	0.0046	0.2319	0.0046	0.0002	0.0264
C_2	0.1724	0.1164	0.1975	0.1164	0.0329	0.2974	0.0329	0.0057	0.1641
C_3	0.1207	0.1756	0.0961	0.1756	0.1156	0.3059	0.1156	0.0530	0.4182
$i\text{-}C_4$	0.0170	0.0351	0.0089	0.0351	0.0350	0.0355	0.0350	0.0268	0.0748
$n\text{-}C_4$	0.0502	0.1142	0.0215	0.1142	0.1244	0.0921	0.1243	0.1055	0.2157
$i\text{-}C_5$	0.0118	0.0323	0.0026	0.0323	0.0416	0.0122	0.0416	0.0433	0.0334
$n\text{-}C_5$	0.0209	0.0590	0.0038	0.0593	0.0776	0.0185	0.0781	0.0832	0.0508
C_6	0.0186	0.0571	0.0014	0.0571	0.0807	0.0060	0.0806	0.0941	0.0156
C_{7+}	0.1034	0.3339	0.0002	0.3336	0.4876	0.0005	0.4873	0.5882	0.0010
	1.0000	1.0000	1.0000	1.0000	1.0000	1.0000	1.0000	1.0000	1.0000
	$\bar{n}_g = 0.6906$			$\bar{n}_g = 0.3156$			$\bar{n}_g = 0.1713$		

The heptanes plus fraction of the primary separator feed has a molecular weight of 171 lb/lb moles and a specific gravity of 0.801. Dew-point pressure of the reservoir gas is 2973 psig at 184°F.

Calculate

a. Primary separator gas-oil ratio, scf/STB

b. Second separator gas-oil ratio, scf/STB

c. Stock-tank gas-oil ratio, scf/STB

d. Total gas-oil ratio, scf/STB

e. Gravity of stock-tank oil, °API

f. Specific gravity of reservoir gas

g. Specific gravity of primary separator gas

h. Specific gravity of second separator gas

i. Specific gravity of stock-tank gas

    j. "Plant products" of primary separator gas, GPM
    k. "Plant products" of second separator gas, GPM
    l. Gross heating value of primary separator gas, BTU/scf
    m. Gross heating value of second separator gas, BTU/scf
    n. Density of primary separator gas at 350 psia and 90°F, lb/cu ft
    o. Liquid recovery, STB/MMscf surface gas
    p. Liquid recovery, STB/MMscf reservoir gas
    q. Wet gas formation volume factor at the dew point, B_{wg},
       bbl res gas/STB

Reference

1. Smith, H.V.: "Oil and Gas Separators," *Petroleum Engineering Handbook,* H.B. Bradley et al. (Eds.), SPE, Richardson, Texas (1987).

14

Equilibrium-Ratio Correlations

The equations for gas-liquid equilibria calculations and the definition of equilibrium ratio (K-factor) were discussed in Chapter 12. This chapter will examine *K-factor correlations*.

The equation which defines K-factor,

$$K_j = \frac{y_j}{x_j}, \qquad (12\text{--}14)$$

and the equation resulting from the combination of Raoult's and Dalton's equations,

$$\frac{y_j}{x_j} = \frac{p_{vj}}{p}, \qquad (12\text{--}2)$$

can be combined to give an equation for calculating the K-factor of a component, j, in a mixture which behaves like an *ideal solution*.

$$K_j = \frac{p_{vj}}{p} \qquad (14\text{--}1)$$

The vapor pressure of component j, p_{vj}, is solely a function of temperature. Thus Equation 14–1 shows that if ideal solution behavior exists, the K-factor of each component simply depends on pressure and temperature.

Remember that one of the principal properties used to define an *ideal solution* is that the intermolecular forces of attraction and repulsion are the same between unlike as between like molecules. This property does not exist in *real solutions*. Molecular behavior in a real solution depends on the types and sizes of the molecules which are interacting.

Therefore, the K-factor for a component of a real solution depends not only on pressure and temperature but also on the types and quantities of other substances present. This means that any correlation of K-factors must be based on at least three quantities: pressure, temperature, and a third property which describes nonideal solution behavior. This property must represent both the types of molecules present and their quantities in the gas and liquid.

Attempts have been made to define this third property in several different ways.[1] We will look only at *convergence pressure*, which appears to be the most convenient for the types of calculations required of petroleum engineers.

A large number of K-factor correlations have been proposed.[1] We will present only two: one graphical, the other in equation form.[2,3] A partial set of K-factor curves is given in Appendix A.[2] Only 14 charts out of a total of 90 of this correlation are given. These should suffice for illustrative exercises in this text.

Notice that the K-factor charts in Appendix A apply to petroleum mixtures which have convergence pressure of 5000 psia. The other charts of this correlation (not reproduced in Appendix A) are for use with mixtures with other convergence pressures. A value of convergence pressure applicable to the fluid of interest must be estimated in order to select the correct set of graphs.

Convergence Pressure

Experimentally determined K-factors normally are plotted against pressure on a log-log scale. Figures 14–1 and 14–2 show equilibrium ratios of two typical petroleum mixtures for several temperatures.

The shapes of the curves in these figures are characteristic of most multicomponent mixtures. At low pressures, the slope of each curve is approximately −1.0. A slope of −1.0 for ideal K-factors is predicted by Equation 14–1. Each curve passes through a value of unity at a pressure very close to the vapor pressure of the component at the temperature of the graph. This also is predicted for ideal-solution behavior.

At higher pressures, the effects of nonideality are seen. The curve for each component departs from a slope of −1.0. The curves tend to converge toward a K-factor of 1.0.

Definition of Convergence Pressure

The value of pressure for which the K-factors appear to converge to unity is known as the *convergence pressure*. If a mixture is at its true critical temperature, the curves in reality will converge to a value of 1.0

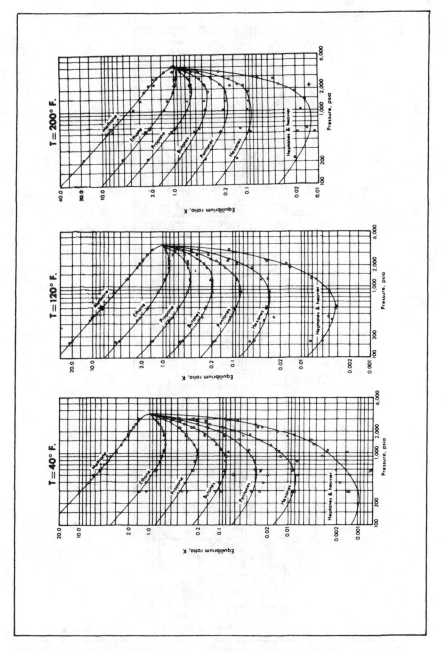

Fig. 14–1. Equilibrium ratios for a retrograde gas at various temperatures and pressures. (Roland, Smith, Kaveler, *Oil and Gas Journal 39,* No. 46, 128, with permission.)

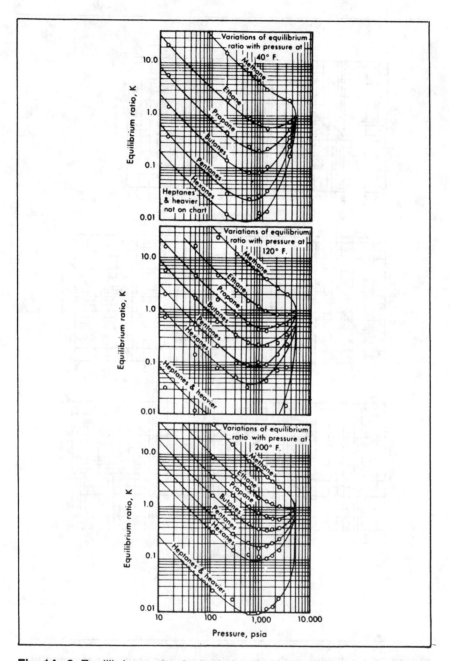

Fig. 14–2. Equilibrium ratios for a crude oil-natural gas mixture at various temperatures and pressures. (Reprinted with permission from *Industrial and Engineering Chemistry 29*, 1072. Copyright 1937 American Chemical Society.)

at the critical pressure. At any temperature other than the critical temperature, the equilibrium-ratio curves actually do not extend past the bubble-point or dew-point pressure of the mixture. However, the curves can be extrapolated to determine the point of *apparent* convergence. Sometimes convergence pressure is called *apparent convergence pressure*.

Although the convergence pressure is not equal to the critical pressure of the mixture, to a certain extent it does characterize the properties of the mixture. Convergence pressure is useful in the correlation of K-factor data.

Estimation of Convergence Pressure

Several methods of estimating convergence pressure have been proposed. These methods have been evaluated using laboratory data of petroleum reservoir samples.[4] The method described below is as accurate as any and is the easiest to apply.

Convergence pressures of binary hydrocarbon mixtures may be estimated from the critical locus curves given in Figure 2–16. A similar curve which includes multicomponent mixtures is presented in Figure 14–3.

This method of estimating convergence pressure involves a trial-and-error procedure in which a first trial value of convergence pressure is used to obtain K-factors. A reasonable estimate of convergence pressure for a first trial value may be obtained from[5]

$$p_k = 60M_{C7+} - 4200, \qquad (14-2)$$

where M_{C7+} is the molecular weight of the heptanes plus fraction.

The compositions and quantities of the equilibrium gas and liquid are computed in the usual manner. Then the liquid is equated to a pseudobinary mixture consisting of the lightest component of the liquid and a hypothetical heavy component.

The lightest component of petroleum mixtures always is taken to be methane. The hypothetical heavy component is represented by a critical temperature and a critical pressure calculated as the weighted average of the true critical properties of all components of the liquid except the lightest.

The critical properties of methane and the weighted-average critical properties of the hypothetical heavy component are plotted on Figure 14–3, and a locus of convergence pressures for that pair is interpolated using the adjacent critical loci as guides.

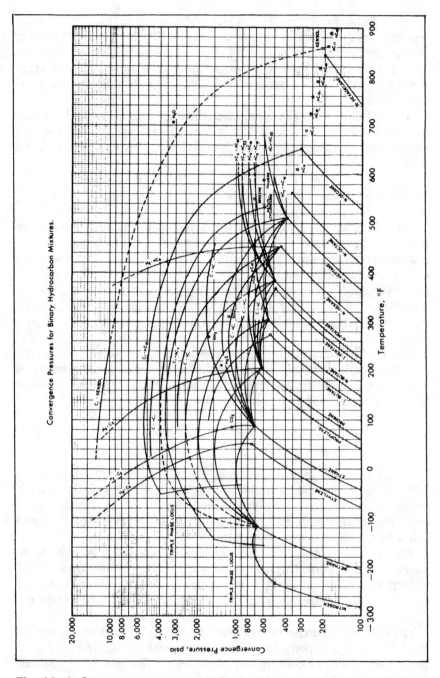

Fig. 14-3. Convergence pressures for hydrocarbons. (*Engineering Data Book,* 10th Ed., GPSA, Tulsa, 1987, with permission.)

The estimate of convergence pressure is taken from the point this locus crosses the temperature at which calculations are desired. If this convergence pressure is not reasonably close to the first trial value of convergence pressure, the procedure should be repeated. The estimated value of convergence pressure can be used as the new trial value.

EXAMPLE 14–1: *The compositions of the gas and liquid from a black oil at 1300 psia and 160°F calculated with Equation 12–17 are given in the table below. A convergence pressure of 5000 psia was used to obtain the K-factors. What value of convergence pressure should have been used for this mixture at 160°F?*

Component	Composition of equilibrium liquid, x_j	Composition of equilibrium gas, y_j
Methane	0.2752	0.8705
Ethane	0.0730	0.0806
Propane	0.0390	0.0217
i-Butane	0.0151	0.0052
n-Butane	0.0442	0.0122
i-Pentane	0.0178	0.0029
n-Pentane	0.0247	0.0033
Hexanes	0.0259	0.0018
Heptanes plus	0.4851	0.0018
	1.0000	1.0000

Properties of heptanes plus
Specific gravity   0.8600
Molecular weight   225 lb/lb mole

Solution

First, the liquid composition must be expressed in weight fraction.

Component	Composition, mole fraction x_j	Molecular weight M_j	$x_j M_j$	Composition, weight fraction $x_j M_j / \sum_j x_j M_j$
C_1	0.2752	16.043	4.415	0.0350
C_2	0.0730	30.070	2.195	0.0174
C_3	0.0390	44.097	1.720	0.0136
i-C_4	0.0151	58.123	0.878	0.0069
n-C_4	0.0442	58.123	2.569	0.0204
i-C_5	0.0178	72.150	1.284	0.0102
n-C_5	0.0247	72.150	1.782	0.0141
C_6	0.0259	86.177	2.230	0.0177
C_{7+}	0.4851	225	109.148	0.8647
	1.0000		$\sum_j x_j M_j = 126.221$	1.0000

Second, delete methane and adjust the weight fraction of the hypothetical heavy component. Then calculate weighted-average critical properties.

Component	Composition excluding C_1, weight fraction w_j	Critical temperature, °R T_{cj}	$w_j T_{cj}$	Critical pressure, psia p_{cj}	$w_j p_{cj}$
C_2	0.0180	549.50	9.9	706.5	12.7
C_3	0.0141	665.64	9.4	616.0	8.7
i-C_4	0.0072	734.04	5.3	527.9	3.8
n-C_4	0.0211	765.20	16.1	550.6	11.6
i-C_5	0.0106	828.68	8.7	490.4	5.2
n-C_5	0.0146	845.38	12.4	488.6	7.1
C_6	0.0183	913.18	16.7	436.9	8.0
C_{7+}	0.8961	*1360	1218.7	*242	216.9
	1.0000	wt avg T_c =1297 °R		wt avg p_c = 274 psia	

*Pseudocritical properties for C_{7+} from Figure 3–10.

Third, plot weight-averaged critical point on Figure 14–3, interpolate a locus of convergence pressures, and read p_k at 160°F.

A convergence pressure of 10,000 psia should have been used for this mixture. See Figure 14–4. Obtain K-factors at p_k = 10,000 and repeat the gas-liquid equilibrium calculation.

When the operating pressure is considerably less than the convergence pressure, an error in the estimate of convergence pressure has little effect on the resulting calculations. As operating pressure approaches convergence pressure, however, equilibrium ratios become very sensitive to the convergence pressure used and care must be taken in the selection of the correct value.

The correlations from which the equilibrium ratio data in Appendix A were taken include charts for convergence pressures of 800, 1000, 1500, 2000, 3000, 5000, and 10,000 psia. When the convergence pressure for the mixture is between the values for which charts are provided, interpolate between charts. Interpolation is necessary when the operating pressure is near the convergence pressure. At low pressure, simply use the chart with convergence pressure nearest the value for the mixture.

Black oils usually have convergence pressures of about 10,000 psia, retrograde gases about 5000 psia, and volatile oils about 7000 psia.

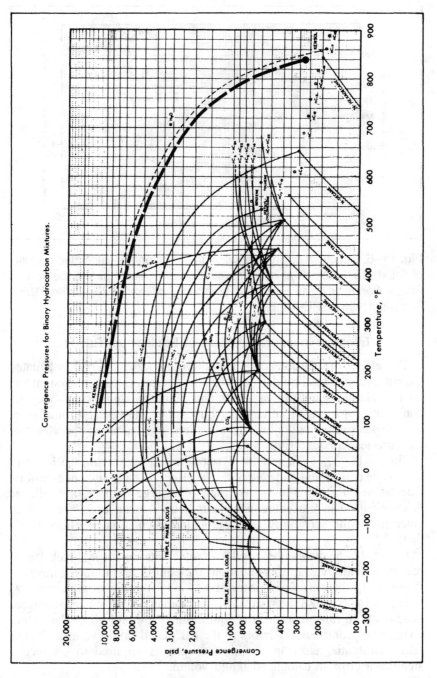

Fig. 14–4. Convergence pressures for hydrocarbons (part of solution to Example 14–1).

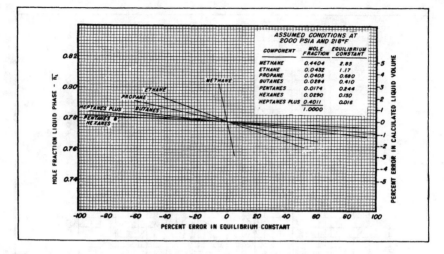

The table shown within the figure reads:

ASSUMED CONDITIONS AT 2000 PSIA AND 218°F		
COMPONENT	MOLE FRACTION	EQUILIBRIUM CONSTANT
METHANE	0.4404	2.85
ETHANE	0.0432	1.17
PROPANE	0.0405	0.680
BUTANES	0.0284	0.410
PENTANES	0.0174	0.244
HEXANES	0.0290	0.150
HEPTANES PLUS	0.4011	0.016
	1.0000	

Fig. 14-5. Effects of errors in equilibrium ratios on the calculated quantity of liquid of a black oil at reservoir conditions. (Standing, *Volumetric and Phase Behavior of oil Field Hydrocarbon Systems,* SPE, Dallas, 1951. Copyright 1951 SPE-AIME.

Effect of Inaccurate Values of Equilibrium Ratios

The effect of inaccuracies in equilibrium ratios on the calculated quantities of gas and liquid depends on the composition of the system and on the relative quantities of vapor and liquid in equilibrium. General comments regarding accuracy are impossible; however, examples of three typical cases will illustrate a method of evaluating the effect of inaccuracies in equilibrium ratios.[5]

Figures 14-5, 14-6, and 14-7 were created by altering the K-factor of a single component and observing the effect on the results of equilibria calculations. Each figure is the result of a large number of equilibria calculations. The procedure is tedious, but it is an excellent way of determining the sensitivity of a particular calculation to errors in K-factor.

Figure 14-5 shows the results of calculations for a black oil at reservoir conditions. In this situation, the equilibrium ratio for methane is obviously the only one that must be known with extreme accuracy.

Figure 14-6 shows the results of calculations for a black oil at surface conditions. In this situation, the equilibrium ratios for all of the lighter components contribute to errors in the calculated liquid volume. Yet, a fairly substantial error in equilibrium ratio is required to produce a significant error in calculated liquid volume.

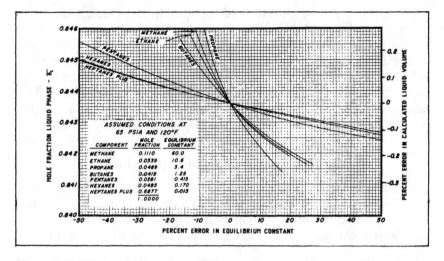

Fig. 14–6. Effects of errors in equilibrium ratios on the calculated quantity of liquid of a black oil at surface conditions. (Standing, *Volumetric and Phase Behavior of Oil Field Hydrocarbon Systems,* SPE, Dallas, 1951. Copyright 1951 SPE-AIME.

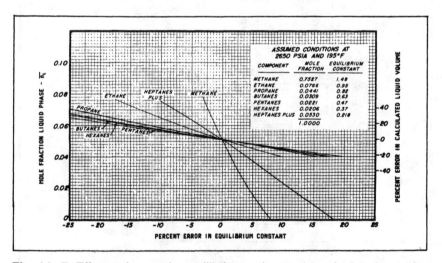

Fig. 14–7. Effects of errors in equilibrium ratios on the calculated quantity of liquid of a retrograde gas at reservoir conditions. (Standing, *Volumetric and Phase Behavior of Oil Field Hydrocarbon Systems,* SPE, Dallas, 1951. Copyright 1951 SPE-AIME.

Figure 14–7 shows the results of calculations for a retrograde gas at reservoir conditions. Inaccuracies in the equilibrium ratios for both the lightest component, methane, and the heaviest components, represented by heptanes plus, result in drastic error in the calculated liquid volume. In fact, an error of only 8% in the equilibrium ratio for methane results in the calculation of a dew point rather than the correct liquid volume. This indicates that calculations near the dew point of a mixture are highly dependent on the accuracy with which equilibrium ratios are known.

Correlating and Smoothing K-Factors

The best source of K-factors is laboratory data on the same or similar petroleum mixture. The nature of the experimental procedures is such that small variations in measurement are magnified when K-factors are calculated from the data. Thus it is desirable to have a method of

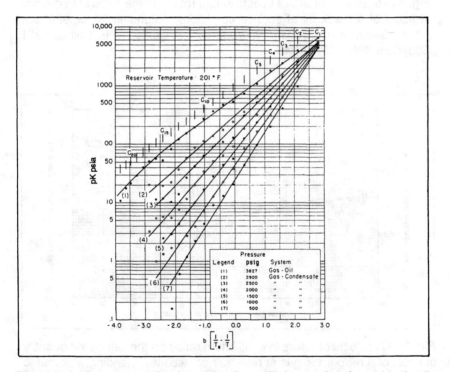

Fig. 14–8. Experimentally determined equilibrium ratios for a black oil (gas-oil) and a retrograde gas (gas-condensate) at 201°F. (Hoffman, Crump, Hocott, *Trans.*, AIME, *198*, 1. Copyright 1953 SPE-AIME.)

evaluating the consistency of the data and of smoothing the data if necessary.

A method has been developed from generalized vapor pressure theory.[6] The logarithm of the product of K-factor and pressure plotted against a component-dependent plotting factor will produce a straight line. See Figure 14–8. The plotting factor for component j is

$$b_j \left[\frac{1}{T_{Bj}} - \frac{1}{T} \right], \qquad (14-3)$$

where T_{Bj} is the normal boiling point of component j and b_j is defined by

$$b_j = \frac{\log p_{cj} - \log 14.7}{\left[\dfrac{1}{T_{Bj}} - \dfrac{1}{T_{cj}} \right]}. \qquad (14-4)$$

All pressures and temperatures are in absolute quantities. Table 14–1 gives values of b_j for paraffin hydrocarbons of interest.

TABLE 14–1: Values of b_j[6]

Hydrocarbon	Smoothed Values b_j
Methane	805
Ethane	1412
Propane	1799
i-Butane	2037
n-Butane	2153
i-Pentane	2368
n-Pentane	2480
n-Hexane	2780
n-Heptane	3068
n-Octane	3335
n-Nonane	3590
n-Decane	3828
n-Undecane	4055
n-Dodecane	4291
n-Tridecane	4500
n-Tetradecane	4715
n-Pentadecane	4919
n-Hexadecane	5105
n-Heptadecane	5290
n-Octadecane	5470
n-Nonadecane	5630
n-Eicosane	5790
n-Heneicosane	5945
n-Docosane	6095

Lines 2 through 7 of Figure 14–8 represent data of one petroleum system at different pressures. Each isobar seems to be straight, within the scatter of the data. The lines appear to converge. If extrapolated, the lines will converge at the convergence pressure of the mixture. The value of K_j for every component is equal to 1.0 at the convergence pressure so that the value of pK_j at the point of convergence is the convergence pressure of the mixture. Figure 14–8 shows a convergence pressure of about 6000 psia for the "gas-condensate" system.

K-factor data are smoothed by drawing a straight line through the data and moving each data point vertically to the line. The straight line may not be a "best fit" line in a statistical sense. One or two data points may be obviously in error and should be ignored in drawing the line.

EXAMPLE 14–2: *The K-factor data for methane through hexanes of line 2, 2900 psig, of Figure 14–8 are given below. Smooth the data.*

Component	K-factor
Methane	1.97
Ethane	0.956
Propane	0.545
i-Butane	0.355
n-Butane	0.376
i-Pentane	0.178
n-Pentane	0.292
Hexanes	0.121

Solution

First, calculate pK_j and $b_j \left[\dfrac{1}{T_{Bj}} - \dfrac{1}{T} \right]$

Component	pK_j, psia	b_j	T_{Bj}, °R	$b_j \left[\dfrac{1}{T_{Bj}} - \dfrac{1}{T} \right]$
C_1	5742	805	201.0	2.79
C_2	2786	1412	332.2	2.11
C_3	1589	1799	416.0	1.60
i-C_4	1035	2037	470.5	1.25
n-C_4	1096	2153	490.8	1.13
i-C_5	519	2368	541.8	0.79
n-C_5	851	2480	556.6	0.70
C_6	353	2780	615.4	0.31

Second, plot pK_j against $b_j \left[\dfrac{1}{T_{Bj}} - \dfrac{1}{T} \right]$, draw a best-fit line, and adjust data points to the line. See Figure 14–9.

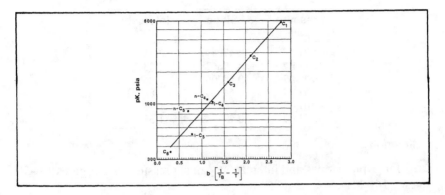

Fig. 14–9. pK vs. b(1/T_B – 1/T) (part of solution to Example 14–2).

Component	pK$_j$, adjusted psia	K$_j$, adjusted
C_1	5844	2.00
C_2	2797	0.960
C_3	1610	0.552
i-C_4	1102	0.378
n-C_4	968	0.332
i-C_5	669	0.230
n-C_5	607	0.208
C_6	398	0.137

The slopes and intercepts of the straight lines through K-factor data at several different pressures, such as lines 2 through 7 on Figure 14–8, can be correlated linearly against pressure.[4] This allows interpolation and some extrapolation of the data. For instance, the upper limit of the data is a horizontal line at pK$_j$ equal to the convergence pressure.

EXAMPLE 14–3: *The slopes and intercepts of straight lines through data sets 2 through 7 of Figure 14–8 are given below. Correlate the slopes and intercepts against pressure so interpolation and limited extrapolation of the data are possible.*

$$\log pK_j = (slope)b_j \left[\frac{1}{T_{Bj}} - \frac{1}{T} \right] + (intercept) \quad (14\text{–}5)$$

finding Convergence Pressure

Pressure, psia	Slope	Intercept
2914.7	0.470497	2.45399
2514.7	0.506678	2.33622
2014.7	0.602067	2.06883
1514.7	0.647047	1.88861
1014.7	0.707713	1.68195
514.7	0.792640	1.38290

Solution

First, plot slopes and intercepts against pressure. See Figures 14–10 and 14–11.

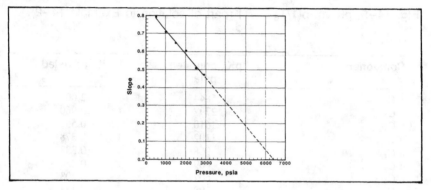

Fig. 14–10. Slopes of pK plots vs. pressure (part of solution to Example 14–3).

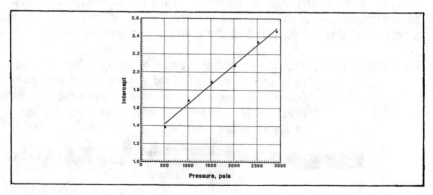

Fig. 14–11. Intercepts of pK plots vs. pressure (part of solution to Example 14–3).

Second, determine equations for the lines

$$\text{slope} = -1.330(10^{-4})p + 0.8536 \qquad (14-6)$$

$$\text{intercept} = 4.418(10^{-4})p + 1.196 \qquad (14-7)$$

Third, remember that at convergence pressure all components have $K = 1.0$. Thus the slope of the pK_j line is zero. Extrapolation of Figure 14–10 to slope = 0 gives a convergence pressure of 6400 psia for this mixture.

The form of the plotting factor, Equation 14–3, indicates the possibility of extrapolation to temperatures other than the temperature at which the data were obtained. Unfortunately, this has not been verified; in fact, limited investigation indicates that extrapolation by 50°F from the experimental temperature could result in errors of as much as 20 percent in individual K-factors.[6]

Exercises

14–1. Gas liquid equilibria calculations at 1500 psia and 160°F have been completed for a black oil. Resulting compositions are given below. A convergence pressure of 5000 psia was used to obtain K-factors. What convergence pressure should have been used for this mixture at 160°F?

Component	Composition of equilibrium liquid, mole fraction	Composition of equilibrium gas, mole fraction
Methane	0.2883	0.8214
Ethane	0.1052	0.1094
Propane	0.0745	0.0403
i-Butane	0.0159	0.0057
n-Butane	0.0439	0.0126
i-Pentane	0.0164	0.0028
n-Pentane	0.0161	0.0023
Hexanes	0.0501	0.0038
Heptanes plus	0.3896	0.0017
	1.0000	1.0000

Properties of heptanes plus		
Specific gravity		0.8515
Molecular weight		218 lb/lb mole

14–2. Gas-liquid equilibrium for a volatile oil has been calculated for a pressure of 1500 psia and a temperature of 150°F. Results are given below. A convergence pressure of 5000 psia was used to obtain K-factors. What convergence pressure should have been used?

Component	Composition of equilibrium liquid, mole fraction	Composition of equilibrium gas, mole fraction
Methane	0.3062	0.8620
Ethane	0.0753	0.0749
Propane	0.0713	0.0370
i-Butane	0.0135	0.0045
n-Butane	0.0416	0.0112
i-Pentane	0.0161	0.0026
n-Pentane	0.0272	0.0036
Hexanes	0.0380	0.0026
Heptanes plus	0.4108	0.0016
	1.0000	1.0000
Properties of heptanes plus		
Specific gravity		0.840
Molecular weight		207 lb/lb mole

14–3. Gas-liquid equilibrium has been calculated for a retrograde gas at 1500 psia and 125°F. A convergence pressure of 5000 psia was used to obtain K-factors. Results are given below. What value of convergence pressure should have been used for this mixture at 125°F?

Component	Composition of equilibrium liquid, mole fraction	Composition of equilibrium gas, mole fraction
Methane	0.3174	0.8681
Ethane	0.0783	0.0708
Propane	0.0669	0.0304
i-Butane	0.0390	0.0110
n-Butane	0.0361	0.0083
i-Pentane	0.0334	0.0044
n-Pentane	0.0261	0.0033
Hexanes	0.0477	0.0027
Heptanes plus	0.3551	0.0010
	1.0000	1.0000
Properties of heptanes plus		
Specific gravity		0.793
Molecular weight		146 lb/lb mole

2 Stage Surface Separation

14-4. Constant volume depletion of a retrograde gas produced the following K-factors at 256°F. $n_{g1}, n_{L1}, x_{j1}, y_{j1}$, Ch. 12,14

Component			Pressure, psig			
	5000	4000	3000	2100	1200	700
Methane	1.527	1.620	2.155	2.917	4.762	6.265
Ethane	1.068	1.067	1.128	1.241	1.703	2.390
Propane	0.826	0.859	0.838	0.822	0.958	1.229
i-Butane	0.613	0.633	0.569	0.541	0.612	0.800
n-Butane	0.571	0.589	0.517	0.483	0.524	0.651
i-Pentane	0.470	0.461	0.399	0.361	0.358	0.407
n-Pentane	0.379	0.432	0.370	0.322	0.319	0.368
Hexanes	0.299	0.299	0.232	0.190	0.172	0.192
Heptanes plus	0.357	0.235	0.104	0.059	0.040	0.038

$n_{g8}, n_{L8}, \ldots$

Smooth the 5000 psig data.

14-5. Smooth the 4000 psig data of Exercise 14-4.

P_{sto}, M_{sto} Ch. 11

14-6. Correlate the slopes and intercepts of the pK_j versus

Calculate ρ_{ob}
Pg. 378

$$b_j \left[\frac{1}{T_{Bj}} - \frac{1}{T} \right]$$

lines of the data of Exercise 14-4.

γ_{sp}, γ_{st} R_{st}

R_{sp} - Pg.377

References

1. Edmister, W.C. and Lee, B.I.: *Applied Hydrocarbon Thermodynamics*, 2nd ed., Gulf Publishing Co., Houston (1984).
2. *Engineering Data Book*, 10th ed., Gas Processors Suppliers Association, Tulsa (1987).
3. Bruno, J.A., Yanosik, J.L., and Tierney, J.W.: "Distillation Calculations with Nonideal Mixtures," *Extractive and Aziotropic Distillation*, Advances in Chemistry Series *115*, ACS, Washington, DC (1972).
4. Wattenbarger, R.C.: *An Investigation of Convergence Pressure Methods,* Master of Science Thesis, Texas A&M University (Dec. 1986).
5. Standing, M.B.: *Volumetric and Phase Behavior of Oil Field Hydrocarbon Systems*, Reinhold Publishing Corp., New York (1952).
6. Hoffmann, A.E., Crump, J.S., and Hocott, C.R.: "Equilibrium Constants for A Gas-Condensate System," *Trans.*, AIME, (1953) *198*, 1–10.

15

Gas-Liquid Equilibria Calculations With Equations of State

Equations of state can be used to calculate gas-liquid equilibria as an alternative to using K-factor correlations. The assumption must be made that the equations of state presented in Chapters 3 and 4 predict pressure-volume-temperature relationships for liquids as well as for gases.

This chapter is an *introduction* to utilizing equations of state in gas-liquid equilibria calculations. It conveys a general understanding of the subject and is as simple and short as possible. Many of the complexities of this subject are not discussed. Thus, the study of this chapter will not result in the ability to apply equations of state to complex petroleum mixtures.

Pure Substances

An equation of state in the spirit of van der Waals will produce isotherms as shown in Figure 15–1 for a pure substance.[1] Notice that the isotherms at and above the critical temperature look very much like the corresponding experimental isotherms of Figure 2–10.

The calculated isotherm of Figure 15–1 for temperature below critical temperature exhibits the "van der Waals loop." At certain temperatures, the calculated pressures in the loop are negative, as shown. This loop does not appear experimentally. For a pure substance a horizontal line connects equilibrium gas and liquid. Again see Figure 2–10.

The van der Waals loop is used to determine the molar volumes of the equilibrium gas and liquid and then is replaced by a tie-line between these two volumes. The basis of this calculation is that the *chemical potential* of the equilibrium gas equals the *chemical potential* of the equilibrium liquid.

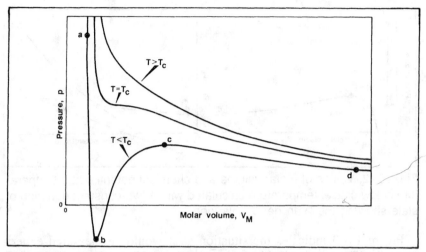

Fig. 15–1. Values of molar volume calculated with a two-constant equation of state.

Chemical Potential

The *chemical potential*, G, of a pure fluid at constant temperature may be calculated as[2]

$$dG = V_M dp . \qquad (15-1)$$

Chemical potential also is called *Gibbs molar free energy*.

Figure 15–2A gives a calculated isotherm below the critical temperature. Point a is selected arbitrarily along the liquid part of the isotherm. Point d is selected arbitrarily along the gas part of the isotherm. Points b and c are minimum and maximum points on the van der Waals loop. Points e represent the points along the loop for which the chemical potentials are equal. Point f is the point along line $\overline{bc}$ which has the same pressure as points e.

Figure 15–2B gives the corresponding chemical potentials calculated as in Equation 15–1. A loop also appears on this figure. The loop is nonexistent physically but can be used analytically. The point of intersection, e, meets the requirements of equilibrium for the gas and liquid of a pure substance. At point e, the pressure of the gas equals the pressure of the liquid, and the chemical potentials of the two phases are equal. Point f has the same pressure as points e but is not an equilibrium point because its chemical potential is higher than that of points e.

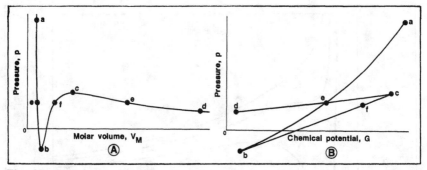

Fig. 15–2. Values of molar volume and chemical potential at a temperature below critical temperature calculated with a two-constant equation of state showing the "van der Waals loop."

Figure 15–3 indicates the situation as it would be measured experimentally. The nonphysical loop $\overline{ebfce}$ in Figure 15–2A has been replaced with the tie-line $\overline{ee}$. Figure 15–3B shows the elimination of the corresponding loop $\overline{ebfce}$. If an equation of state could be devised that would exactly reproduce isotherm $\overline{aeed}$ on Figure 15–3A, the calculated chemical potentials of all states along tie-line $\overline{ee}$ would be identical.

The point e to the left on Figure 15–3A is the position of the bubble point at the temperature of the isotherm, and the point e to the right is the position of the dew point. If the above analysis were performed at various temperatures below the critical temperature, the phase envelope would be defined. Figure 15–4 shows the position of the phase envelope along with three isotherms.

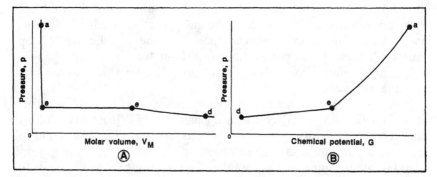

Fig. 15–3. Values of molar volume and chemical potential at a temperature below critical temperature calculated with a two-constant equation of state with the "van der Waals loop" removed.

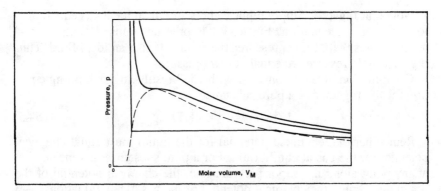

Fig. 15–4. Values of molar volume calculated with two-constant equation of state with "van der Waals loop" removed from the isotherm below the critical temperature.

The calculation of the properties of equilibrium gas and liquid resolves into the calculation of the chemical potentials of the two phases.

An equation of state is required to evaluate Equation 15-1. For an ideal gas

$$V_M = \frac{RT}{p} \tag{3-13}$$

Thus for a pure ideal gas

$$dG = (RT/p)dp = RT\, d(\ln p). \tag{15-2}$$

The Fugacity

The chemical potential of a real fluid can be expressed by replacing pressure in Equation 15–2 with a property called *fugacity,* f.

$$dG = RT\, d(\ln f) \tag{15-3}$$

Since Equation 15–3 defines fugacity in differential form, a reference value is required.

$$\lim_{p \to 0} f = p \tag{15-4}$$

Note that fugacity simply replaces pressure in an ideal gas equation to form a real gas equation. Fugacity has pressure units. Equation 15–4 merely states that at low pressures the fluid acts like an ideal fluid. Thus fugacity and pressure are equal at low pressures.

Combination of Equation 15–1 with 15–3 results in the defining equation for the fugacity of a pure substance.

$$d \, (\ln f) = (V_M/RT) \, dp \qquad (15-5)$$

Remember that chemical potential for the liquid must equal chemical potential for the gas at equilibrium. For a pure substance this means that at any point along the vapor pressure line, the chemical potential of the liquid must equal the chemical potential of the gas. Thus Equation 15–3 shows that the fugacity of the liquid must equal the fugacity of the gas at equilibrium on the vapor pressure line. So gas-liquid equilibria can be calculated under the condition that

$$f_g = f_L . \qquad (15-6)$$

Fugacity Coefficient

The following expression can be derived from Equation 15–5 under the constraint of Equation 15–4.[2]

$$\ln \frac{f}{p} = z - 1 - \ln z + \frac{1}{RT} \int_{\infty}^{V_M} \left(\frac{RT}{V_M} - p \right) dV_M \qquad (15-7)$$

where z is the z-factor as defined by Equations 3–39 and 3–40.

For a pure substance, the ratio of fugacity to pressure, f/p, is called *fugacity coefficient*.

Example of State Calculation for a Pure Substance

The Peng-Robinson equation of state, Equation 4–35, will be used with Equation 15–7 to develop a procedure for calculating of the vapor pressure of a pure substance.[3] The vapor pressure is simply the pressure, points e on Figure 15–2, for which the fugacity of the liquid equals the fugacity of the gas.

The application here of the Peng-Robinson equation of state does not mean that it is the best equation of state. This equation was merely selected to illustrate the application of a typical two-constant, cubic equation of state to gas-liquid equilibria calculations.

The Peng-Robinson equation of state

$$\left[p + \frac{a_T}{V_M (V_M+b) + b(V_M-b)} \right] (V_M-b) = RT \tag{4-35}$$

can be arranged into cubic form,

$$V_M^3 - \left(\frac{RT}{p} - b \right) V_M^2 + \left(\frac{a_T}{p} - \frac{2bRT}{p} - 3b^2 \right) V_M$$

$$- b \left(\frac{a_T}{p} - \frac{bRT}{p} - b^2 \right) = 0 . \tag{15-8}$$

Application of Equations 4–8 at the critical point results in

$$b = 0.07780 \frac{RT_c}{p_c} \tag{15-9}$$

and

$$a_c = 0.45724 \frac{R^2 T_c^2}{p_c} . \tag{15-10}$$

The term b is a constant. The term a_T varies with temperature; a_c is its value at the critical temperature. The temperature variation of term a_T resides in α, that is

$$a_T = a_c \alpha , \tag{15-11}$$

where α is determined as

$$\alpha^{1/2} = 1 + (0.37464 + 1.54226\omega - 0.26992\omega^2)(1 - T_r^{1/2}) . \tag{15-12}$$

The acentric factor, ω, is a constant. Each pure substance has a different value of acentric factor.

The coefficients in Equations 15–9 and 15–10 often are assigned symbols as follows.

$$\Omega_b = 0.07780 \quad \text{and} \quad \Omega_a = 0.45724 \tag{15-13}$$

Substitution of

$$V_M = zRT/p \qquad (3-39)$$

into the Peng-Robinson equation of state gives

$$z^3 - (1 - B)z^2 + (A - 2B - 3B^2)z$$

$$- (AB - B^2 - B^3) = 0, \qquad (15-14)$$

where

$$A = \frac{a_T p}{R^2 T^2} \qquad (15-15)$$

and

$$B = \frac{bp}{RT}. \qquad (15-16)$$

Equation 15–14 is a cubic equation with real coefficients. Thus, three values of z-factor cause the equation to equal zero. These three "roots" are all real when pressure and temperature are on the vapor pressure line—that is, when liquid and gas are present. One real root and two complex roots exist when the temperature is above the critical temperature.

Figure 15–5 gives the shape of an isotherm calculated with Equation 15–14 at a temperature below the critical temperature. Points a through f are equivalent to points a through f on Figures 15–2 and 15–3. Points e are the values of z-factor that would be measured experimentally. Point f is a nonphysical solution.

As before, curve $\overline{ebfce}$ is eliminated. See Figure 15–6. The upper point e is the z-factor of the equilibrium gas, and the lower point e is the z-factor of the equilibrium liquid. The dotted line connecting these two points has no physical meaning. The dashed curve represents the complete phase envelope. Notice the similarity of the isotherm on Figure 15–6 to the experimental 104°F isotherm of Figure 3–4.

Thus, Equation 15–14 is solved for its three roots. If there is only one real root, temperature is above the critical temperature. If there are three real roots, the largest is the z-factor of the equilibrium gas and the

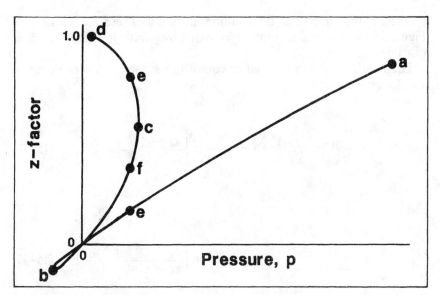

Fig. 15–5. Values of z-factor at a temperature below critical temperature calculated with a two-constant equation of state showing "van der Waals loop."

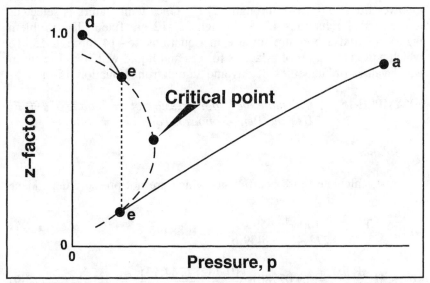

Fig. 15–6. Values of z-factor at a temperature below critical temperature calculated with a two-constant equation of state with "van der Waals loop" removed.

smallest is the z-factor of the equilibrium liquid. These are points e on Figure 15–5. The middle root, represented by point f in Figure 15–5, is discarded.

Equation 15–14 can be used to complete the integration of Equation 15–7, resulting in[2]

$$\ln\left(\frac{f_g}{p}\right) \doteq z_g - 1 - \ln(z_g - B) - \frac{A}{2^{1.5}B} \ln\left[\frac{z_g + (2^{1/2} + 1)B}{z_g - (2^{1/2} - 1)B}\right] \qquad (15\text{--}17)$$

and

$$\ln\left(\frac{f_L}{p}\right) = z_L - 1 - \ln(z_L - B) - \frac{A}{2^{1.5}B} \ln\left[\frac{z_L + (2^{1/2} + 1)B}{z_L - (2^{1/2} - 1)B}\right]. \qquad (15\text{--}17)$$

Equation 15–17 is applied twice: once with the liquid z-factor to calculate the fugacity of the liquid and again with the gas z-factor to calculate the fugacity of the gas.

The procedure to calculate the vapor pressure of a pure substance involves Equations 15–9 through 15–17. Once temperature is selected, the results of Equations 15–9 through 15–12 are fixed. The problem then is to find a pressure for use in Equations 15–14 through 15–16 which will give values of z-factors for gas and liquid which will result in equal values of fugacities of gas and liquid from Equation 15–17.

EXAMPLE 15–1: *Calculate the vapor pressure of iso-butane at 190°F. Use the Peng-Robinson equation of state.*

Solution

First, calculate those coefficients which are not pressure dependent

$$T_r = \frac{T}{T_c} = \frac{190 + 459.67}{274.46 + 459.67} = 0.88495 \qquad (3\text{--}41)$$

$$b = \Omega_b \frac{RT_c}{p_c} = (0.07780)\frac{(10.732)\,(734.13)}{(527.9)} = 1.1611 \qquad (15\text{--}9)$$

$$a_c = \Omega_a \frac{R^2 T^2_c}{p_c} = (0.45724) \frac{(10.732)^2 (734.13)^2}{(527.9)} = 53,765 \quad (15\text{--}10)$$

$$\alpha^{1/2} = 1 + (0.37464 + 1.54226\omega - 0.26992\omega^2) (1 - T_r^{1/2}) \quad (15\text{--}12)$$

$$\alpha = 1.0787 \text{ where } \omega = 0.1852$$

$$a_T = a_c\alpha = (53,765)(1.0787) = 57,995 \quad (15\text{--}11)$$

Second, by trial and error, find a pressure which causes the fugacity of the liquid calculated with Equations 15–14, 15–15, 15–16, and 15–17 to equal the fugacity of the gas calculated with the same equations. Only the last calculation with a final trial value of p = 228.79 psia will be shown.

$$A = \frac{a_T p}{(RT)^2} = \frac{(57995)\,(228.79)}{(10.732)^2\,(649.67)^2} = 0.27295 \quad (15\text{--}15)$$

$$B = \frac{bp}{RT} = \frac{(1.1611)\,(228.79)}{(10.732)(649.67)} = 0.038101 \quad (15\text{--}16)$$

The three roots which solve

$$z^3 - (1 - B)z^2 + (A - 2B - 3B^2)z$$
$$- (AB - B^2 - B^3) = 0 \quad (15\text{--}14)$$

are 0.067258, 0.18678, and 0.70786.

$$z_L = 0.067258 \text{ and } z_g = 0.70786$$

Then

$$\ln\left(\frac{f_g}{p}\right) = z_g - 1 - \ln(z_g - B) - \frac{A}{2^{1.5}B} \ln\left(\frac{z_g + (2^{1/2} + 1)B}{z_g - (2^{1/2} - 1)B}\right) \quad (15\text{--}17)$$

$$f_g = 176.79 \text{ psia}$$

and

$$\ln\left(\frac{f_L}{p}\right) = z_L - 1 - \ln(z_L - B) - \frac{A}{2^{1.5}B} \ln\left(\frac{z_L + (2^{1/2} + 1)B}{z_L - (2^{1/2} - 1)B}\right) \quad (15\text{--}17)$$

$$f_L = 176.79 \text{ psia}$$

$f_L = f_g$, thus the trial value of p, 228.79 psia, is the vapor pressure of iso-butane at 190°F.

Notice that the molar volumes can be calculated easily.

$$V_{Mg} = \frac{z_g RT}{p} = \frac{(0.70786)(10.732)(649.67)}{(228.79)}$$

$$= 21.57 \text{ cu ft/lb mole} \quad (3\text{--}39)$$

$$V_{ML} = \frac{z_L RT}{p} = \frac{(0.067258)(10.732)(649.67)}{(228.79)}$$

$$= 2.050 \text{ cu ft/lb mole} \quad (3\text{--}39)$$

Compare results with experimental data.[5]

	Calculated	Experimental
Vapor pressure, psia	228.8	228.3
Saturated liquid molar volume, cu ft/lb mole	2.050	2.035
Saturated vapor Molar volume, cu ft/lb mole	21.57	21.68
Liquid z-factor	0.0673	0.0666
Gas z-factor	0.7078	0.7101

The results do not always agree this closely with experimental observations.

The trial-and-error process illustrated in Example 15–1 is rather tedious. Several methods have been proposed to speed convergence to the correct solution. These methods can be grouped into two general types: successive substitutions and Newton-Raphson. These techniques will not be discussed in this text.

Mixtures

The situation with regard to mixtures is somewhat more difficult to visualize. However, equilibrium is attained when the chemical potential of each component in the liquid equals the chemical potential of that component in the gas.

Chemical Potential

The chemical potential of a *component* of a *mixture* may be calculated as

$$dG_j = RT \, d(\ln f_j) . \tag{15–18}$$

The reference value for fugacity in this equation is

$$\lim_{p \to 0} f_j = y_j p = p_j . \tag{15–19}$$

That is, as pressure approaches zero the fluid approaches ideal behavior and the fugacity of a component approaches the partial pressure of that component.

Remember that the *chemical potential*, G_j, for a component of a mixture at equilibrium must be the same in both the gas and the liquid. Thus Equation 15–18 shows that at equilibrium the *fugacities* of a component must be equal in both the gas and the liquid. So gas-liquid equilibria can be calculated under the condition that

$$f_{gj} = f_{Lj} \tag{15–20}$$

for all components j. This is analogous to the development of the equations for ideal solutions, early in Chapter 12, in which the partial pressure of the liquid (Dalton's equation) was set equal to the partial pressure of the gas (Raoult's equation).

Values of the fugacity for each component are calculated with an equation of state. Any equation of state can be used for these calculations. Later, as an example of the procedure, we develop equations using the Peng-Robinson equation of state.

Fugacity Coefficient

Another useful term is *fugacity coefficient*. Fugacity coefficient for each component of a mixture is defined as the ratio of fugacity to partial pressure.

$$\phi_j = \frac{f_j}{y_j p} \qquad (15\text{--}21)$$

Fugacity coefficient may be calculated as[2]

$$\ln \phi_j = \frac{1}{RT} \int_{\infty}^{V} \left[\frac{RT}{V} - \left(\frac{\partial p}{\partial n_j} \right)_{T,V,n_i} \right] dV - \ln z, \quad (15\text{--}22)$$

where z is z-factor as defined in Chapter 3.

Further, the ratio of the fugacity coefficients can be used to calculate K-factor.[2]

$$K_j = \frac{\phi_{Lj}}{\phi_{gj}} = \frac{\dfrac{f_{Lj}}{x_j p}}{\dfrac{f_{gj}}{y_j p}} = \frac{y_j}{x_j}, \qquad (15\text{--}23)$$

where $f_{Lj} = f_{gj}$ at equilibrium.

Example of State Calculation for Mixtures

The Peng-Robinson equation of state is

$$p = \frac{RT}{V_M - b} - \frac{a_T}{V_M(V_M + b) + b(V_M - b)}. \qquad (4\text{--}35)$$

Mixture rules are

$$b = \sum_j y_j b_j \tag{4-38}$$

and

$$a_T = \sum_i \sum_j y_i y_j (a_{Ti} a_{Tj})^{\frac{1}{2}} (1 - \delta_{ij}), \tag{4-40}$$

where subscripts j and i refer to components. Also

$$\delta_{ii} = \delta_{jj} = 0 \tag{15-24}$$

and

$$\delta_{ij} = \delta_{ji} \tag{15-25}$$

Values of the coefficients for the individual components are calculated as

$$b_j = 0.07780 \frac{RT_{cj}}{p_{cj}} \tag{15-9}$$

and

$$a_{Tj} = a_{cj} \alpha_j , \tag{15-11}$$

where

$$a_{cj} = 0.45724 \frac{R^2 T_{cj}^2}{p_{cj}} . \tag{15-10}$$

and

$$\alpha_j^{\frac{1}{2}} = 1 + (0.37464 + 1.54226\omega_j - 0.26992\omega_j^2)(1 - T_{rj}^{\frac{1}{2}}) . \tag{15-12}$$

As before, the Peng-Robinson equation can be written as

$$z^3 - (1 - B)z^2 + (A - 2B - 3B^2)z$$
$$- (AB - B^2 - B^3) = 0, \qquad (15-14)$$

where

$$A = \frac{a_T p}{R^2 T^2} \qquad (15-15)$$

and

$$B = \frac{bp}{RT} . \qquad (15-16)$$

Equation 15–14 is cubic in z-factor. Thus, three values of z-factor cause the equation to equal zero. These three "roots" are all real when pressure and temperature are such that the mixture is two phase. There will be one real root and two complex roots when the mixture is single phase.

When three roots are obtained, the lowest root is the z-factor of the liquid. The highest root is the z-factor of the gas, and the middle root is discarded. This is analogous to eliminating point f on Figure 15–5.

Combining the Peng-Robinson equation of state and Equation 15–22 results in an equation for fugacity coefficient of each component.[3]

$$\ln \phi_j = -\ln (z - B) + (z - 1)B'_j - \frac{A}{2^{1.5}B} (A'_j - B'_j)$$

$$\ln \left[\frac{z + (2^{1/2} + 1)B}{z - (2^{1/2} - 1)B} \right], \qquad (15-26)$$

where

$$B'_j = \frac{b_j}{b} \qquad (15-27)$$

and

$$A'_j = \frac{1}{a_T} \left[2a_{Tj}^{1/2} \sum_i y_i a_{Ti}^{1/2} (1 - \delta_{ij}) \right]. \qquad (15\text{-}28)$$

Equations 15–26 through 15–28 are written twice: once for values of z, b, and a_T of the gas and again for values of z, b, and a_T of the liquid.

The procedure for calculating gas-liquid equilibria at given temperature and pressure is as follows.

Values of a_{Tj} and b_j for each component of the mixture are obtained with Equations 15–9 through 15–12 from a knowledge of the critical properties and acentric factors of the pure components.

A first trial set of K-factors is obtained. For instance, the K-factor equation given in Appendix B can be applied to get the first trial set of K-factors. These are used in a gas-liquid equilibria calculation, as described in Chapter 12, to determine the compositions of the gas and liquid. The remaining equations are solved twice, once for the liquid and once for the gas.

Values of a_T and b are calculated from Equations 4–38 and 4–40 with the compositions determined above. When the composition of the liquid is used, the values are a_{TL} and b_L. When the composition of the gas is used, the values are a_{Tg} and b_g. Values of binary interaction coefficients, δ_{ij}, can be included in Equation 4–40 if they are known. If unknown, the values of δ_{ij} can be set equal to zero. Values of A and B for the liquid, A_L and B_L, are calculated with Equations 15–15 and 15–16 using a_{TL} and b_L. When a_{Tg} and b_g are used in Equations 15–15 and 15–16, A_g and B_g result.

Also, B'_j and A'_j must be calculated for each component j. B'_{Lj} results when b_L is used in Equation 15–27, and B'_{gj} results when b_g is used. A'_{Lj} and A'_{gj} result similarly from Equation 15–28.

The smallest root of Equation 15–14 is z_L when A_L and B_L are used. The largest root of Equation 15–14 is z_g when A_g and B_g are used.

Equation 15–26 is solved for the fugacity coefficients of the components of the liquid, ϕ_{Lj}, using values of A_L, B_L, z_L, A'_{Lj}, and B'_{Lj}. Values of ϕ_{gj} result when the corresponding gas coefficients and z-factor are used in Equation 15–26.

Then values of liquid fugacity and gas fugacity for each component are obtained from

$$f_{Lj} = x_j p \phi_{Lj} \qquad (15\text{-}21)$$

and

$$f_{gj} = y_j p \phi_{gj} \qquad (15\text{--}21)$$

Equilibrium is obtained and the calculation is complete when all

$$f_{gj} = f_{Lj} \qquad (15\text{--}20)$$

There are as many Equations 15–20 as there are components. All these equations cannot be satisfied simultaneously. Thus some sort of error function based on Equation 15–20 must be devised. One approach is[4]

$$\epsilon_j = f_{Lj} - f_{gj} , \qquad (15\text{--}29)$$

where a solution is obtained when the Euclidean norm of the ϵ_j

$$\sqrt{\sum_j \epsilon_j^2} \qquad (15\text{--}30)$$

is less than some selected tolerance.

Another error function used in converging on a correct solution by a method of successive substitution involves K-factors. The K_j for the mixture are determined from the fugacity coefficients with Equation 15–23. Then

$$\epsilon_j = \frac{(K_j^T - K_j^C)^2}{K_j^T K_j^C} , \qquad (15\text{--}31)$$

where K_j^C are the K-factors just calculated and the K_j^T are the trial values of K-factors.

Convergence on a correct solution is obtained when the sum of the error functions is less than some selected tolerance. If the sum of the error functions is greater than the tolerance, the K_j^C are used as new trial values of K_j, and the process is repeated.

EXAMPLE 15–2: *Calculate the compositions of the gas and liquid when the mixture with composition given below is brought to equilibrium at 160°F and 1000 psia. Use the Peng-Robinson equation of state. Use binary interaction coefficients of 0.02 for methane-n-butane, 0.04 for methane-n-decane, and 0.0 for n-butane-n-decane.*

Component	Composition, mole fraction
Methane	0.5301
n–Butane	0.1055
n–Decane	0.3644
	1.0000

Solution

First, calculate the coefficients of the components of the mixture.

$$\alpha_j^{\frac{1}{2}} = 1 + (0.37464 + 1.54226\omega_j - 0.26992\omega_j^2)\,(1 - T_{rj}^{\frac{1}{2}}) \qquad (15\text{–}12)$$

$$a_{cj} = 0.45724\,\frac{R^2 T_{cj}^2}{P_{cj}} \qquad (15\text{–}10)$$

$$a_{Tj} = a_{cj}\alpha_j \qquad (15\text{–}11)$$

$$b_j = 0.07780\,\frac{RT_{cj}}{P_{cj}} \qquad (15\text{–}9)$$

Component	T_{cj} °R	p_{cj} psia	ω_{cj}	α_j	a_{cj}	a_{Tj}	b_j
C_1	343.0	666.4	0.0104	0.7481	9,297	6,956	0.4297
$n-C_4$	765.3	550.6	0.1995	1.1394	56,017	63,827	1.1604
$n-C_{10}$	1111.7	305.2	0.4898	1.6139	213,240	344,149	3.0411

Second, select trial values of K-factors and calculate trial compositions of equilibrium gas and liquid. Only the final trial, with K-factors as given below, is shown.

$$\sum_j x_j = \sum_j \frac{z_j}{1 + \bar{n}_g (K_j - 1)} \qquad (12-17)$$

This calculation requires trial and error; only the final trial with $\bar{n}_g = 0.4015$ is shown.

Component	K_j	z_j	x_j	y_j
C_1	3.992	0.5301	0.2408	0.9613
$n-C_4$	0.2413	0.1055	0.1517	0.0366
$n-C_{10}$	0.00340	0.3644	0.6075	0.0021
		1.0000	1.0000	1.0000

Third, calculate the composition dependent coefficients necessary for z-factor calculations for both liquid and gas.

$$a_T = \sum_i \sum_j y_i y_j (a_{Ti} a_{Tj})^{1/2} (1 - \delta_{ij}) \qquad (4-40)$$

$$b = \sum_j y_j b_j \qquad (4-38)$$

$$A = \frac{a_T p}{R^2 T^2} \qquad (15\text{--}15)$$

$$B = \frac{bp}{RT} \qquad (15\text{--}16)$$

Phase	a_T	b	A	B
Liquid	171,446	2.1270	3.8766	0.31983
Gas	8,177	0.4619	0.1849	0.06945

Fourth, calculate z-factors of liquid and gas.

$$z^3 - (1 - B)z^2 + (A - 2B - 3B^2)z$$
$$- (AB - B^2 - B^3) = 0, \qquad (15\text{--}14)$$

$$z_L = 0.3922 \text{ and } z_g = 0.9051$$

Fifth, calculate the composition dependent coefficients necessary for calculating fugacity coefficients for both liquid and gas.

$$A'_j = \frac{1}{a_T} \left[2a_{Tj}^{1/2} \sum_i y_i a_{Ti}^{1/2} (1 - \delta_{ij}) \right] \qquad (15\text{--}28)$$

$$B'_j = \frac{b_j}{b} \qquad (15\text{--}27)$$

Component	Liquid		Gas	
	A'_j	B'_j	A'_j	B'_j
C_1	0.38893	0.20204	1.8440	0.93042
$n-C_4$	1.22127	0.54559	5.5014	2.51253
$n-C_{10}$	2.83309	1.42979	12.5445	6.58435

Sixth, calculate the fugacity coefficients of the components of liquid and gas.

$$\ln \phi_j = -\ln (z - B) + (z - 1)B_j' - \frac{A}{2^{1.5}B} (A'_j - B'_j)$$

$$\ln \left[\frac{z + (2^{1/2} + 1)B}{z - (2^{1/2} - 1)B} \right], \qquad (15-26)$$

Component	ϕ_{Lj}	ϕ_{gj}
C_1	3.67552	0.92065
$n-C_4$	0.12878	0.53373
$n-C_{10}$	0.000699	0.20600

Seventh, calculate the K-factors of the components and the error functions.

$$K_j = \frac{\phi_{Lj}}{\phi_{gj}} \qquad (15-23)$$

$$\epsilon_j = \frac{(K_j^T - K_j^C)^2}{K_j^T K_j^C}, \qquad (15-31)$$

Component	K_j^C	ϵ_j
C_1	3.992	0.000
$n-C_4$	0.2413	0.000
$n-C_{10}$	0.00340	0.000
		$\sum_i \epsilon_j = 0.000$

The sum of the error functions is less than a tolerance of 0.001, so the set of trial values of K-factors was correct and the calculated values of liquid and gas compositions are correct.

Compare results with experimental data.[5]

	Calculated		Experimental	
Component	x_j	y_j	x_j	y_j
C_1	0.2408	0.9613	0.242	0.963
$n-C_4$	0.1517	0.0366	0.152	0.036
$n-C_{10}$	0.6075	0.0021	0.606	0.0021
	1.0000	1.0000	1.000	1.0011

Also, the molar volumes and densities can be calculated.

$$V_M = \frac{zRT}{p} \tag{3-39}$$

$$\rho = \frac{pM}{zRT} \tag{3-39}$$

Phase	z	V_M cu ft/lb mole	M_a lb/lb mole	ρ lb/cu ft
Liquid	0.3922	2.61	99.12	38.00
Gas	0.9051	6.02	17.84	2.96

Much more sophisticated and powerful methods of converging on a solution are available. They will not be discussed here.

These equations can be used also to calculate the bubble points and dew points of mixtures. The solution techniques in these applications differ from those used in Example 15–2.

Exercises

15–1. Use the Peng-Robinson equation of state to calculate the vapor pressure of ethane at 32°F. Also, calculate the densities of the liquid and gas at 32°F. Compare your answers with values from Figures 2–7, 2–12, and 3–3.

15–2. Use the Peng-Robinson equation of state to calculate the vapor pressure of propane at 104°F. Also, calculate the densities of the liquid and gas at 104°F. Compare your answers with values from Figures 2–7, 2–12, and 3–4.

15–3. Use the Peng-Robinson equation of state to calculate the compositions and densities of the equilibrium liquid and gas of the mixture given below at 160°F and 2000 psia. Use binary interaction coefficients of 0.02 for methane-n-butane, 0.035 for methane-n-decane, and 0.0 for n-butane-n-decane.

Component	Composition, mole fraction
Methane	0.5532
n–Butane	0.3630
n–Decane	0.0838
	1.0000

Compare your answer with experimental results shown below.[5]

Component	Composition, mole fraction	
	liquid	gas
Methane	0.485	0.826
n–Butane	0.412	0.167
n–Decane	0.103	0.0063
	1.000	0.9993

15–4. Use the Peng-Robinson equation of state to calculate the compositions, densities, and quantities (lb moles) of the equilibrium liquid and gas of the mixture given below at 160°F and 500 psia. Use binary interaction coefficients of 0.0 for methane-propane, 0.02 for methane-n-pentane, and 0.01 for propane-n-pentane.

Component	Composition, mole fraction
Methane	0.500
Propane	0.150
n–Pentane	0.350
	1.000

Compare your answer with the results of Example 2–8.

References

1. Fussell, L.T.: "A Technique for Calculating Multiphase Equilibria," *Soc. Pet. Eng. J.* (Aug. 1979) *19* 203–210.
2. Edmister, W.C. and Lee, B.I.: *Applied Hydrocarbon Thermodynamics*, 2nd ed., Gulf Publishing Co., Houston (1984).
3. Peng, D.Y. and Robinson, D.B.: "A New Two-Constant Equation of State," *I.&E.C. Fundamentals* (1976) *15*, No. 1, 59–64.
4. Fussell, D.D. and Yanosik, J.L.: "An Iterative Sequence for Phase Equilibria Calculations Incorporating the Redlich-Kwong Equation of State," *Soc. Pet. Eng. J.* (June 1978) *18*, 173–182.
5. Sage, B.H. and Lacey, W.N.: "Thermodynamic Properties of the Lighter Paraffin Hydrocarbons and Nitrogen," *Monograph on API Research Project 37*, API, New York (1950).

General References

Phase Behavior, SPE Reprint Series *15*, SPE, Dallas (1981).

Chao, K.C. and Robinson, R.L., Jr. (eds.): *Equations of State in Engineering and Research*, Advances in Chemistry Series *182*, ACS, Washington (1979).

Properties of Oilfield Waters

16

Water invariably occurs with petroleum deposits. Thus, a knowledge of the properties of this *connate*, or *interstitial*, or *formation* water is important to petroleum engineers. In this chapter, we examine the composition of oilfield water; water density, compressibility, formation volume factor and viscosity; solubility of hydrocarbons in water and solubility of water in both liquid and gaseous hydrocarbons; and, finally, water-hydrocarbon interfacial tension. An unusual process called *hydrate formation* in which water and natural gas combine to form a solid at temperatures above the freezing point of water is discussed in Chapter 17.

The properties of oilfield waters have not been studied as carefully and systematically as the properties of other fluids of interest to petroleum engineers. Therefore, many of the correlations discussed in this chapter are based on limited data.

Composition of Oilfield Water

All formation waters contain dissolved solids, primarily sodium chloride. The water sometimes is called *brine* or *salt water*. However, oilfield brines bear no relationship to seawater, either in the concentration of solids or in the distribution of the ions present. Generally, oilfield waters contain much higher concentrations of solids than seawater does. Formation waters have been reported with total solid concentrations ranging from as little as 200 ppm to saturation, which is approximately 300,000 ppm. Seawater contains about 35,000 ppm total solids.

The dissolved cations commonly found in oilfield waters are Na^+, Ca^{++}, and Mg^{++}. Occasionally K^+, Ba^{++}, Li^+, Fe^{++}, and Sr^{++} are present. The most common anions are Cl^-, $SO_4^=$, and HCO_3^-. Also, $CO_3^=$, NO_3^-, Br^-, I^-, $BO_3^=$, and $S^=$ are often present. Trace quantities of as many as 30 to 40 other ions frequently occur in these very complex brines.

Current techniques in analytical chemistry permit quantitative analysis of all the cations listed above. Previously, analysis of the sodium ion was tedious. Sodium seldom was measured directly but was calculated as the difference in the reacting values of the cations and anions which were measured. Thus, many of the older analyses report only calcium, magnesium, and sodium, with potassium and other cations included in the value for sodium.

In addition to dissolved salts, microorganisms of different species and strains are usually present in oilfield brines. The origin of these organisms is not really known; they could be present in untapped reservoirs or could be introduced during drilling. These organisms contribute to corrosion in the wellbore and to plugging of the formation during water-injection operations. The very hardy anaerobic sulfate-reducing bacteria are considered to be the most troublesome organisms commonly found in formation brines.

The variations in concentrations and ionic compositions of formation waters may be due to many factors. Among the explanations that have been proposed are the nonmarine origin of some sediments, dilution by ground water, concentration through evaporation by migrating gas, sulfate reduction by anaerobic bacteria or petroleum constituents, adsorption and base exchange of cations with clay minerals, dissolution of mineral salts by migrating formation water, exchange of magnesium and calcium ions during dolomitization, salt sieving by shales, precipitation of magnesium and calcium sulfates and carbonates, and chemical reaction with constituents of the confining sediments. Any one or more of these theories might apply to a particular formation water.

The important point is that the water contained in a producing formation has different composition than any other brine, even those in the immediate vicinity of that formation.

The concentrations of the solids present in brines are reported in several different ways. Among these are *parts per million, milligrams per liter*, and *weight percent solids*. Parts per million refers to grams of solids per one million grams of brine. Milligrams per liter refers to milligrams of solid per liter of brine.

Parts per million is multiplied by the density of the brine at standard conditions in grams per cubic centimeter to obtain milligrams per liter. Often, parts per million and milligrams per liter are used interchangeably. This is correct only if the density of the brine at standard conditions can be assumed to be 1.0 g/cc.

The units of weight percent solids may be obtained by dividing parts per million by 10,000.

Another set of units is *milliequivalents per liter*. Milligrams per liter can be converted to milliequivalents per liter by dividing by the *equivalent weight*. For ionization reactions, equivalent weight is obtained by dividing the atomic weight of the ion by its valence.

Table 16–1 gives definitions of the several different ways of expressing the concentration of dissolved solids in formation water.

TABLE 16–1

Summary of nomenclature and units for concentration of dissolved solids in formation waters (adapted from Monograph Series, SPE 9, 38)

Term	Symbol	Definition	Equations
Molality	C_m	$\dfrac{\text{g mole solid}}{1000 \text{ g pure water}}$	
Molarity	C_M	$\dfrac{\text{g mole solid}}{1000 \text{ ml brine}}$	
Normality	C_N	$\dfrac{\text{eq wt solid}}{1000 \text{ ml brine}}$	
Milliequivalents per liter	$C_{meq/l}$	$\dfrac{\text{meq solid}}{1000 \text{ ml brine}}$	$C_{meq/l} = 1000 \times C_N = C_{mg/l}/\text{eq wt}$
Weight percent solids	C_W	$\dfrac{\text{g solid}}{100 \text{ g brine}}$	$C_W = C_{ppm} \times 10^{-4}$
Parts per million	C_{ppm}	$\dfrac{\text{g solid}}{10^6 \text{ g brine}}$	$C_{ppm} = C_W \times 10^4 = C_{mg/l}/\rho_w$
Milligrams per liter	$C_{mg/l}$	$\dfrac{\text{g solid}}{10^6 \text{ ml brine}}$	$C_{mg/l} = \rho_w \times C_{ppm} = \rho_w \times C_W \times 10^4$
Grains per gallon	$C_{gr/gal}$	$\dfrac{\text{grains solid}}{\text{gal brine}}$	$C_{gr/gal} = 17.1 \times C_{mg/l} = 17.1 \times \rho_w \times C_{ppm}$

where ρ_w is in g/cc at standard conditions

EXAMPLE 16–1: *Your black oil reservoir is producing water with the oil. The reservoir is 8421 feet subsea. Initial conditions are 165°F and 3916 psig. The black oil has a bubble-point pressure of 3161 psig at 165°F. The reservoir produces 44.7°API stock-tank oil and 0.847 specific gravity gas at 1345 scf/STB. The composition of the produced water is given as follows.*

| | Ion concentration, | | Ion concentration, | |
Cation	ppm	Anion	ppm
Sodium	23,806	Chloride	41,312
Calcium	1,906	Bicarbonate	51.5
Magnesium	375	Sulfate	283
Iron	297	Carbonate	0
Barium	0		

Total dissolved solids is 68,030 ppm. Convert the analysis to milligrams per liter.

Solution

ρ_w = 65.40 lb/cu ft = 1.0476 gm/cc at 6.8% solids, Figure 16–8
Multiply parts per million by brine density to obtain mg/l.

| | Ion concentration, | | Ion concentration, | |
Cation	mg/l	Anion	mg/l
Sodium	24,939	Chloride	43,278
Calcium	1,997	Bicarbonate	54
Magnesium	393	Sulfate	296
Iron	311	Carbonate	0
Barium	0		

EXAMPLE 16–2: *Continue Example 16–1. Convert milligrams per liter to milliequivalents per liter.*

Solution

The atomic weight of sodium is 22.99 g/g mole, and the valence of sodium is 1 eq wt/g mole, so the equivalent weight of sodium is

$$\frac{22.99 \text{ g/g mole}}{1 \text{ eq wt/g mole}} = 22.99 \text{ g/eq wt} = 22.99 \text{ mg/meq}$$

Thus, the milliequivalents per liter of sodium in this analysis are

$$\frac{24,939 \text{ mg/liter}}{22.99 \text{ mg/meq}} = 1,084.8 \text{ meq/l}$$

The ionic weight of the calcium ion is 40.08 g/g mole, and the valence is 2 eq wt/g mole, so that the equivalent weight is

$$\frac{40.08 \text{ g/g mole}}{2 \text{ eq wt/g mole}} = 20.04 \text{ g/eq wt} = 20.04 \text{ mg/meq}$$

and the milliequivalents per liter of calcium are

$$\frac{1997 \text{ mg/l}}{20.04 \text{ mg/meq}} = 99.6 \text{ meq/l}$$

Thus,

	Ion concentration,		Ion concentration,	
Cation	meq/l	Anion		meq/l
Sodium	1084.8	Chloride		1220.6
Calcium	99.6	Bicarbonate		0.9
Magnesium	32.3	Sulfate		6.2
Iron	11.1	Carbonate		0
Barium	0			

Water analyses contain from 6 to 10 or more numbers representing the compositions of the various ions. Therefore, the data are best viewed in the form of a pattern. Differences between waters are seen easily by visually comparing patterns. Figure 16–1 shows one of the more commonly used methods of presenting water-analysis data.[1] The scale used for the pattern is given in the center of the figure. Positive ions are plotted to the left and negative ions are plotted to the right from a central vertical zero line. The number immediately beneath the identification of each ion is the scale in milliequivalents per liter.

Some oilfield waters can be plotted on a scale where the maximum lengths of the horizontal lines represent 100 meq of sodium and chloride ions and 10 meq of each of the other ions. For highly concentrated

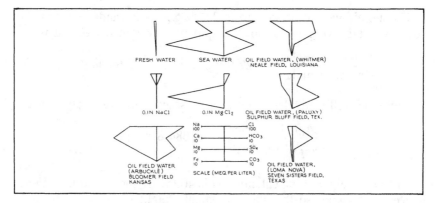

Fig. 16–1. Common water patterns. (Stiff. *Trans.*, AIME, *192*, 377. Copyright 1951 SPE-AIME.)

brines, 1000 meq of sodium and chloride and 100 meq of each of the other ions may be used. Sometimes a logarithmic scale is used for ion concentration.

EXAMPLE 16–3: *Continue Example 16–1. Draw a Stiff diagram for the brine.*

Solution

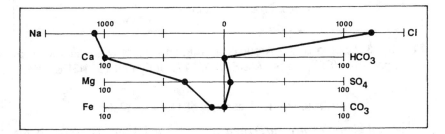

Fig. 16–2. Stiff diagram. Solution to Example 16–3.

There are many applications of the pattern method of presentation of water-analysis data. Figure 16–3 shows the correlation of the Arbuckle formation through four Kansas counties by comparisons of the patterns of the water from four fields completed in the Arbuckle. Other uses include determination of the source of contamination of fresh water, confirmation of the zone in which a new well is to be completed, and detection of incursion of foreign water into an existing well due to improper cementing of casing or leaks in the casing.

Geologists use radical changes in the characteristics of a series of brines going toward deeper formations to indicate a different geologic environment. Also, increases in the salinity of shallow ground waters have aided in the search for buried salt domes.

Mixing of unlike brines can cause the precipitation of sulfates of barium, strontium, and calcium. This *hard scale*, which is a nuisance in oilfield flow lines, often can be explained by leaky casing. *Soft scale*, composed of calcium carbonate, usually is caused by the loss of dissolved carbon dioxide from the brine during pressure reduction.

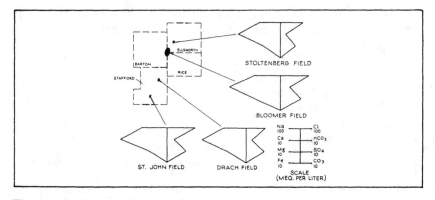

Fig. 16–3. Course of the Arbuckle formation through Kansas shown by water patterns. (Stiff, *Trans.*, AIME, *192*, 377. Copyright 1951 SPE-AIME.)

Bubble-Point Pressure of Oilfield Water

The *bubble-point pressure* of a gas-saturated brine is equal to the bubble-point pressure of the coexisting oil. The two are equal because of thermodynamic equilibrium between the brine and the oil.

When reservoir pressure drops below the bubble-point pressure of the oil, the brine releases some of its dissolved gas. Therefore the saturation pressure of the brine equals reservoir pressure. This is analogous to the oil which is saturated at all pressures below bubble-point pressure.

The brine in a gas reservoir is considered saturated at all reservoir pressures. Thus the bubble-point pressure of the brine in contact with gas is equal to the initial reservoir pressure.

EXAMPLE 16–4: *Continue Example 16–1. Estimate the bubble-point pressure of the brine.*

Solution

$$p_b \text{ of brine } = 3161 \text{ psig at } 165°F$$

Formation Volume Factor of Oilfield Water

The *water formation volume factor* represents the change in volume of the brine as it is transported from reservoir conditions to surface conditions. The units are reservoir barrels per surface barrel at standard conditions, res bbl/STB. As with the oil formation volume factor, three effects are involved:

the evolution of dissolved gas from the brine as pressure is reduced,
the expansion of the brine as pressure is reduced,
and
the contraction of the brine as temperature is reduced.

Remember that the most important contribution to the oil formation volume factor is the amount of gas evolved from the liquid. The solubility of hydrocarbon gas in water is considerably less than the amount of gas held by oil so that gas solubility has little effect on the water formation-volume factor. The contraction and expansion due to reduction in temperature and pressure are small and offsetting so the formation volume factor of water is numerically small, rarely larger than 1.06 res bbl/STB.

Figure 16–4 gives the typical relationship of water formation volume factor to pressure. The figure shows the initial reservoir pressure to be above the bubble-point pressure of the water. As reservoir pressure is reduced from initial pressure to bubble-point pressure, the formation volume factor increases because of the expansion of water in the reservoir.

A reduction in reservoir pressure below bubble-point pressure results in the evolution of gas from the water into the pore spaces of the reservoir. The loss in liquid volume due to evolution of gas only partially offsets the expansion of the water due to reduction in pressure. So the formation volume factor continues to increase as reservoir pressure is reduced.

If reservoir pressure is reduced to atmospheric pressure, the maximum value of formation volume factor is reached. At this point temperature

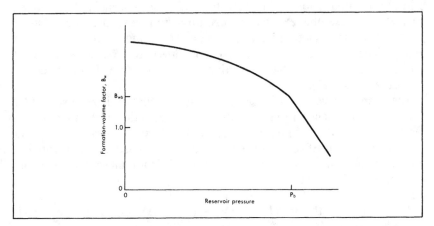

Fig. 16–4. Typical shape of water formation volume factor as a function of pressure at constant reservoir temperature.

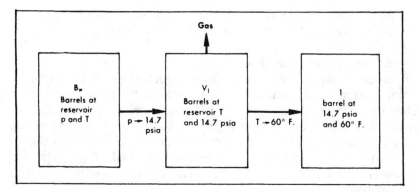

Fig. 16-5. Stepwise change in water volume from reservoir conditions to standard conditions.

must be reduced to 60°F to bring the formation volume factor to exactly 1.0 res bbl/STB.

Note that water formation volume factor can have values less than 1.0 res bbl/STB. This occurs at high reservoir pressures when the brine expansion caused by pressure decrease during the trip to the surface is greater than the brine contraction due to temperature drop and loss of gas.

The formation volume factors of gas-saturated brines have been correlated in a manner similar to an early oil formation volume factor correlation.[2] Figure 16-5 illustrates the correlation procedure. A volume of B_w barrels of brine at reservoir conditions is converted to V_1 barrels of brine by the reduction in pressure to 14.7 psia. Normally, V_1 is larger than B_w because the expansion resulting from pressure reduction is greater than the reduction in liquid volume resulting from the evolution of gas. The V_1 barrels of brine are converted to 1 barrel of brine at surface conditions by the reduction in temperature from reservoir temperature to 60°F.

The change in volume during the pressure reduction is represented by ΔV_{wp}, and the change in volume due to the reduction in temperature is represented by ΔV_{wT}. Figures 16-6 and 16-7 give values of ΔV_{wp} and ΔV_{wT} as functions of reservoir temperature and pressure. The formation volume factor of water may be computed from these values using Equation 16-1.

$$B_w = (1 + \Delta V_{wp})(1 + \Delta V_{wT}) \qquad (16-1)$$

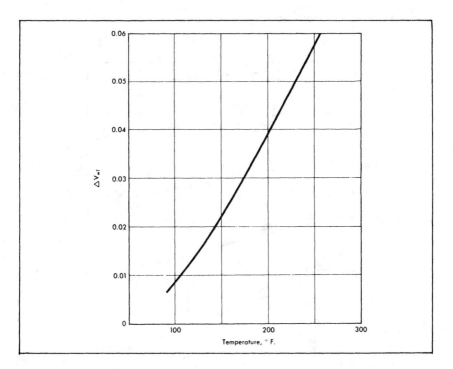

Fig. 16–6. ΔV_{wT} as a function of reservoir temperature. (Based on data from *International Critical Tables* II, 79, 1929; *Chemical Engineer's Handbook,* McGraw-Hill, 1950, 157.)

Values of formation volume factor of water estimated using this correlation agree with the limited published experimental data to within one percent.

Surprisingly, this correlation is valid for oilfield waters with widely varying brine concentrations. An increase in brine concentration causes a slight increase in the coefficient of thermal expansion of the water. This causes a slight increase in ΔV_{wT}. An increase in brine concentration also causes a decrease in the solubility of gas in the water. This causes a slight decrease in ΔV_{wp}. These changes in ΔV_{wT} and ΔV_{wp} are offsetting to within one percent throughout a range of brine concentration from zero to 30 percent.

EXAMPLE 16–5: *Continue Example 16–1. Estimate the formation volume factor of brine at 3161 psig and 165°F.*

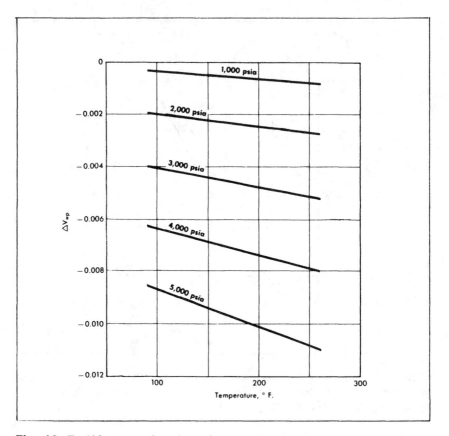

Fig. 16–7. ΔV_{wp} as a function of reservoir pressure and temperature. (Based on data from Dodson and Standing, *Drill. and Prod. Prac.,* API, 1944, 176, McCain and Stacy, *SPEJ 11,* 4; Frick, *Petroleum Production Handbook,* McGraw-Hill, 1962, 22–23.)

Solution

$$\Delta V_{wT} = 0.027 \text{ at } 165°F, \text{ Figure } 16–6$$

$$\Delta V_{wp} = -0.0047 \text{ at } 3176 \text{ psia and } 165°F, \text{ Figure } 16–7$$

$$B_w = (1 + \Delta V_{wp}) (1 + \Delta V_{wT}) \qquad\qquad (16–1)$$

$$B_w = (1 + 0.0272)(1 - 0.0049)$$

$$B_w = 1.022 \text{ res bbl/STB at } 3176 \text{ psia and } 165°F$$

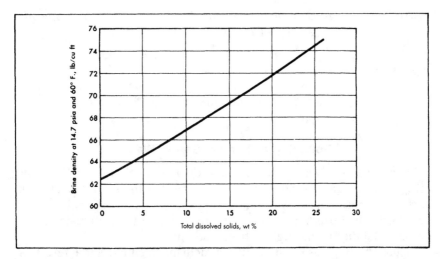

Fig. 16–8. Effect of salinity on the density of brine. (Data from *International Critical Tables II*, 1927, 79.)

Density of Oilfield Water

Figure 16–8 gives the *density of oilfield brine at standard conditions* as a function of total dissolved solids. The *density at reservoir conditions* is determined by dividing the density at standard conditions by the formation volume factor of the reservoir water at reservoir conditions.

EXAMPLE 16–6: *Continue Example 16–1. Estimate the density of the brine at 3161 psig and 165°F.*

Solution

ρ_w = 65.40 lb/cu ft at 14.65 psia and 60°F, 6.8% solids, Figure 16–8

B_w = 1.022 res bbl/STB at 3176 psia and 165°F, Example 16–5.

$$\rho_w = \frac{65.40 \text{ lb/ST cu ft}}{1.022 \text{ res cu ft/ST cu ft}} = 64.0 \text{ lb/cu ft at 3176 psia and 165°F}$$

The specific gravity of brine, γ_w, is defined as the ratio of the density of the brine to the density of pure water, both taken at the same pressure and temperature (usually atmospheric pressure and 60°F.)

EXAMPLE 16–7: *Continue Example 16–1. Calculate the specific gravity of the brine.*

Solution

$$\gamma_w = 65.40 lb/cu\ ft/62.368\ lb/cu\ ft = 1.049$$

Solubility of Natural Gas in Water

Figure 16–9 gives a comparison of the solubilities of various components of natural gas in water.[3] Notice that the solubility of each paraffin hydrocarbon is two to three times less than the solubility of the next lighter hydrocarbon. Also, pressure seems to have a greater effect on the solubility of the lighter hydrocarbons.

Figure 16–10 gives the solubility of methane in pure water at reservoir conditions.[4] This figure may be used with accuracy to within about five percent to estimate the solubility of natural gas in pure water. Figure 16–11 can be used to adjust the solubility to account for varying degrees of salinity of the water.[5]

EXAMPLE 16–8: *Continue Example 16–1. Estimate the solution gas-water ratio of the brine at 3161 psig and 165°F.*

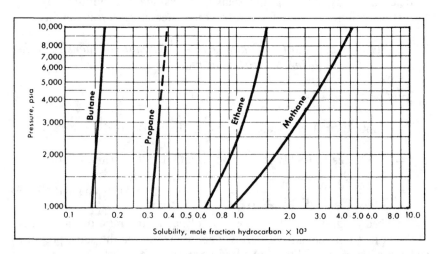

Fig. 16–9. Solubility of pure hydrocarbons in pure water at 200°F. (Brooks et al., *Pet. Refiner 30.* 120, with permission.)

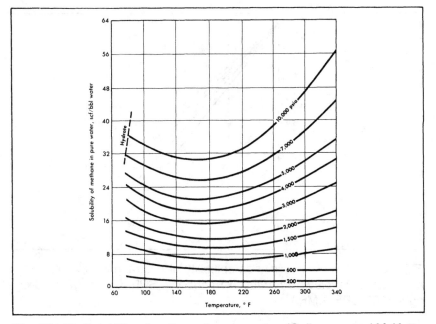

Fig. 16–10. Solubility of methane in pure water. (Culberson and McKetta, *Trans.*, AIME *192*, 226. Copyright 1951 SPE-AIME.)

Solution

R_{sw} = 15.7 scf/STB of pure water at 3176 psia and 165°F, Figure 16–10

Adjustment = 0.74 for 6.8% brine, Figure 16–11

R_{sw} = (15.7)(0.74) = 11.6 scf/STB at 3161 psig and 165°F

The *solution gas-water ratio*, R_{sw}, changes as reservoir pressure changes in the same way as the oil solution gas-oil ratio. However, R_{sw} is very much smaller than R_s. Figure 8–2 illustrates the typical shape of R_{sw}.

The Coefficient of Isothermal Compressibility of Water

We saw in Chapter 8 that the coefficient of isothermal compressibility of oil has a discontinuity at the bubble point. The *coefficient of isothermal compressibility of water* has the same discontinuity for the same reason. Figure 8–7 is typical of the relationship between *water*

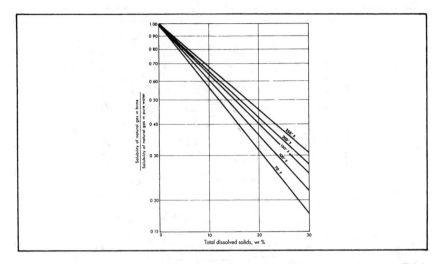

Fig. 16–11. Effect of salinity on solubility of natural gas in water. (Frick and Taylor, *Petroleum Production Handbook,* SPE, Dallas, 1962. Copyright 1962 SPE-AIME.)

compressibility and reservoir pressure. But water compressibility values are somewhat lower than those of oil compressibility.

The Coefficient of Isothermal Compressibility of Water at Pressures Above the Bubble Point

At pressures above the bubble point the compressibility of water is defined as

$$c_w = -\frac{1}{V_w}\left(\frac{\partial V_w}{\partial p}\right)_T, \; c_w = -\frac{1}{B_w}\left(\frac{\partial B_w}{\partial p}\right)_T, \; c_w = \frac{1}{\rho_w}\left(\frac{\partial \rho_w}{\partial p}\right)_T.$$

$$(16-2)$$

Figure 16–12 gives the coefficient of isothermal compressibility of pure water. The figure is a combination of two sets of data which are in fair agreement in the region of overlap.[6,7]

Figure 16–13 gives a correction factor to be used to adjust the compressibility for the effect of dissolved solids. Surprisingly, the amount of dissolved gas has no effect on the compressibility of water.[7]

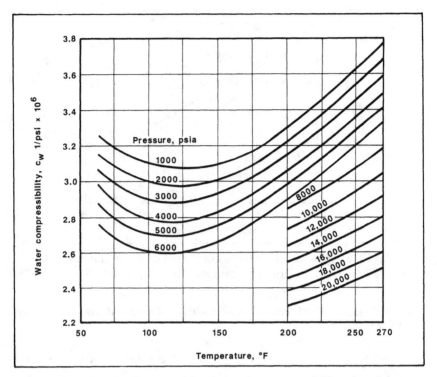

Fig. 16-12. The coefficient of isothermal compressibility of water.

EXAMPLE 16-9: *Continue Example 16-1. Estimate the coefficient of isothermal compressibility of the brine at 3500 psia and 165°F.*

Solution

$$c_w = 2.94 \times 10^{-6} \, psi^{-1} \text{ for pure water at 3500 psia and 165°F,}$$
Figure 16-12

Adjustment $= 0.89$ at 6.8 wt % solids and 3500 psia, Figure 16-13

$$c_w = (2.94 \times 10^{-6})(0.89)$$
$$= 2.62 \times 10^{-6} \, psi^{-1} \text{ at 3500 psia and 165°F}$$

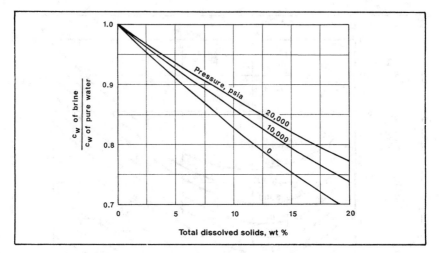

Fig. 16–13. Effect of salinity on the coefficient of isothermal compressibility of water. (From an equation by Osif, SPE *Res. Eng. 3,* 1988, 175.)

The Coefficient of Isothermal Compressibility of Water at Pressures Below the Bubble Point

At pressures below the bubble point, the compressibility of water is defined as[8]

$$c_w = - \frac{1}{B_w} \left[\left(\frac{\partial B_w}{\partial p} \right)_T - B_g \left(\frac{\partial R_{sw}}{\partial p} \right)_T \right]. \qquad (16-3)$$

This equation is analogous to Equation 8–24 for c_o. Equation 16–3 is different from Equation 16–2 for the same reasons that Equation 8–24 is different from Equations 8–7.

Equation 16–3 can be written as

$$c_w = - \frac{1}{B_w} \left(\frac{\partial B_w}{\partial p} \right)_T + \frac{B_g}{B_w} \left(\frac{\partial R_{sw}}{\partial p} \right)_T. \qquad (16-4)$$

The first term on the right side of the Equation 16–4 is related to c_w at pressures above the bubble point and is calculated using Figures 16–12 and 16–13.

The second term on the right side of the equation is estimated by separating it into three parts.

The formation volume factor of water, B_w, is computed with Equation 16–1 and Figures 16–6 and 16–7.

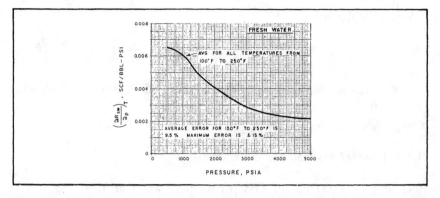

Fig. 16–14. Derivative of solution gas-water ratio with respect to pressure. (Ramey, *Trans.*, AIME *231*, 447. Copyright 1964 SPE-AIME.)

The formation-volume factor of gas, B_g, is calculated using Equation 6–3. Use a value of 0.63 for the specific gravity of the gas evolved from the water to determine a z-factor for Equation 6–3. This value is based on limited data and its accuracy is unknown; however, it gives values which appear reasonable.

The derivative $(\partial R_{sw}/\partial p)_T$ is estimated from Figure 16–14.[8] Multiply values of the derivative from Figure 16–14 by the adjustment factor from Figure 16–11 to account for the effect of dissolved solids.

EXAMPLE 16–10: *Continue Example 16–1. Calculate the coefficient of isothermal compressibility of the brine at 3000 psia and 165°F.*

Solution

First, estimate c_w at 3000 psia and 165°F as if pressure were above bubble point.

$$c_w = 2.98 \times 10^{-6} \text{ psi}^{-1} \text{ for pure water at 3000}$$
$$\text{psia and 165°F, Figure 16–12}$$

Adjustment = 0.89 at 6.8 wt % solids and 3000 psia, Figure 16–13

$$c_w = (2.98 \times 10^{-6})(0.89)$$
$$= 2.65 \times 10^{-6} \text{ psi}^{-1} \text{ at 3000 psia and 165°F}$$

Second, calculate c_w

$$c_w = -\frac{1}{B_w}\left(\frac{\partial B_w}{\partial p}\right)_T + \frac{B_g}{B_w}\left(\frac{\partial R_{sw}}{\partial p}\right)_T \qquad (16\text{-}4)$$

B_w = 1.023 res bbl/STB at 3000 psia and 165°F, Figures 16–6 and 16–7

z = 0.884, Figure 3–7 at γ_g = 0.63

B_g = 0.00092 res bbl/scf, Equation 6–3

$(\partial R_{sw}/\partial p)_T$ = 0.00285 scf/STB psi for pure water at 3000 psia, Figure 16–14.

Adjustment = 0.74 for 6.8 wt % brine, Figure 16–11

$(\partial R_{sw}/\partial p)_T$ = (0.00285)(0.74) = 0.00211 scf/STB psi

$$c_w = 2.65 \times 10^{-6} \text{ psi}^{-1} + \left(\frac{0.00092 \text{ res bbl/scf}}{1.023 \text{ res bbl/STB}}\right)(0.00211 \text{ scf/STB psi})$$
$$= 4.55 \times 10^{-6} \text{ psi}^{-1}$$

The Coefficient of Viscosity of Oilfield Water

Water viscosity is a measure of the resistance to flow exerted by water. Units of centipoise are used most often by petroleum engineers. The definitions of other viscosity units are given in Chapter 6 in connection with gas viscosity.

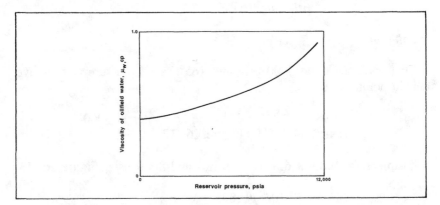

Fig. 16–15. Typical shape of brine viscosity as a function of pressure at constant reservoir temperature.

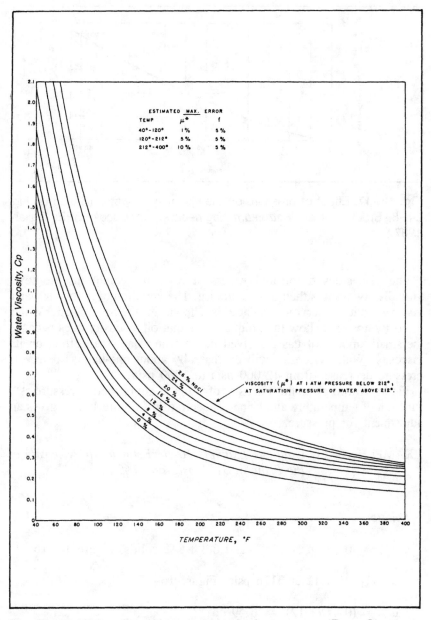

Fig. 16–16. Water viscosities at atmospheric pressure. (From Chestnut, unpublished, Shell Development Co.)

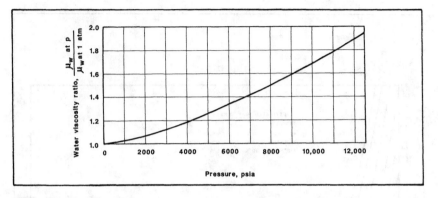

Fig. 16-17. Effect of pressure on the viscosity of water. (Based on Fig. 24-8, Bradley et al., *Petroleum Engineering Handbook,* SPE, Dallas, 1987.)

The viscosities of oilfield waters at reservoir conditions are low, virtually always less than one centipoise. The variation of reservoir water viscosity with reservoir pressure is shown in Figure 16-15. Water viscosity does not show the unique shape that oil viscosity does because the small amount of gas dissolved in the water has small effect on its viscosity. Water viscosity will decrease by about one-half as reservoir pressure decreases from 12,000 psia to 1000 psia.

Figure 16-16 gives the viscosity of water at atmospheric pressure as a function of temperature and brine concentration.[9] Figure 16-17 gives an adjustment for pressure.

EXAMPLE 16-11: *Continue Example 16-1. Estimate the viscosity of the brine at 3161 psig and 165°F.*

Solution

μ_{w1} = 0.44 cp at 165°F and 6.8 wt % solids, Figure 16-16

μ_w/μ_{w1} = 1.12 at 3176 psia, Figure 16-17

μ_w = (0.44)(1.12) = 0.49 cp

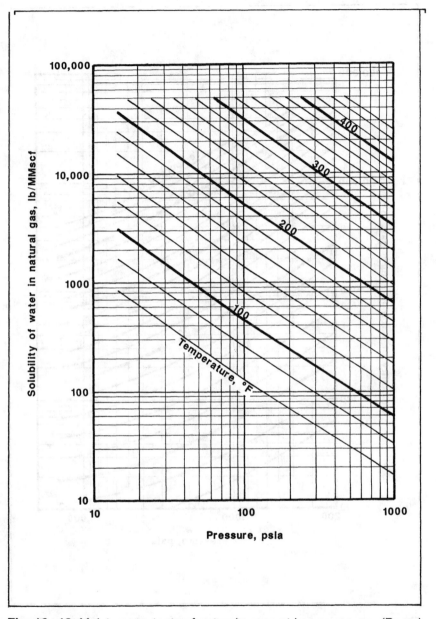

Fig. 16–18. Moisture contents of natural gases at low pressures. (Based on equations of Bukacek, *Equilibrium Moisture Content of Natural Gases*, Res. Bull., IGT, Chicago, 1955.)

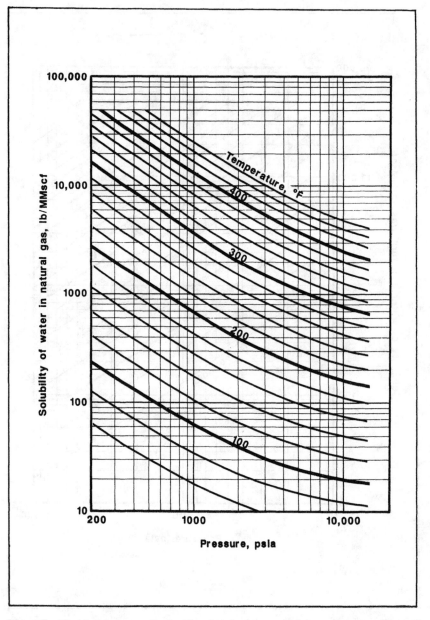

Fig. 16–19. Moisture contents of natural gases at high pressures. (Based on equations of Bukacek, *Equilibrium Moisture Content of Natural Gases*, Res. Bull., IGT, Chicago, 1955.)

Solubility of Water in Natural Gas

Figures 16–18 and 16–19 give the dew-point water-vapor content of a nitrogen-free natural gas in equilibrium with liquid water.

The *water contents* obtained from the graph at temperatures below hydrate-formation conditions represent dew-point formation under metastable equilibrium between gas and liquid water rather than between gas and solid hydrates. The water contents of natural gases in equilibrium with hydrates are significantly lower than the water contents given in Figure 16–18, especially at lower temperatures.

The upper ends of the isobars on Figures 16–18 and 16–19 terminate at the vapor pressure of pure water at each temperature. Liquid water ceases to exist at pressures below the end of each line.

The correlation on which these figures are based is good to within five percent. This is about as accurate as water content can be measured. The correlation was developed for dry gases. The presence of heavier hydrocarbons in wet gases and retrograde gases increases the water content by as much as 10 percent at 1000 psia and 20 percent at 10,000 psia.

Natural gas which contains substantial amounts of carbon dioxide or hydrogen sulfide will contain more water due to the affinity of these substances for water.[10] No quantitative data are available; however, the water content could be as much as five percent greater than predicted by the figures.

Appreciable quantities of nitrogen or helium in the natural gas should lower the saturated water content of the gas.[11] Again, limited data are available, but the decrease could be as much as five percent on Figure 16–18 and ten percent on Figure 16–19.

Dissolved solids in the water lower the partial pressure of the water, thereby reducing the water content of the gas. Figure 16–20 gives an adjustment to be applied to the moisture content from Figure 16–18 or 16–19 to approximate the effect of the brine content of the oilfield water.[10]

EXAMPLE 16–12: *Your dry gas reservoir is producing gas of 0.700 specific gravity. The reservoir water is a 20.0 weight percent brine; however, no reservoir brine is produced with the gas. Reservoir conditions are 5000 psia and 200°F. Estimate the quantity of water vapor in the reservoir gas. Estimate the quantity of water vapor in the surface gas at 1000 psia and 80°F. The difference is the quantity of liquid water which will arrive at surface.*

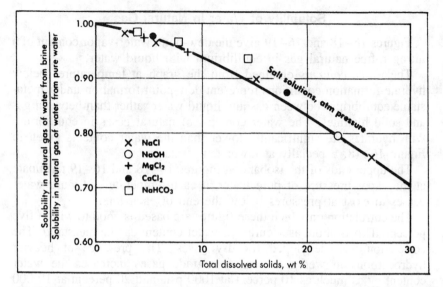

Fig. 16–20. Effect of salinity on the moisture contents of natural gases. (From *Handbook of Natural Gas Engineering* by Katz et al. Copyright 1959 by McGraw-Hill Book Co. Used with permission of McGraw-Hill Book Co.)

Solution

First, calculate water content at reservoir conditions

Water content = 210 lb/MMscf from pure water
                                    at 5000 psia and 200°F, Figure 16–19

Adjustment = 0.83 at 20 wt % solids, Figure 16–20

Water content = (210)(0.83) = 174 lb/MMscf at 5000 psia and 200°F

Second, calculate water content at surface conditions

Water content = 33 lb/MMscf at 1000 psia and 80°F, Figure 16–18

No adjustment, surface water is condensed from the gas, thus pure water.

Third, difference is quantity of liquid at surface

Surface water = 174 − 33 = 141 lb/MMscf

$$\text{Surface water} = \frac{141 \text{ lb/MMscf}}{62.37 \text{ lb/cu ft}} = 2.26 \text{ cu ft/MMscf} = 0.40 \text{ bbl/MMscf}$$

Solubility of Water in Hydrocarbon Liquid

The mutual attraction between water and hydrocarbons is extremely small. Therefore, the solubility of water in hydrocarbon liquid is very limited. Published data are insufficient to develop a correlation of the solubility of water in liquid hydrocarbon at reservoir pressures and temperatures.

Figure 16–21 shows the solubility of water in some hydrocarbon liquids as a function of temperature.[12] The data are for saturation of the hydrocarbon with water at a given temperature under the equilibrium pressure of the existing water, hydrocarbon liquid, and hydrocarbon gas phases. The conditions correspond to the equilibrium line $\overline{Q_2C}$ on Figure 17–2.

These data are given primarily to show the limited solubility of water in hydrocarbon liquid. The data show that the type of hydrocarbon has

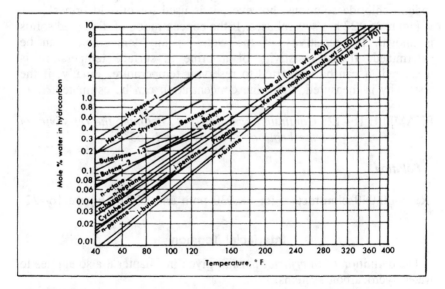

Fig. 16–21. Solubility of water in hydrocarbon liquids at three-phase equilibria. (Hoot et al., *Pet. Refiner 36,* 256, with permission.)

considerably more effect on solubility than does the molecular weight of the hydrocarbon.

Resistivity of Oilfield Water

Pure water does not conduct electricity; however, water containing dissolved salts will conduct electricity. Electrical current is carried in water by the movement of ions from the dissolved salts. The conductivity of brine usually is measured in terms of its reciprocal, *resistivity*.

Resistivity of brine is defined as

$$R_w = \frac{rA}{L}, \tag{16-5}$$

where r is the resistance in ohms, A is cross-sectional area perpendicular to the current path in square meters, and L is the length of the current path in meters. The units of resistivity are ohm-meters.

Since the current is conducted by the ions present in the water, resistivity must be related inversely to the concentration of dissolved solids. Figure 16–22 gives the relationships between resistivity, dissolved solids, and temperature.[13] This graph was developed from data for sodium chloride solutions; however, it is used for oilfield brines.

Figure 16–22 has several uses. If the concentration of dissolved solids is known, the resistivity of the water at any temperature can be determined. If the resistivity of a brine at surface temperature is measured, it can be converted to reservoir temperature. Finally, if the resistivity is measured, the brine concentration can be estimated.

EXAMPLE 16–12: *Continue Example 16–1. Estimate the resistivity of the brine at 3161 psig and 165°F.*

Solution

R_w = 0.049 ohm-meters for 68,030 ppm and 165°F, Figure 16–22

Interfacial Tension

The definition of *interfacial tension* given in Chapter 8 also applies to water-hydrocarbon systems.

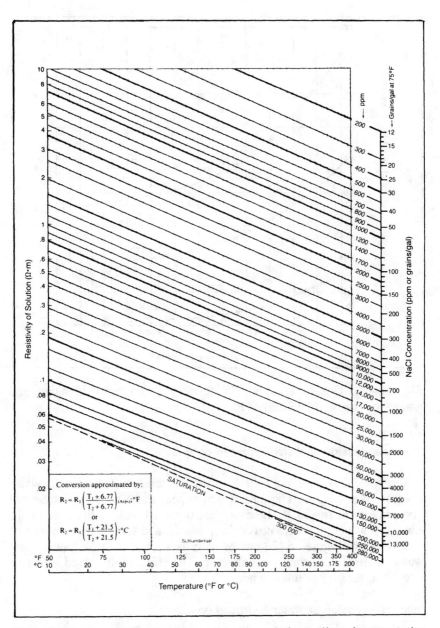

Fig. 16–22. Resistivity of sodium chloride solutions. (*Log Interpretation Charts,* Schlumberger Well Services, Houston, with permission.)

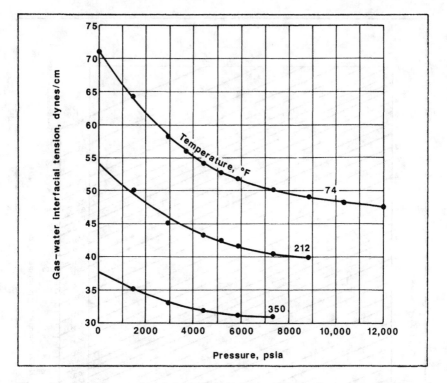

Fig. 16–23. Gas-water interfacial tension. (Data from Jennings and Newman, *Trans.*, AIME *251*, 171.)

Interfacial Tension of Water-Hydrocarbon Gas

The several sets of published interfacial tension data for gas-water systems do not agree. The data which appear to be most consistent are given in Figure 16–23.[14] The data were obtained with methane and pure water. However, the data cover the pressure and temperature ranges of usual interest. Figure 16–23 can be used as a correlation for natural gas-water systems; its accuracy is unknown. The effect of dissolved solids on interfacial tension is unknown.

Interfacial Tension of Water-Hydrocarbon Liquid

Published data of interfacial tension of reservoir oil-water systems is limited. Figure 16–24 can be used to get an order-of-magnitude estimate of oil-water interfacial tension.[15]

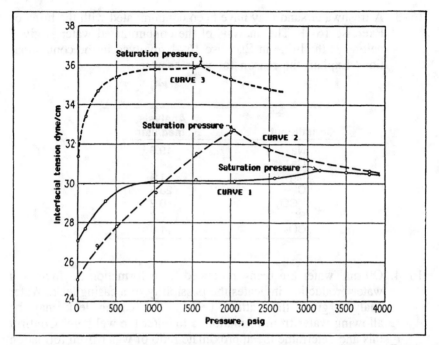

Fig. 16–24. Interfacial tension between oil and water. (Hocott, *Trans.,* AIME *132*, 184. Copyright 1939 SPE-AIME.)

The effect of dissolved solids on interfacial tension is unknown.

Exercises

16–1. Convert the concentrations of solids for the brine given below to milligrams per liter, percent solids, milliequivalents per liter, and grains per gallon.

Constituent	Analysis, ppm
Na	7,365
Ca	1,582
Mg	305
SO$_4$	521
Cl	14,162
CO$_3$	705
HCO$_3$	0

16–2. Draw a pattern of the brine given in Exercise 16–1.

16–3. A freshwater sand may have been contaminated with the brine of
Exercise 16–1. The analysis of the contaminated water is given
below. Is the brine in Exercise 16–1 involved in the contamination? Explain the reason for your answer.

Constituent	Analysis, meq/liter
Na	19.3
Ca	4.2
Mg	1.4
Cl	23.3
HCO_3	0.2
SO_4	0.7
CO_3	1.0

16–4. Oil and water are being produced from formation A. Increased
water production indicates the possibility of a casing leak. Water
analysis gives the pattern below. The casing leak may be
allowing water from formation B to enter the wellbore. Confirm
this and determine the approximate ratio of water from formation
B to water from formation A by comparing patterns of mixtures
of water A and water B with the pattern of the produced water.

Constituent	Analysis	
	Formation A water, meq/liter	Formation B water, meq/liter
Na	230.8	250.0
Ca	3.8	46.5
Mg	1.0	25.0
Fe	0	0
Cl	211.5	267.3
HCO_3	26.0	9.2
SO_4	1.5	20.2
CO_3	0	0

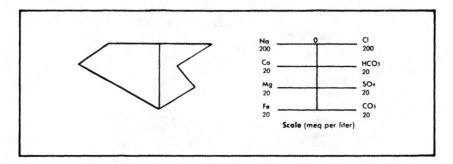

Fig. 16–25. Stiff diagram of produced water of Exercise 16–4.

16–5. Analysis of brine produced along with oil from your reservoir is given below.

Constituent	Ion concentration, mg/l
Sodium	14,400
Calcium	1,900
Magnesium	190
Iron	110
Chloride	24,500
Bicarbonate	2,000
Sulfate	850
Carbonate	0

Total dissolved solids = 43,950 mg/l,
resistivity = 0.1695 ohm-meters at 76.3°F,
specific gravity = 1.0310.

    a. convert ion concentrations to meq/l.
    b. Convert total dissolved solids to ppm.
    c. Plot a Stiff diagram for this brine.
    d. Farmer Brown claims that your brine has contaminated his freshwater well. The analysis of his contaminated well water is given as follows. Are you guilty? Document your answer.

Constituent	Ion concentration, meq/l
Sodium	68.9
Calcium	13.2
Magnesium	3.4
Iron	0.4
Chloride	76.5
Bicarbonate	4.5
Sulfate	2.8
Carbonate	0.2

16–6. Brine with composition given below is produced with a black oil. Reservoir conditions are 140°F and 2300 psig. Initial production is 675 scf/STB of 0.788 specific gravity gas and 41.3°API oil.

Cation	Ion concentration, mg/l	Anion	Ion concentration, mg/l
Sodium	75,000	Chloride	146,200
Calcium	14,300	Bicarbonate	100
Magnesium	1,900	Sulfate	180
Iron	100	Carbonate	0
Barium	0	Hydroxide	0

Total dissolved solids 237,780 mg/l. Convert the analysis to parts per million and percent solids.

16–7. Initial production from a black oil reservoir is 150 STB/d and 10 barrels of water per day. The stock-tank oil is 41.0°API. The produced gas has specific gravity of 0.715. Initial producing gas-oil ratio is 1100 scf/STB. Initial reservoir conditions are 7000 psig and 260°F. Bubble-point pressure of the reservoir oil is estimated to be 4850 psia at 260°F. The composition of the produced water is given below.

Cation	Ion concentration, mg/l	Anion	Ion concentration, mg/l
Sodium	24,200	Chloride	60,000
Calcium	12,100	Bicarbonate	180
Magnesium	500	Sulfate	50
Iron	350	Carbonate	0
Barium	20	Hydroxide	0

Convert the analysis to parts per million and percent solids.

16–8. Estimate the bubble-point pressure of the reservoir water of Exercise 16–6.

16-9. Estimate the bubble-point pressure of the reservoir water of Exercise 16-7.

16-10. Estimate the formation volume factor of the reservoir water of Exercise 16-6 at 2000 psia and 140°F.

16-11. Estimate the formation volume factor of the reservoir water of Exercise 16-7 at 4850 psia and 260°F.

16-12. Estimate the formation volume factor of the reservoir water of Exercise 16-7 at 2000 psia and 260°F.

16-13. Estimate the density of the reservoir water of Exercise 16-6 at 2000 psig and 140°F.

16-14. Estimate the density of the reservoir water of Exercise 16-7 at 4850 psia and 260°F.

16-15. Estimate the density of the reservoir water of Exercise 16-7 at 2000 psia and 260°F.

16-16. Estimate the solubility of reservoir gas in the reservoir water of Exercise 16-6 at 2000 psia and 140°F.

16-17. Estimate the solubility of reservoir gas in the reservoir water of Exercise 16-7 at 4850 psia and 260°F.

16-18. Estimate the solubility of reservoir gas in the reservoir water of Exercise 16-7 at 2000 psia and 260°F.

16-19. Estimate a value of the coefficient of isothermal compressibility of the reservoir water of Exercise 16-6 at 2300 psig and 140°F.

16-20. Estimate a value of the coefficient of isothermal compressibility of the reservoir water of Exercise 16-6 at 1000 psia and 140°F.

16-21. Estimate a value of the coefficient of isothermal compressibility of the reservoir water of Exercise 16-7 at 7000 psig and 260°F.

16-22. Estimate a value of the coefficient of isothermal compressibility of the reservoir water of Exercise 16-7 at 3000 psia and 260°F.

16-23. Estimate a value of formation volume factor of the reservoir water of Exercise 16-7 at 6000 psia and 260°F.

16-24. Estimate a value of viscosity for the reservoir water of Exercise 16-6 at 2000 psia and 140°F.

16-25. Estimate a value of viscosity for the reservoir water of Exercise 16-7 at 4850 psia and 260°F.

16-26. Calculate the volume of water condensed from production of 12 MMscfd of 0.68 specific gravity gas. Reservoir conditions are 4640 psia and 203°F. The reservoir water is a 10 wt % brine. Surface conditions are 600 psia and 60°F.

16-27. How much water must be removed from the surface gas of Exercise 16-26 to meet pipeline specifications of 6 lb/MMscf?

16-28. A gas well completed in a reservoir at 5600 psia and 225°F is producing 10 MMscfd of 0.710 specific gravity gas and 8

barrels of water per day. Surface conditions are 700 psia and
75°F. How much of the produced water is condensed from the
gas and how much is formation water? The reservoir water is a
62,000 ppm brine.

16–29. Estimate the resistivity of the formation water of Exercise
16–28 at 225°F.

16–30. Brine produced from an oil reservoir is sampled. The resistivity
is measured as 0.306 ohm-meters at 81°F. What is the resistivity
of the brine at reservoir conditions of 2500 psia and 180°F?
What is the concentration of solids in the brine?

References

1. Stiff, H.A., Jr.: "The Interpretation of Chemical Water Analysis by
   Means of Patterns," *Trans.*, AIME (1951) *192*, 376–379.

2. Katz, D.L.: "Prediction of Shrinkage of Crude Oils," *Drill. and
   Prod. Prac.*, API (1942) 137–147.

3. Brooks, W.B., Gibbs, G.B. and McKetta, J.J., Jr.: "Mutual
   Solubility of Light Hydrocarbon-Water Systems," *Pet. Refiner*
   (1951) *30*, No. 10, 118–120.

4. Culberson, O.L. and McKetta, J.J., Jr.: "Phase Equilibria in
   Hydrocarbon-Water Systems III—The Solubility of Methane in
   Water at Pressures to 10,000 psia," *Trans.*, AIME (1951) *192*, 223-
   226.

5. McKetta, J.J. and Wehe, A.H.: "Hydrocarbon-Water and Formation Water Correlations," *Petroleum Production Handbook*, Vol. II,
   T.C. Frick and R.W. Taylor (eds.), SPE, Dallas (1962) 22–13.

6. Dodson, C.R. and Standing, M.B.: "Pressure, Volume, Temperature and Solubility Relations for Natural Gas-Water Mixtures,"
   Drill. and Prod. Prac., API (1944) 173–179.

7. Osif, T.L.: "The Effects of Salt, Gas, Temperature, and Pressure on
   the Compressibility of Water," *SPE Res. Eng.* (Feb. 1988) *3*, No. 1,
   175–181.

8. Ramey, H.J., Jr.: "Rapid Methods for Estimating Reservoir
   Compressibilities," *Trans.*, AIME (1964) *231*, 447–454.

9. Matthews, C.S. and Russell, D.G.: *Pressure Buildup and Flow
   Tests in Wells*, Monograph Series, SPE, Dallas (1967) *1*.

10. Katz, D.L., et al.: *Handbook of Natural Gas Engineering*, McGraw-
    Hill Book Co., Inc., New York City (1959).

11. Deaton, W.M. and Frost, E.M., Jr.: *Gas Hydrates and Their
    Relation to the Operation of Natural Gas Pipelines*, Monograph
    Series, USBM, (1946) *8*.

12. Hoot, W.F., Azarnoosh, A. and McKetta, J.J., Jr.: "Solubility of Water in Hydrocarbons," *Pet. Refiner* (May 1957) *36*, 255-256.
13. *Log Interpretation Charts*, Schlumberger Well Services, Houston (1986) 5.
14. Jennings, H.Y., Jr. and Newman, G.H.: "The Effect of Temperature and Pressure on the Interfacial Tension of Water Against Methane-Normal Decane Mixtures," *Trans.*, AIME (1971) *251*, 171–175.
15. Hocott, C.R.: "Interfacial Tension between Water and Oil Under Reservoir Conditions," *Trans.*, AIME (1939) *132*, 184–190.

General References

Patton, C.C.: *Oil Field Water Systems*, Campbell Petroleum Series, Norman, Oklahoma (1981).

Case, L.C.: *Water Problems in Oil Production*, 2d Ed., PennWell Publishing Co., Tulsa (1970).

Levorsen, A.I.: *Geology of Petroleum*, 2d Ed., W.H. Freeman and Co., San Francisco (1967).

Collins, A.G.: "Properties of Produced Waters," *Petroleum Engineering Handbook*, H.B. Bradley et al. (eds.), SPE, Richardson, TX (1987) 24-1–24-23.

Kinghorn, R.R.F.: *An Introduction to the Physics and Chemistry of Petroleum*, John Wiley and Sons, New York (1983).

Gas Hydrates

17

Hydrocarbon gas and liquid water combine to form solids resembling wet snow at temperatures somewhat above the temperature at which water solidifies. These solids are called *gas hydrates*. They are one of a form of complexes known as *clathrates*. This phenomenon particularly interests those in the petroleum industry because these solids can form at temperatures and pressures normally encountered in producing and transporting natural gases.

Gas Hydrate Formation

Gas hydrates behave as solutions of gases in crystalline solids rather than as chemical compounds. The main framework of the hydrate crystal is formed with water molecules. The hydrocarbon molecules occupy void spaces within the lattice of water molecules.

Hydrate formation is physical rather than chemical in nature. Apparently, no strong chemical bonds are formed between the hydrocarbon and water molecules. Actually, the hydrocarbon molecules are free to rotate within the void spaces.

The water framework seems ice-like because of comparable heats of formation. However, the crystal lattice is somewhat different than ice since an ice lattice provides no space for even the smallest hydrocarbon molecule.

Two types of hydrate crystal lattices are known. Each contains void spaces of two different sizes. One lattice has voids sized to accept small molecules such as methane and larger molecules such as propane at a ratio of about two small molecules to one large molecule. The other lattice accepts methane molecules and medium-sized molecules—such as ethane—at a ratio of about three mediums to one small.

Although gas hydrates appear to be solid solutions rather than chemical compounds, a specific number of water molecules is associated

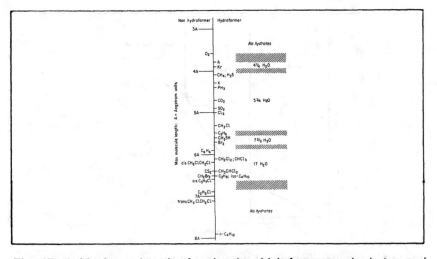

Fig. 17–1. Maximum length of molecule which forms gas hydrates and the water-to-gas ratio in the hydrate crystal. Several molecules of the same maximum dimension which do not form hydrates are also given. (From *Handbook of Natural Gas Engineering* by Katz, et al. Copyright 1959 by McGraw-Hill Book Company. Used with permission of McGraw-Hill Book Company.)

with each gas molecule. This is due to the framework of the crystal. The ratio depends primarily on the size of the gas molecule.

Figure 17–1 shows a number of gases which form hydrates, along with the maximum length of each molecule in angstrom units and the approximate ratio of water molecules to gas molecules in the resulting crystals.[1] Not only the size of the gas molecule but also its isomeric configuration control whether or not hydrate will form. For example, iso-butane readily forms hydrate, but the longer n-butane molecule forms hydrate only at temperatures slightly above the freezing point of water. Yet, in the presence of other gases with smaller molecules, the n-butane molecules readily enter the hydrate. This indicates that n-butane molecules occupy the larger voids in the crystal lattice once hydrate formation begins.

Conditions For Gas Hydrate Formation

The most important consideration in hydrate formation is that liquid water must be present for hydrate to form. Even with liquid water present a metastable equilibrium can exist between water and gas at conditions of

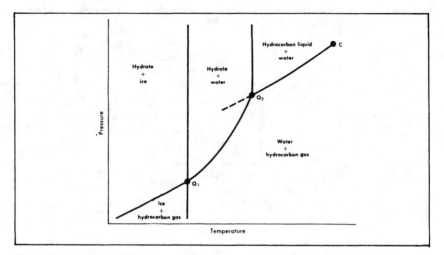

Fig. 17–2. Hydrate portion of the phase diagram of a typical mixture of water and a light hydrocarbon.

pressure and temperature for which hydrate formation could occur. But, once seed crystals are formed, hydration occurs readily.

Seed crystals begin to form at temperatures 3 to 10°F lower than the hydrate-forming temperatures discussed later in this chapter.[2] Or, at a given temperature, seed crystals start forming at 300 or more psi above hydrate-forming pressure. However, dust or rust particles may act like seed crystals in initiating hydrate formation.

A portion of the phase diagram for a mixture of water and a light hydrocarbon is given in Figure 17–2. The line $\overline{Q_2C}$ separates the region in which liquid water and hydrocarbon gas exist from the region in which liquid water and hydrocarbon liquid exist. None of the phases is pure; all contain slight amounts of the other substance according to their mutual solubility.

Line $\overline{Q_2C}$ is parallel to and slightly above the vapor-pressure line for the pure hydrocarbon. The dashed line to the left of point Q_2 is simply an extension of the vapor-pressure line of the hydrocarbon. Point C is the three-phase critical point at which the properties of the hydrocarbon gas and liquid merge to form a single hydrocarbon phase in equilibrium with liquid water.

The line $\overline{Q_1Q_2}$ separates the area in which liquid water and hydrocarbon gas exist from the area in which liquid water and hydrate exist. This line represents the conditions at which gas and liquid water combine to form hydrate.

Point Q_2 is a quadruple point. At Q_2, four phases are in equilibrium: liquid water, hydrocarbon liquid, hydrocarbon gas, and solid hydrate. The almost vertical line extending from point Q_2 separates the area of liquid water and hydrocarbon liquid from the area of liquid water and hydrate.

Q_1, which occurs at approximately 32°F, is also a quadruple point representing the point at which ice, hydrate, liquid water, and hydrocarbon gas exist in equilibrium. The vertical line extending from point Q_1 separates the area for hydrate and liquid water from the area for hydrate and ice.

The line of major interest on this phase diagram is the line $\overline{Q_1Q_2}$, which represents the equilibrium between hydrocarbon gas, liquid water, and hydrate.

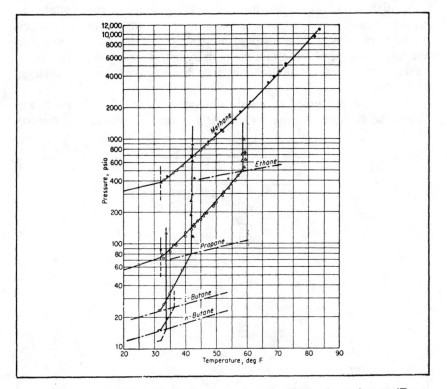

Fig. 17–3. Hydrate-forming conditions for paraffin hydrocarbons. (From *Handbook of Natural Gas Engineering* by Katz et al. Copyright 1959 by McGraw-Hill Book Company. Used with permission of McGraw-Hill Book Company.)

Figure 17–3 shows the $\overline{Q_1Q_2}$ lines for the five hydrate-forming hydrocarbon constituents of natural gas.[1] Portions of the vapor-pressure lines of the hydrocarbons and parts of the vertical lines from Q_1 and Q_2 also are shown. Pressures and temperatures along the $\overline{Q_1Q_2}$ line for each hydrocarbon indicate conditions for which hydrate formation occurs for mixtures of that hydrocarbon with liquid water. Methane forms hydrate at much higher temperatures than the larger hydrocarbons. However, the pressure required for hydrate formation becomes significantly lower as molecular size increases.

Figure 17–4 gives a comparison of the hydrate-formation lines of methane and propane with the hydrate-formation lines for mixtures of these two hydrocarbons.[3] Note that very small quantities of the larger hydrocarbon in the mixtures cause large reductions in the pressures required to initiate hydrate formation.

A mixture of 40% methane and 60% propane forms hydrate at pressures nearly as low as pure propane, but hydrate forms at much higher temperatures than for pure propane. Thus, mixtures of methane and larger hydrocarbons retain the high hydrate-forming temperatures of methane and approach the lower hydrate-forming pressures of the larger molecules.

This fact is confirmed by the data given in Figure 17–5.[1] This figure shows the hydrate-forming conditions for several natural gases along

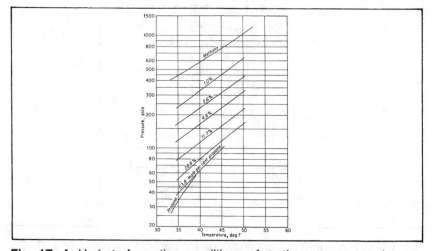

Fig. 17–4. Hydrate-formation conditions of methane-propane mixtures. (Deaton and Frost, *U.S. Bureau of Mines, Monograph 8*, 1946, 20. Courtesy, Bureau of Mines, U.S. Department of the Interior.)

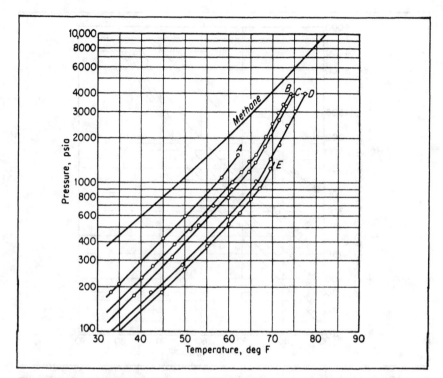

Fig. 17–5. Hydrate-forming conditions for natural gases. Symbols A through E represent gases of increasing specific gravity. (From *Handbook of Natural Gas Engineering* by Katz et al. Copyright 1959 by McGraw-Hill Book Company. Used with permission of McGraw-Hill Book Company.)

with the hydrate-formation line for methane. The pressures at which hydrate formation occurs are significantly lower for the natural gases than for methane. The natural gases with higher specific gravities (that is, gases with more of the heavier hydrocarbons) form hydrates at much lower pressures than the lighter gases.

The data are incomplete at the higher temperatures, but apparently all the gases form hydrates at temperatures somewhat above 70°F.

Figure 17–6 was developed from the data of Figure 17–5.[4] Figure 17–6 is a correlation of hydrate-forming conditions for natural gases with various specific gravities. This figure can be used to estimate the conditions under which hydrates will form. The resulting hydrate formation conditions must be used with caution because there is a great discrepancy between the limited published data and the correlation shown in Figure 17–6.[5] The differences in hydrate-forming pressures

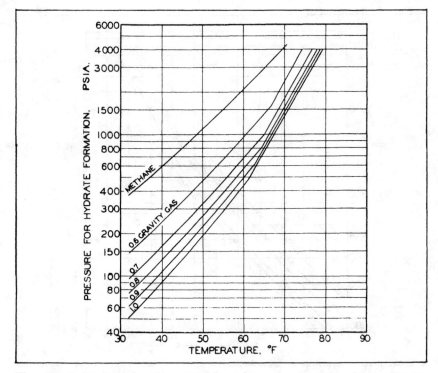

Fig. 17–6. Hydrate-forming conditions for natural gases. (Katz, *Trans.*, AIME, *160*, 140. copyright 1945 SPE-AIME.)

between the available data and the correlation in Figure 17–6 range from about 10% at the higher specific gravities to as much as 35% at the lower specific gravities.[5] Given a value of pressure, the error in the estimate of hydrate-forming temperature could be 5°F or more.

Figure 17–6 is for sweet gases. The presence of hydrogen sulfide and carbon dioxide shifts the lines to the right. This results in an increase in the hydrate temperature for a given pressure or a decrease in hydrate pressure for a given temperature.

EXAMPLE 17–1: *A 0.7 specific gravity natural gas is to be compressed to 500 psia and cooled to 50°F. Is there a possibility of hydrate formation under these conditions?*

Solution

Yes. Figure 17–6 indicates that a 0.7 specific gravity gas forms hydrates at temperatures below 56°F at a pressure of 500 psia if liquid water is present.

Inhibition of Gas Hydrate Formation

The presence of dissolved solids in the water reduces the temperatures at which natural gases form hydrates. Figure 17–7 gives values to reduce the temperatures estimated from Figure 17–6 to account for the effect of dissolved solids in the water.[1]

The water which condenses from natural gas at surface conditions is pure water. Often an inhibitor is added to this water to lower the hydrate-

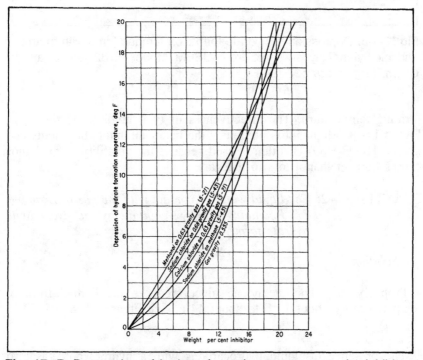

Fig. 17–7. Depression of hydrate-formation temperatures by inhibitors. (From *Handbook of Natural Gas Engineering* by Katz et al. Copyright 1959 by McGraw-Hill Book Company. Used with permission of McGraw-Hill Book Company.)

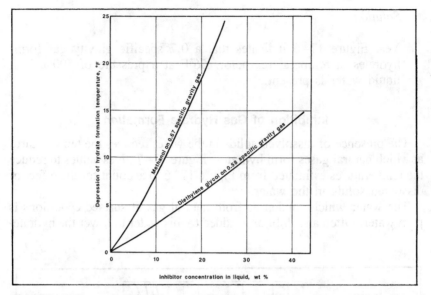

Fig. 17–8. Depression of hydrate-formation temperatures with methanol and diethylene glycol. (Data from USBM *Mono. 8,* 32 and Scauzillo, *Chem. Eng. Prog. 52,* 324.)

forming temperature. The effectiveness of two inhibitors is shown in Figure 17–8. Methanol is obviously the better inhibitor, but diethylene glycol, $HO(C_2H_4O)_2H$, often is used because the volatility of methanol causes high methanol loss to the gas.

EXAMPLE 17–2: *The liquid water in contact with the gas of Example 17–1 contains 35% by weight diethylene glycol. Will hydrate form?*

Solution

Probably not. The diethylene glycol reduces the hydrate-forming temperature by about 12°F to about 44°F. See Figure 17–8.

The presence of liquid hydrocarbons with a hydrocarbon gas also lowers hydrate-forming temperatures.[2] Figure 17–9 indicates this depression in hydrate-forming temperatures. The figure represents a gas with specific gravity less than 0.6 and is probably not sufficiently

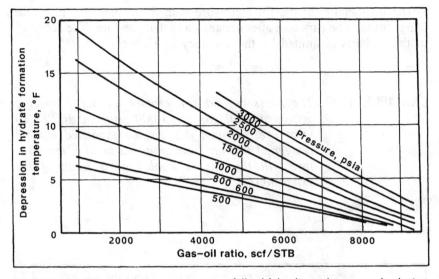

Fig. 17–9. Effects of the presence of liquid hydrocarbons on hydrate-forming temperatures. (Adapted from Scauzillo, *Chem. Eng. Prog. 52*, 1956, 324.)

accurate for use as a general correlation. However, it does give an indication of the inhibiting effect of the presence of hydrocarbon liquid.

Gas Hydrate Formation Caused by Reduction of Pressure

Reducing pressure at normal surface conditions, such as across a choke, causes a reduction in the temperature of the gas. This temperature reduction could cause the condensation of water vapor from the gas. It also could bring the mixture of gas and liquid water to the conditions necessary for hydrate formation.

The temperature reductions accompanying pressure reductions have been calculated for typical natural gases. These results combined with the hydrate-formation conditions given in Figure 17–6 provide calculation charts giving the maximum reduction in pressure to which a natural gas can be subjected prior to the onset of hydrate formation.[4] Figures 17–10 through 17–14 give these calculation charts for natural gases of various specific gravities. Their use is relatively simple.

The charts are entered at the intersection of the initial pressure and the initial temperature. The lowest pressure to which the gas can be expanded without danger of hydrate formation is obtained from the abscissa directly below this intersection.

The intersection of the final pressure with the dashed line gives the temperature to be expected after expansion to that pressure. The accuracy of these charts is limited by the accuracy of Figure 17–6.

EXAMPLE 17–3: *The pressure and temperature on a 0.7 specific gravity natural gas are 3,000 psia and 160°F. To what pressure can this gas be expanded without danger of hydrate formation and to what temperature will the gas be cooled as a result of the expansion?*

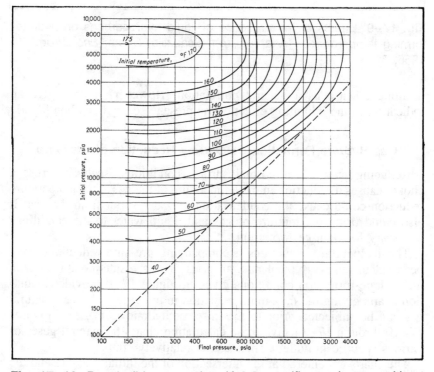

Fig. 17–10. Permissible expansion of 0.6 specific gravity gas without hydrate formation. (Katz, *Trans.*, AIME, *160*, 140. Copyright 1945 SPE-AIME.)

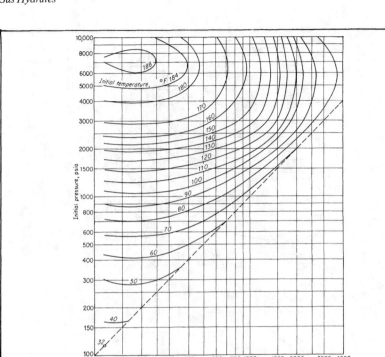

Fig. 17–11. Permissible expansion of 0.7 specific gravity gas without hydrate formation. (Katz, *Trans.*, AIME, *160*, 140. Copyright 1945 SPE-AIME.)

Solution

Enter the ordinate of Figure 17–11 with an initial pressure of 3000 psia. Move horizontally to the 160°F isotherm. Move vertically to the abscissa, 580 psia, which is the lowest final pressure precluding hydrate formation. The intersection of a final pressure of 580 psia with the dashed line gives 59°F, which is the temperature after expansion to 580 psia.

Exercises

17–1. Methane and liquid water are in contact at 4000 psia and 60°F. Do you expect hydrate to form? Explain the reason for your answer.

17–2. Suppose the gas of Exercise 17–1 contains 10 mole % propane. Do you expect hydrate to form? Explain the reason for your answer.

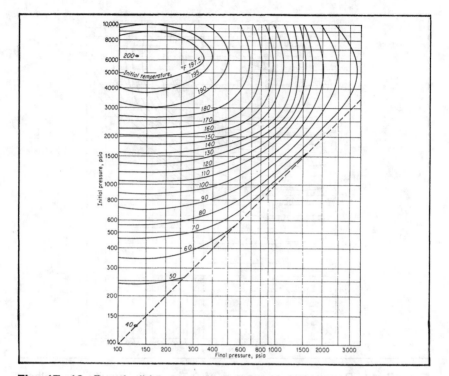

Fig. 17–12. Permissible expansion of 0.8 specific gravity gas without hydrate formation. (Katz, *Trans.*, AIME, *160*, 140. Copyright 1945 SPE-AIME.)

17–3. Suppose the liquid water of Exercise 17–1 contains 20 wt % solids. Will hydrate form? Explain the reason for your answer.

17–4. Methane and liquid water are in contact at 1500 psia and 60°F. Do you expect hydrate to form? Explain the reason for your answer.

17–5. Suppose the methane of Exercise 17–4 contains 5 mole % propane. Will hydrate form? Explain the reason for your answer.

17–6. A 0.7 specific gravity gas is handled in a gas gathering system. Some liquid water is present. At what pressures is hydrate formation possible when the temperature is 70°F?

17–7. The liquid water of Exercise 17–6 contains 10 wt % solids. At what pressures is hydrate formation possible when the temperature is 70°F?

17–8. A 0.75 specific gravity gas is collected in a gathering system. The pressure in the gathering system is 1000 psia. The liquid water in the gathering system condensed from the produced gas (0%

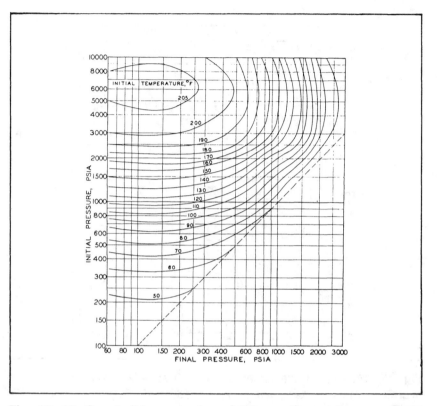

Fig. 17–13. Permissible expansion of 0.9 specific gravity gas without hydrate formation. (Katz, *Trans.*, AIME, *160*, 140. Copyright 1945 SPE-AIME.)

solids). At what temperatures is hydrate formation possible?

17–9. The gathering system of Exercise 17–8 must be protected from hydrate formation to a temperature of 40°F. Methanol will be added to the liquid water. What concentration must the methanol reach to inhibit hydrate formation?

17–10. What concentration of diethylene glycol is required to inhibit hydrate formation to 40°F in Exercise 17–8?

17–11. Tubinghead conditions of a dry gas well are 2600 psia and 110°F. The gas has specific gravity of 0.600. Some water condenses from the gas during production. Gas pressure is reduced through a choke to 1200 psia. Is there a possibility of hydrate formation?

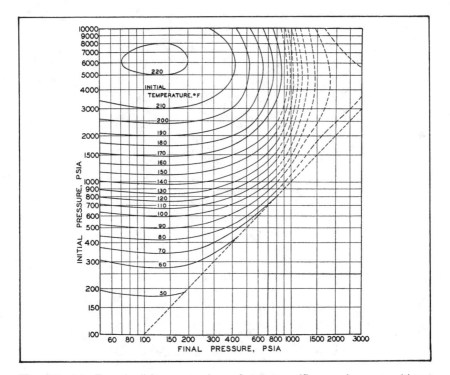

Fig. 17-14. Permissible expansion of 1.0 specific gravity gas without hydrate formation. (Katz, *Trans.*, AIME, *160*, 140. Copyright 1945 SPE-AIME.)

17-12. A 0.70 specific gravity gas is expanded through a choke. Upstream conditions are 3000 psia and 150°F. What is the lowest pressure to which the gas can be expanded prior to onset of hydrate formation? What is the temperature at this pressure?

References

1. Katz, D.L., et al.*: Handbook of Natural Gas Engineering,* McGraw-Hill Book Co., Inc., New York City (1959).
2. Scauzillo, F.R.: "Inhibiting Hydrate Formations in Hydrocarbon Gases," *Chem. Eng. Prog.* (Aug. 1956) *52,* 324–328.
3. Deaton, W.M. and Frost, E.M., Jr.: *Gas Hydrates and Their Relation to the Operation of Natural Gas Pipelines,* Monograph Series, USBM, (1946) *8.*

4. Katz, D.L.: "Prediction of Conditions for Hydrate Formation in Natural Gases," *Trans.*, AIME (1945) *160*, 141-149.
5. Sloan, E.D.: "Phase Equilibria of Natural Gas Hydrates," paper presented at 63rd Annual GPA Convention, New Orleans, March 19–21, 1984.

General Reference

Kobayashi, R., Kyoo, Y.S. and Sloan, E.D.: "Phase Behavior of Water/ Hydrocarbon Systems," *Petroleum Engineering Handbook*, H.B. Bradley et al. (eds.), SPE, Richardson, TX (1987).

Appendix A

Table A–1

Physical Constants

*See the Table of Notes and References.

Number	Compound	Formula	A. Molar mass (molecular weight)	B. Boiling point, °F 14.696 psia	Vapor pressure, psia 100 °F	C. Freezing point, °F 14.696 psia	D. Refractive index, n_D 60 °F	Critical constants Pressure, psia	Temperature, °F	Volume, ft³/lbm	Number
1	Methane	CH_4	16.043	-258.73	(5000)•	-296.44•	1.00042•	666.4	-116.67	0.0988	1
2	Ethane	C_2H_6	30.070	-127.49	(800)•	-297.04•	1.20971•	706.5	89.92	0.0783	2
3	Propane	C_3H_8	44.097	-43.75	188.64	-305.73•	1.29480•	616.0	206.06	0.0727	3
4	Isobutane	C_4H_{10}	58.123	10.78	72.581	-255.28	1.3245•	527.9	274.46	0.0714	4
5	n-Butane	C_4H_{10}	58.123	31.08	51.706	-217.05	1.33588•	550.6	305.62	0.0703	5
6	Isopentane	C_5H_{12}	72.150	82.12	20.445	-255.82	1.35631	490.4	369.10	0.0679	6
7	n-Pentane	C_5H_{12}	72.150	96.92	15.574	-201.51	1.35992	488.6	385.8	0.0675	7
8	Neopentane	C_5H_{12}	72.150	49.10	36.69	2.17	1.342•	464.0	321.13	0.0673	8
9	n-Hexane	C_6H_{14}	86.177	155.72	4.9597	-139.58	1.37708	436.9	453.8	0.0688	9
10	2-Methylpentane	C_6H_{14}	86.177	140.47	6.769	-244.62	1.37387	436.6	435.83	0.0682	10
11	3-Methylpentane	C_6H_{14}	86.177	145.89	6.103	———	1.37888	453.1	448.4	0.0682	11
12	Neohexane	C_6H_{14}	86.177	121.52	9.859	-147.72	1.37128	446.8	420.13	0.0667	12
13	2,3-Dimethylbutane	C_6H_{14}	86.177	136.36	7.406	-199.38	1.37730	453.5	440.29	0.0665	13
14	n-Heptane	C_7H_{16}	100.204	209.16	1.620	-131.05	1.38989	396.8	512.7	0.0691	14
15	2-Methylhexane	C_7H_{16}	100.204	194.09	2.272	-180.89	1.38714	396.5	495.00	0.0673	15
16	3-Methylhexane	C_7H_{16}	100.204	197.33	2.131	———	1.39091	408.1	503.80	0.0646	16
17	3-Ethylpentane	C_7H_{16}	100.204	200.25	2.013	-181.48	1.39566	419.3	513.39	0.0665	17
18	2,2-Dimethylpentane	C_7H_{16}	100.204	174.54	3.494	-190.86	1.38446	402.2	477.23	0.0665	18
19	2,4-Dimethylpentane	C_7H_{16}	100.204	176.89	3.293	-182.63	1.38379	396.9	475.95	0.0668	19
20	3,3-Dimethylpentane	C_7H_{16}	100.204	186.91	2.774	-210.01	1.38564	427.2	505.87	0.0662	20
21	Triptane	C_7H_{16}	100.204	177.58	3.375	-12.81	1.39168	428.4	496.44	0.0636	21
22	n-Octane	C_8H_{18}	114.231	258.21	0.53694	-70.18	1.39956	360.7	564.22	0.0690	22
23	Diisobutyl	C_8H_{18}	114.231	228.39	1.102	-132.11	1.39461	360.6	530.44	0.0676	23
24	Isooctane	C_8H_{18}	114.231	210.63	1.709	-161.27	1.38624	372.4	519.46	0.0656	24
25	n-Nonane	C_9H_{20}	128.258	303.47	0.17953	-64.28	1.40746	331.8	610.68	0.0684	25
26	n-Decane	$C_{10}H_{22}$	142.285	345.48	0.06088	-21.36	1.41385	305.2	652.0	0.0679	26
27	Cyclopentane	C_5H_{10}	70.134	120.65	9.915	-136.91	1.40896	653.8	461.2	0.0594	27
28	Methylcyclopentane	C_6H_{12}	84.161	161.25	4.503	-224.40	1.41210	548.9	499.35	0.0607	28
29	Cyclohexane	C_6H_{12}	84.161	177.29	3.266	43.77	1.42862	590.8	536.6	0.0586	29
30	Methylcyclohexane	C_7H_{14}	98.188	213.68	1.609	-195.87	1.42538	503.5	570.27	0.0600	30
31	Ethene(Ethylene)	C_2H_4	28.054	-154.73	(1400)•	-272.47•	(1.228)•	731.0	48.54	0.0746	31
32	Propene(Propylene)	C_3H_6	42.081	-53.84	227.7	-301.45•	(1.3130)•	668.6	197.17	0.0689	32
33	1-Butene(Butylene)	C_4H_8	56.108	20.79	62.10	-301.63•	1.3494•	583.5	295.48	0.0685	33
34	cis-2-Butene	C_4H_8	56.108	38.69	45.95	-218.06	1.3665•	612.1	324.37	0.0668	34
35	trans-2-Butene	C_4H_8	56.108	33.58	49.87	-157.96	1.3563•	587.4	311.86	0.0679	35
36	Isobutene	C_4H_8	56.108	19.59	63.02	-220.65	1.3512•	580.2	292.55	0.0682	36
37	1-Pentene	C_5H_{10}	70.134	85.93	19.12	-265.39	1.37426	511.8	376.93	0.0675	37
38	1,2-Butadiene	C_4H_6	54.092	51.53	36.53	-213.16	———	(653.)•	(340.)•	(0.065)•	38
39	1,3-Butadiene	C_4H_6	54.092	24.06	59.46	-164.02	1.3975•	627.5	305.	0.0654	39
40	Isoprene	C_5H_8	68.119	93.31	16.68	-230.73	1.42498	(558.)•	(412.)•	(0.065)•	40
41	Acetylene	C_2H_2	26.038	-120.49•	———	-114.5•	———	890.4	95.34	0.0695	41
42	Benzene	C_6H_6	78.114	176.18	3.225	41.95	1.50396	710.4	552.22	0.0531	42
43	Toluene	C_7H_8	92.141	231.13	1.033	-139.00	1.49942	595.5	605.57	0.0550	43
44	Ethylbenzene	C_8H_{10}	106.167	277.16	0.3716	-138.966	1.49826	523.0	651.29	0.0565	44
45	o-Xylene	C_8H_{10}	106.167	291.97	0.2643	-13.59	1.50767	541.6	674.92	0.0557	45
46	m-Xylene	C_8H_{10}	106.167	282.41	0.3265	-54.18	1.49951	512.9	651.02	0.0567	46
47	p-Xylene	C_8H_{10}	106.167	281.07	0.3424	55.83	1.49810	509.2	649.54	0.0570	47
48	Styrene	C_8H_8	104.152	293.25	0.2582	-23.10	1.54937	587.8	(703.)•	0.0534	48
49	Isopropylbenzene	C_9H_{12}	120.194	306.34	0.1884	-140.814	1.49372	465.4	676.3	0.0572	49
50	Methyl alcohol	CH_4O	32.042	148.44	4.629	-143.79	1.33034	1174.	463.08	0.0590	50
51	Ethyl alcohol	C_2H_6O	46.069	172.90	2.312	-173.4	1.36346	890.1	465.39	0.0581	51
52	Carbon monoxide	CO	28.010	-312.68	———	-337.00•	1.00036•	507.5	-220.43	0.0532	52
53	Carbon dioxide	CO_2	44.010	-109.257•	———	-69.83•	1.00048•	1071.	87.91	0.0344	53
54	Hydrogen sulfide	H_2S	34.08	-76.497	394.59	-121.88•	1.00060•	1300.	212.45	0.0461	54
55	Sulfur dioxide	SO_2	64.06	14.11	85.46	-103.86	1.00062•	1143.	315.8	0.0305	55
56	Ammonia	NH_3	17.0305	-27.99	211.9	-107.88•	1.00036•	1646.	270.2	0.0681	56
57	Air	N_2+O_2	28.9625	-317.8	———	———	1.00028•	546.9	-221.31	0.0517	57
58	Hydrogen	H_2	2.0159	-422.955•	———	-435.26•	1.00013•	188.1	-399.9	0.5185	58
59	Oxygen	O_2	31.9988	-297.332•	———	-361.820•	1.00027•	731.4	-181.43	0.0367	59
60	Nitrogen	N_2	28.0134	-320.451	———	-346.00•	1.00028•	493.1	-232.51	0.0510	60
61	Chlorine	Cl_2	70.906	-29.13	157.3	-149.73•	1.3878•	1157.	290.75	0.0280	61
62	Water	H_2O	18.0153	212.000•	0.9501	32.00	1.33335	3198.8	705.16	0.0497	62
63	Helium	He	4.0026	-452.09	———	———	1.00003•	32.99	-450.31	0.2300	63
64	Hydrogen chloride	HCl	36.461	-121.27	906.71	-173.52•	1.00042•	1205.	124.77	0.0356	64

12/12/86

NOTE: Numbers in this table do not have accuracies greater than 1 part in 1000; in some cases extra digits have been added to calculated values to achieve consistency or to permit recalculation of experimental values.

Physical properties (from *Engineering Data Book*, GPSA, 1987, with permission).

Table A-1 (cont.)

Physical Constants

*See the Table of Notes and References.

Number	E. Density of liquid 14.696 psia, 60°F — Relative density (specific gravity) 60°F/60°F	lbm/gal	gal./lb mole	F. Temperature coefficient of density, 1/°F	G. Acentric factor, ω	H. Compressibility factor, Z of real gas, 14.696 psia, 60°F	I. Ideal gas 14.696 psia, 60°F — Relative density (specific gravity) Air = 1	ft³ gas/lbm	ft³ gas/gal. liquid	J. Specific Heat 60°F 14.696 psia Btu/(lbm·°F) — Cp Ideal gas	Cp Liquid	Number
1	(0.3)•	(2.5)•	(6.4172)•	——	0.0104	0.9980	0.5539	23.654	(59.135)•	0.52669	——	1
2	0.35619•	2.9696•	10.126•	——	0.0979	0.9919	1.0382	12.620	37.476•	0.40782	0.97225	2
3	0.50699•	4.2268•	10.433•	-0.00162•	0.1522	0.9825	1.5226	8.6059	36.375•	0.38852	0.61996	3
4	0.56287•	4.6927•	12.386•	-0.00119•	0.1852	0.9711	2.0068	6.5291	30.639•	0.38669	0.57066	4
5	0.58401•	4.8690•	11.937•	-0.00106•	0.1995	0.9667	2.0068	6.5291	31.790•	0.39499	0.57272	5
6	0.62470	5.2082	13.853	-0.00090	0.2280	——	2.4912	5.2596	27.393	0.38440	0.53331	6
7	0.63112	5.2617	13.712	-0.00086	0.2514	——	2.4912	5.2596	27.674	0.38825	0.54363	7
8	0.59666•	4.9744•	14.504•	-0.00106•	0.1963	0.9582	2.4912	5.2596	26.163•	0.39038	0.55021	8
9	0.66383	5.5344	15.571	-0.00075	0.2994	——	2.9755	4.4035	24.371	0.38628	0.53327	9
10	0.65785	5.4846	15.713	-0.00076	0.2780	——	2.9755	4.4035	24.152	0.38526	0.52732	10
11	0.66901	5.5776	15.451	-0.00076	0.2732	——	2.9755	4.4035	24.561	0.37902	0.51876	11
12	0.65385	5.4512	15.809	-0.00076	0.2326	——	2.9755	4.4035	24.005	0.38231	0.51367	12
13	0.66631	5.5551	15.513	-0.00076	0.2469	——	2.9755	4.4035	24.462	0.37762	0.51308	13
14	0.68820	5.7376	17.464	-0.00068	0.3494	——	3.4598	3.7872	21.729	0.38447	0.52802	14
15	0.68310	5.6951	17.595	-0.00070	0.3298	——	3.4598	3.7872	21.568	0.38041	0.52199	15
16	0.69165	5.7664	17.377	-0.00070	0.3232	——	3.4598	3.7872	21.838	0.37882	0.51019	16
17	0.70276	5.8590	17.103	-0.00069	0.3105	——	3.4598	3.7872	22.189	0.38646	0.51410	17
18	0.67829	5.6550	17.720	-0.00070	0.2871	——	3.4598	3.7872	21.416	0.38594	0.51678	18
19	0.67733	5.6470	17.745	-0.00073	0.3026	——	3.4598	3.7872	21.386	0.39414	0.52440	19
20	0.69772	5.8170	17.226	-0.00067	0.2674	——	3.4598	3.7872	22.030	0.38306	0.50138	20
21	0.69457	5.7907	17.304	-0.00068	0.2503	——	3.4598	3.7872	21.930	0.37724	0.49920	21
22	0.70696	5.8940	19.381	-0.00064	0.3977	——	3.9441	3.3220	19.580	0.38331	0.52406	22
23	0.69793	5.8187	19.632	-0.00067	0.3564	——	3.9441	3.3220	19.330	0.37571	0.51130	23
24	0.69624	5.8046	19.679	-0.00065	0.3035	——	3.9441	3.3220	19.283	0.38222	0.48951	24
25	0.72187	6.0183	21.311	-0.00051	0.4445	——	4.4284	2.9588	17.607	0.38246	0.52244	25
26	0.73421	6.1212	23.245	-0.00057	0.4898	——	4.9127	2.6671	16.326	0.38179	0.52103	26
27	0.75050	6.2570	11.209	-0.00073	0.1950	——	4.2215	5.4110	33.856	0.27199	0.42182	27
28	0.75349	6.2819	13.397	-0.00069	0.2302	——	2.0959	4.5090	28.325	0.30100	0.44126	28
29	0.78347	6.5319	12.885	-0.00065	0.2096	——	2.0959	4.5090	29.452	0.28817	0.43584	29
30	0.77400	6.4529	15.216	-0.00062	0.2358	——	3.3902	3.8649	24.940	0.31700	0.44012	30
31					0.0865	0.9936	0.9686	13.527	——	0.35697	——	31
32	0.52095•	4.3432•	9.6889•	-0.00173•	0.1356	0.9844	1.4529	9.0179	39.167•	0.35714	0.57116	32
33	0.60107•	5.0112•	11.197•	-0.00112•	0.1941	0.9699	1.9373	6.7636	33.894•	0.35446	0.54533	33
34	0.62717•	5.2288•	10.731•	-0.00105•	0.2029	0.9665	1.9373	6.7636	35.366•	0.33754	0.52980	34
35	0.60996•	5.0853•	11.033•	-0.00106•	0.2128	0.9667	1.9373	6.7636	34.395•	0.35574	0.54215	35
36	0.60040•	5.0056•	11.209•	-0.00117•	0.1999	0.9700	1.9373	6.7636	37.680•	0.35839	0.54839	36
37	0.64571	5.3834	13.028	-0.00089	0.2333	——	2.4215	5.4110	29.129	0.36351	0.51782	37
38	0.65799•	5.4857•	9.8605•	-0.00101•	0.2540	(0.969)	1.8677	7.0156	38.485•	0.34347	0.54029	38
39	0.62723•	5.2293•	10.344•	-0.00110•	0.2007	(0.965)	1.8677	7.0156	36.687•	0.34120	0.53447	39
40	0.68615	5.7205	11.908	-0.00082	0.1568	——	2.3520	5.5710	31.869	0.35072	0.51933	40
41	(0.41796)•	(3.3740)•	(7.473)•	——	0.1949	0.9930	0.8990	14.574	——	0.39754	——	41
42	0.88448	7.3740	10.593	-0.00067	0.2093	——	2.6971	4.8581	35.824	0.24296	0.40989	42
43	0.87190	7.2691	12.676	-0.00059	0.2633	——	3.1814	4.1184	29.937	0.26370	0.40095	43
44	0.87168	7.2673	14.609	-0.00056	0.3027	——	3.6657	3.5744	25.976	0.27792	0.41139	44
45	0.88467	7.3756	14.394	-0.00052	0.3942	——	3.6657	3.5744	26.363	0.28964	0.41620	45
46	0.86875	7.2429	14.658	-0.00053	0.3257	——	3.6657	3.5744	25.889	0.27427	0.40545	46
47	0.86578	7.2181	14.708	-0.00056	0.3216	——	3.6657	3.5744	25.800	0.27471	0.40255	47
48	0.91108	7.5958	13.712	-0.00053	(0.2412)	——	3.5961	3.6435	27.675	0.27110	0.41220	48
49	0.86634	7.2228	16.641	-0.00055	0.3260	——	4.1500	3.1573	22.804	0.29170	0.42053	49
50	0.79626	6.6385	4.8267	-0.00066	0.5649	——	1.1063	11.843	78.622	0.32316	0.59187	50
51	0.79399	6.6196	6.9595	-0.00058	0.6438	——	1.5906	8.2372	54.527	0.33222	0.56610	51
52	0.78939•	6.5812•	4.2561•	——	0.0484	0.9959	0.9671	13.548	89.163•	0.24847	——	52
53	0.81802•	6.8199•	4.6532•	-0.00583•	0.2667	0.9943	1.5196	8.6229	58.807•	0.19911	——	53
54	0.80144•	6.6817•	5.1005•	-0.00157•	0.0948	0.9846	1.1767	11.135	74.401•	0.23827	0.50418	54
55	1.3974•	11.650•	5.4987•	——	0.2548	0.9802	2.2118	5.9238	69.012•	0.14804	0.32460	55
56	0.61832•	5.1550•	3.3037•	——	0.2557	0.9877	0.5880	22.283	114.87•	0.49677	1.1209	56
57	0.87476•	7.2930•	3.9713•	——		1.0000	1.0000	13.103	95.557•	0.23988	——	57
58	0.071070•	0.59252•	3.4022•	——	-0.2202	1.0006	0.06960•	188.25	111.54•	0.24258	——	58
59	1.1421•	9.5221•	3.3605•	——	0.0216	0.9992	1.1048	11.859	112.93•	0.21892	——	59
60	0.80940•	6.7481•	4.1513•	——	0.0372	0.9997	0.9672	13.546	91.413•	0.24828	——	60
61	1.4244•	11.875•	5.9710•	——	0.0878	(0.9875)	2.4482	5.3519	63.554•	0.11377	——	61
62	1.00000	8.33712	2.1609	-0.00009	0.3443	——	0.62202	21.065	175.62	0.44457	0.99974	62
63	0.12510•	1.0430•	3.8376•	——	0.	1.0006	0.1382	94.814	98.891•	1.2404	——	63
64	0.85129•	7.0973•	5.1375•	-0.00300•	0.1259	0.9923	2.1589	10.408	73.869•	0.19086	——	64

12/12/86

NOTE: Numbers in this table do not have accuracies greater than 1 part in 1000; in some cases extra digits have been added to calculated values to achieve consistency or to permit recalculation of experimental values.

Table A-1 (cont.)

Physical Constants

*See the Table of Notes and References.

		K. Heating value, 60°F					L.	M.	Flammability limits, vol % in air mixture		ASTM octane number		
Number	Compound	Net Btu/ft³ Ideal gas 14.696 psia	Btu/lbm Liquid	Gross Btu/ft³ Ideal gas 14.696 psia	Btu/lbm Liquid	Btu/gal. Liquid	Heat of vaporization 14.696 psia at boiling point, Btu/lbm	Air required for combustion Ideal gas ft³(air)/ft³(gas)	Lower	Higher	Motor method D-357	Research method D-908	Number
1	Methane	909.4		1010.0			219.45	9.548	5.0	15.0			1
2	Ethane	1618.7	20277.•	1769.6	22181.•	65869.•	211.14	16.710	2.9	13.0	+0.05	+1.6•	2
3	Propane	2314.9	19757.•	2516.1	21489.•	90830.•	183.01	23.871	2.0	9.5	97.1	+1.8•	3
4	Isobutane	3000.4	19437.•	3251.9	21079.•	98917.•	157.23	31.032	1.8	8.5	97.6	+0.1•	4
5	n-Butane	3010.8	19494.•	3262.3	21136.•	102911.•	165.93	31.032	1.5	9.0	89.6•	93.8•	5
6	Isopentane	3699.0	19303.	4000.9	20891.	108805.	147.12	38.193	1.3	8.0	90.3	92.3	6
7	n-Pentane	3706.9	19335.	4008.9	20923.	110091.	153.57	38.193	1.4	8.3	62.6•	61.7•	7
8	Neopentane	3682.9	19235.•	3984.7	20822.•	103577.•	135.58	38.193	1.3	7.5	80.2	85.5	8
9	n-Hexane	4403.8	19232.	4755.9	20783.	115021.	143.94	45.355	1.1	7.7	26.0	24.8	9
10	2-Methylpentane	4395.2	19202.	4747.3	20753.	113822.	138.45	45.355	1.18	7.0	73.5	73.4	10
11	3-Methylpentane	4398.2	19213.	4750.3	20764.	115813.	140.05	45.355	1.2	7.7	74.3	74.5	11
12	Neohexane	4384.0	19163.	4736.2	20714.	112918.	131.23	45.355	1.2	7.0	93.4	91.8	12
13	2,3-Dimethylbutane	4392.9	19195.	4745.0	20746.	115246.	136.07	45.355	1.2	7.0	94.3	+0.3	13
14	n-Heptane	5100.0	19155.	5502.5	20679.	118648.	136.00	52.516	1.0	7.0	0.0	0.0	14
15	2-Methylhexane	5092.2	19133.	5494.6	20657.	117644.	131.58	52.516	1.0	7.0	46.4	42.4	15
16	3-Methylhexane	5096.0	19146.	5498.6	20671.	119197.	132.10	52.516	(1.01)	(6.8)	55.8	52.0	16
17	3-Ethylpentane	5098.3	19154.	5500.7	20679.	121158.	132.82	52.516	(1.00)	(6.5)	69.3	65.0	17
18	2,2-Dimethylpentane	5079.6	19095.	5481.9	20620.	116606.	125.12	52.516	(1.09)	(6.8)	95.6	92.8	18
19	2,4-Dimethylpentane	5084.2	19111.	5486.7	20635.	116526.	126.57	52.516	(1.08)	(6.8)	83.8	83.1	19
20	3,3-Dimethylpentane	5086.4	19119.	5488.8	20643.	120080.	127.20	52.516	(1.04)	(7.0)	86.6	80.8	20
21	Triptane	5081.2	19103.	5483.5	20628.	119451.	124.21	52.516	(1.08)	(6.8)	+0.1	+1.8	21
22	n-Octane	5796.1	19096.	6248.9	20601.	121422.	129.52	59.677	0.8	6.5			22
23	Diisobutyl	5780.5	19047.	6233.5	20552.	119586.	122.83	59.677	(0.92)	(6.3)	55.7	55.2	23
24	Isooctane	5778.8	19063.	6231.7	20568.	119389.	112.94	59.677	0.95	6.0	100.0	100.0	24
25	n-Nonane	6493.2	19054.	6996.5	20543.	123634.	124.36	66.839	0.7	5.6			25
26	n-Decane	7189.6	19018.	7742.9	20494.	125448.	119.65	74.000	0.7	5.4			26
27	Cyclopentane	3512.1	18825.	3763.7	20186.	126304.	167.33	35.806	(1.48)	(8.3)	84.9•	+0.1	27
28	Methylcyclopentane	4199.4	18771.	4501.2	20132.	126467.	148.54	42.968	1.0	8.35	80.0	91.3	28
29	Cyclohexane	4179.7	18675.	4481.7	20036.	130873.	153.03	42.968	1.3	8.35	77.2	83.0	29
30	Methylcyclohexane	4863.6	18640.	5215.9	20002.	129071.	136.30	50.129	1.1	6.7	71.1	75.8	30
31	Ethene(Ethylene)	1499.1		1599.8			207.41	14.323	2.7	36.0	75.6	+0.03	31
32	Propene(Propylene)	2181.8	19858.	2332.7	(21208.)	(92113.)	188.19	21.484	2.0	11.7	84.9	+0.2	32
33	1-Butene(Butylene)	2878.7	19309.•	3079.0	20670.•	103582.•	167.96	28.645	1.6	10.	80.8•	97.4	33
34	cis-2-Butene	2871.0	19241.•	3072.2	20602.•	107724.•	178.89	28.645	1.6	10.	83.5	100.0	34
35	trans-2-Butene	2866.8	19221.•	3068.0	20582.•	104866.•	174.37	28.645	1.6	10.			35
36	Isobutene	2859.9	19182.•	3061.1	20543.•	102830.•	169.47	28.645	1.8	10.			36
37	1-Pentene	3575.0	19184.	3826.5	20545.	110602.	154.48	35.806	1.5				37
38	1,2-Butadiene	2789.0	19378.•	2939.2	20437.•	112115.•	191.88	26.258	(1.62)	(10.3)	77.1	90.9	38
39	1,3-Butadiene	2729.0	18967.•	2879.9	20025.•	104717.•	185.29	26.258	2.0	12.5			39
40	Isoprene	3410.8	18832.	3612.1	19953.	114141.	163.48	33.419	(1.12)	(8.5)	81.0	99.1	40
41	Acetylene	1423.2	(20887.)	1473.5	(21613.)	(75204.)	151.90	11.935	1.5	100.			41
42	Benzene	3590.9	17256.	3741.8	17989.	132651.	169.24	35.806	1.2	8.0	+2.8		42
43	Toluene	4273.6	17421.	4475.0	18250.	132681.	154.83	42.968	1.2	7.1	+0.3	+5.8	43
44	Ethylbenzene	4970.5	17593.	5222.2	18492.	134387.	144.02	50.129	1.0	8.0	97.9	+0.8	44
45	o-Xylene	4958.2	17544.	5209.9	18444.	136036.	149.10	50.129	1.0	7.6	100.0		45
46	m-Xylene	4956.3	17541.	5207.9	18440.	133559.	147.24	50.129	1.0	7.0	+2.8	+4.0	46
47	p-Xylene	4957.1	17545.	5208.8	18444.	133131.	145.71	50.129	1.0	7.0	+1.2	+3.4	47
48	Styrene	4829.8	17414.	5031.1	18147.	137841.	152.85	47.742	1.1	8.0	+0.2	>+3.•	48
49	Isopropylbenzene	5660.9	17709.	5962.8	18662.	134792.	134.24	57.290	0.8	6.5	99.3	+2.1	49
50	Methyl alcohol	766.1	8559.	866.7	9751.	64731.	462.58	7.161	5.5	44.0			50
51	Ethyl alcohol	1448.1	11530.	1599.1	12770.	84539.	359.07	14.323	3.28	19.0			51
52	Carbon monoxide	320.5		320.5			92.77	2.387	12.50	74.20			52
53	Carbon dioxide	0.0		0.0			246.47•						53
54	Hydrogen sulfide	586.8	6337.•	637.1	6897.•	46086.•	235.63	7.161	4.30	45.50			54
55	Sulfur dioxide	0.0		0.0			167.22						55
56	Ammonia	359.0		434.4			589.48	3.581	15.50	27.00			56
57	Air	0.0		0.0			88.20						57
58	Hydrogen	273.8		324.2			192.74	2.387	4.00	74.20			58
59	Oxygen	0.0		0.0			91.59						59
60	Nitrogen	0.0		0.0			85.59						60
61	Chlorine						123.75						61
62	Water	0.0	•	•	0.0	0.0	970.18						62
63	Helium	0.0		0.0									63
64	Hydrogen chloride						190.43						64

13/13/66

NOTE: Numbers in this table do not have accuracies greater than 1 part in 1000; in some cases extra digits have been added to calculated values to achieve consistency or to permit recalculation of experimental values.

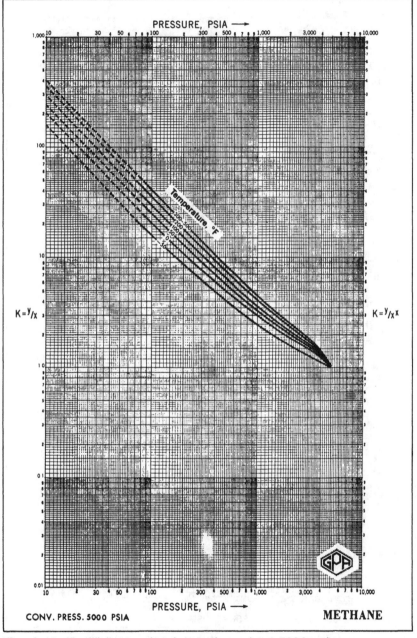

Fig. A–1. Equilibrium ratios for methane in a 5000 psia convergence pressure system. (*Engineering Data Book*, 10th Ed., GPSA, Tulsa, 1987, with permission.)

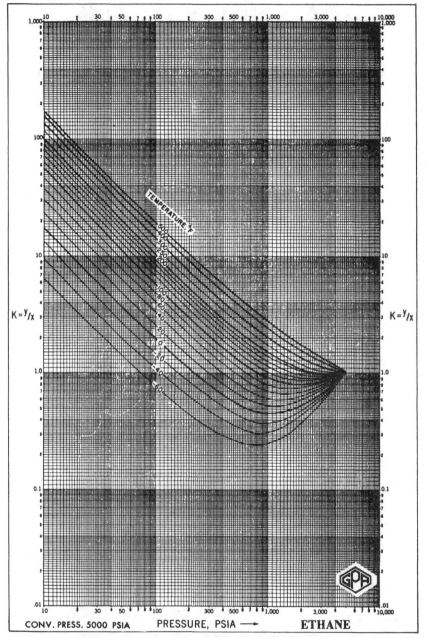

Fig. A–2. Equilibrium ratios for ethane in a 5000 psia convergence pressure system. (*Engineering Data Book*, 10th Ed., GPSA, Tulsa, 1987, with permission.)

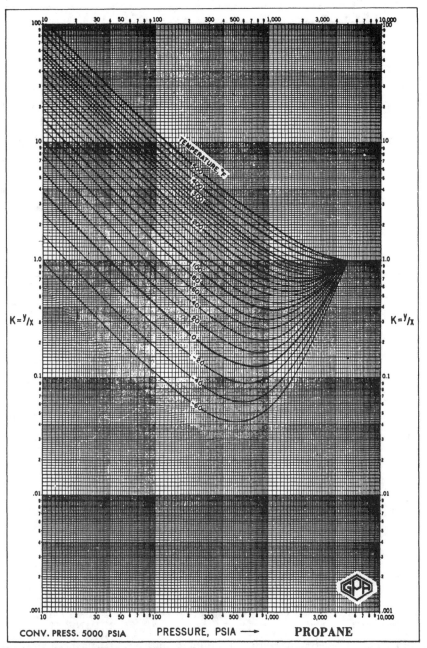

Fig. A–3. Equilibrium ratios for propane in a 5000 psia convergence pressure system. (*Engineering Data Book*, 10th Ed., GPSA, Tulsa, 1987, with permission.)

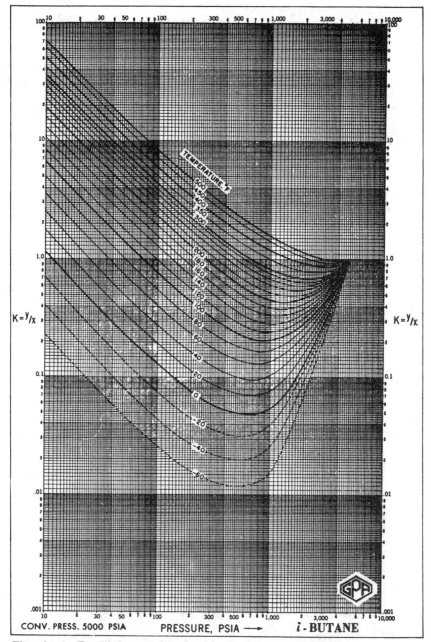

Fig. A–4. Equilibrium ratios for i-butane in a 5000 psia convergence pressure system. (*Engineering Data Book*, 10th Ed., GPSA, Tulsa, 1987, with permission.)

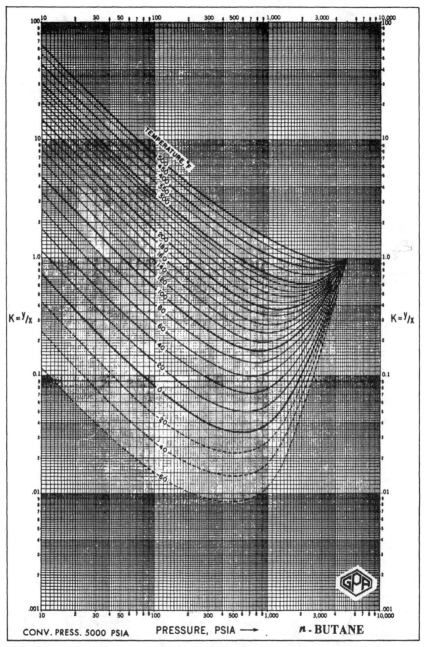

Fig. A–5. Equilibrium ratios for n-butane in a 5000 psia convergence pressure system. (*Engineering Data Book*, 10th Ed., GPSA, Tulsa, 1987, with permission.)

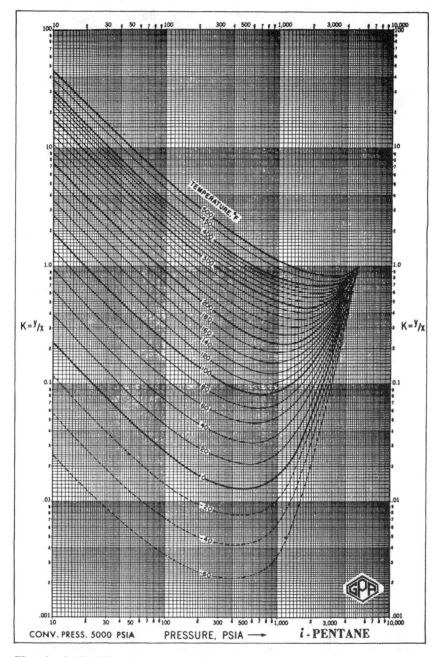

Fig. A–6. Equilibrium ratios for i-pentane in a 5000 psia convergence pressure system. (*Engineering Data Book*, 10th Ed., GPSA, Tulsa, 1987, with permission.)

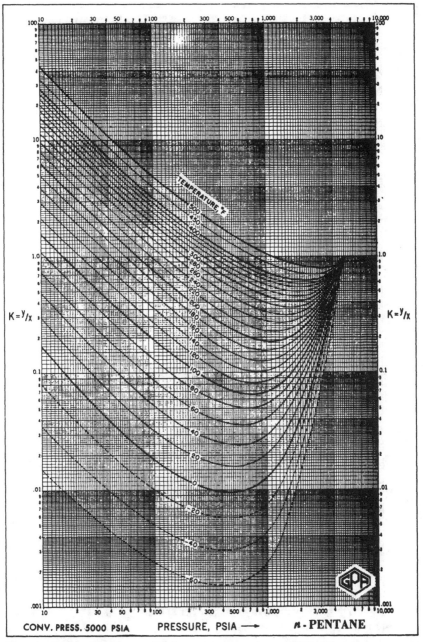

Fig. A–7. Equilibrium ratios for n-pentane in a 5000 psia convergence pressure system. (*Engineering Data Book*, 10th Ed., GPSA, Tulsa, 1987, with permission.)

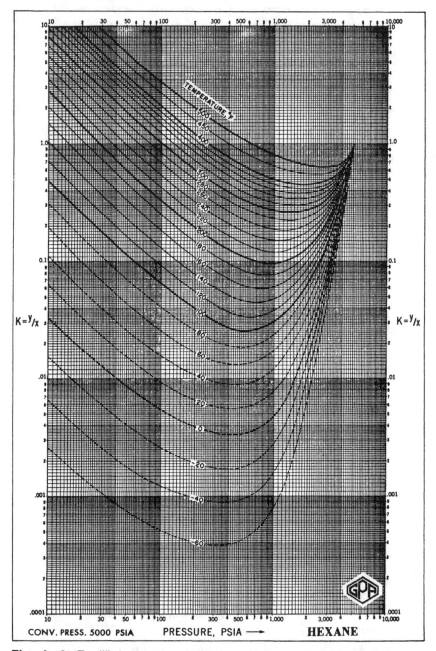

Fig. A–8. Equilibrium ratios for hexane in a 5000 psia convergence pressure system. (*Engineering Data Book*, 10th Ed., GPSA, Tulsa, 1987, with permission.)

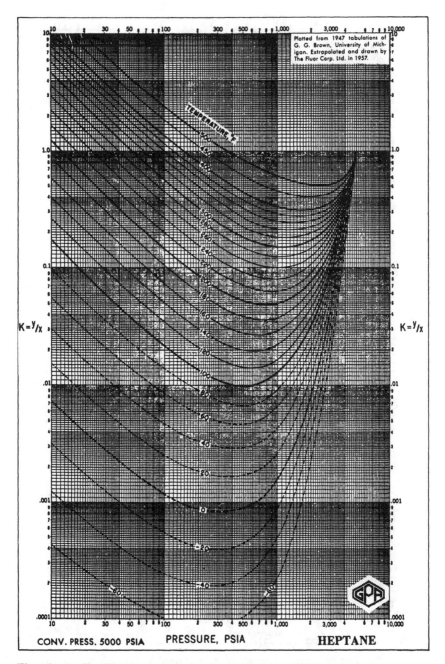

Fig. A–9. Equilibrium ratios for heptane in a 5000 psia convergence pressure system. (*Engineering Data Book*, 10th Ed., GPSA, Tulsa, 1987, with permission.)

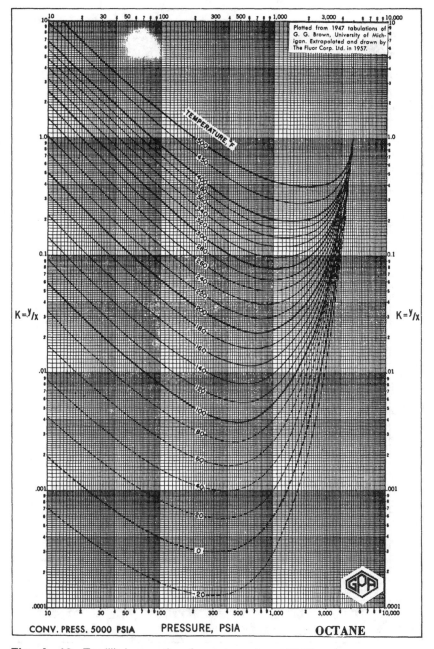

Fig. A–10. Equilibrium ratios for octane in a 5000 psia convergence pressure system. (*Engineering Data Book*, 10th Ed., GPSA, Tulsa, 1987, with permission.)

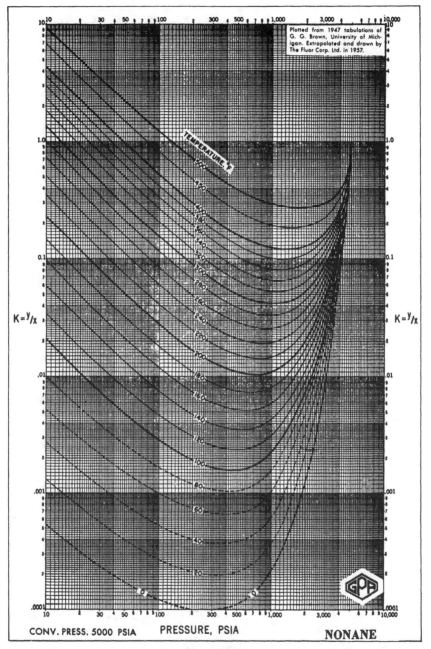

Fig. A–11. Equilibrium ratios for nonane in a 5000 psia convergence pressure system. (*Engineering Data Book*, 10th Ed., GPSA, Tulsa, 1987, with permission.)

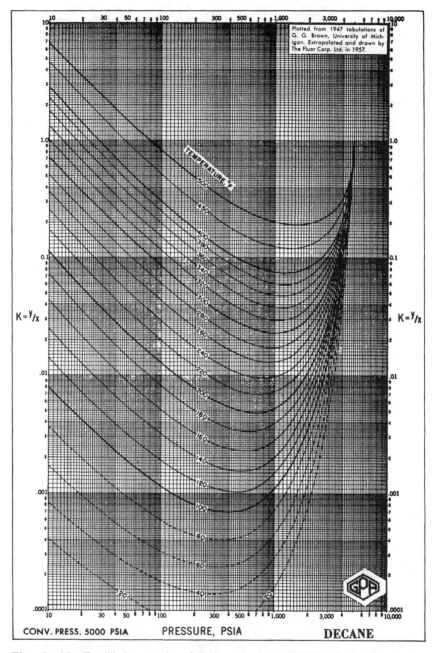

Fig. A-12. Equilibrium ratios for decane in a 5000 psia convergence pressure system. (*Engineering Data Book*, 10th Ed., GPSA, Tulsa, 1987, with permission.)

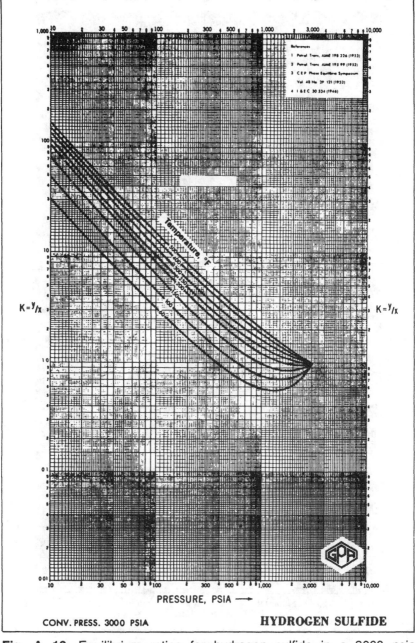

Fig. A–13. Equilibrium ratios for hydrogen sulfide in a 3000 psia convergence pressure system. (*Engineering Data Book*, 10th Ed., GPSA, Tulsa, 1987, with permission.)

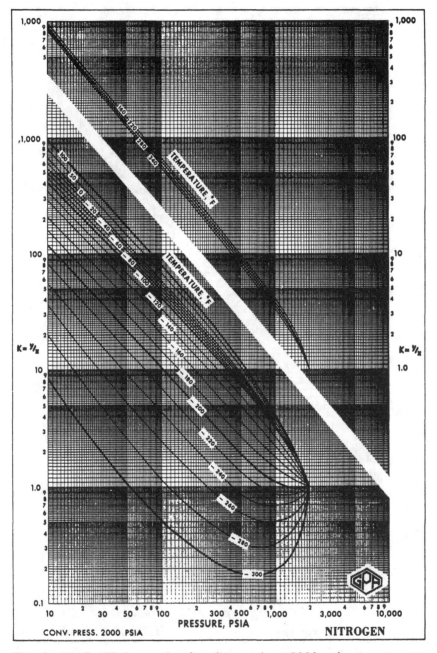

Fig. A–14. Equilibrium ratios for nitrogen in a 2000 psia convergence pressure system. (*Engineering Data Book*, 10th Ed., GPSA, Tulsa, 1987, with permission.)

Appendix B

This appendix presents equations which underlie the calculational charts in the book. Also, several alternate calculational procedures are given. The nomenclature used here is consistent with the nomenclature in the rest of the book. Units are defined with each set of equations as needed.

Equation 3–42, page 111, Pseudocritical Properties of Gas Mixtures

An alternative to Kay's method for calculating pseudocritical properties when gas composition is known is given below. This method provides pseudocritical properties which give more accurate values of z-factors when used with Figure 3–7 or 3–8.

Kay's method results in reasonable values of z-factor at pressures below about 3000 psia and for gases with specific gravities less than 0.75. The Stewart, Burkhardt, Voo method as modified by Sutton should be used at higher pressures.[1,2]

$$T_{pc} = (K')^2/J', \qquad (B-1)$$

$$p_{pc} = T_{pc}/J', \qquad (B-2)$$

where

$$J' = J - \epsilon_J, \qquad (B-3)$$

$$K' = K - \epsilon_K, \qquad (B-4)$$

and

$$J = (\tfrac{1}{3})\sum_j y_j(T_c/p_c)_j + (\tfrac{2}{3})\left[\sum_j y_j(T_c/p_c)_j^{1/2}\right]^2, \qquad (B-5)$$

$$K = \sum_j y_j(T_c/p_c^{1/2})_j, \qquad (B-6)$$

where

$$\epsilon_J = 0.6081F_J + 1.1325F_J^2 - 14.004F_J y_{C7+} + 64.434F_J y_{C7+}^2 \quad (B-7)$$

$$\epsilon_K = [(T_c/p_c^{1/2})_{C7+}]\left[0.3129y_{C7+} - 4.8156y_{C7+}^2 + 27.3751y_{C7+}^3\right], \quad (B-8)$$

and

$$F_J = (1/3)\left[y(T_c/p_c)\right]_{C7+} + (2/3)\left[y(T_c/p_c)^{1/2}\right]_{C7+}^2. \quad (B-9)$$

The units of the critical and pseudocritical properties are psia and °R.

Figures 3–7 and 3–8, pages 112-113, Compressibility Factors of Natural Gases

A review by Takas of several numerical representations of the Standing Katz z-factor chart indicates that the Dranchuk Abou-Kassem equations duplicate Figure 3–7 with an average absolute error of 0.6 percent.[3,4] The equations fit Figure 3–8 almost as well, with the accuracy deteriorating as pressure and temperature increase. The results are three percent high at $p_{pr} = 30$ and $T_{pr} = 2.8$.

$$z = 1 + (A_1 + A_2/T_{pr} + A_3/T_{pr}^3 + A_4/T_{pr}^4 + A_5/T_{pr}^5)\rho_{pr}$$

$$+ (A_6 + A_7/T_{pr} + A_8/T_{pr}^2)\rho_{pr}^2$$

$$- A_9(A_7/T_{pr} + A_8/T_{pr}^2)\rho_{pr}^5$$

$$+ A_{10}(1 + A_{11}\rho_{pr}^2)(\rho_{pr}^2/T_{pr}^3) \, EXP \, (-A_{11}\rho_{pr}^2), \quad (B-10)$$

where

$$\rho_{pr} = 0.27\left[p_{pr}/(zT_{pr})\right]. \quad (B-11)$$

The constants are as follows: $A_1 = 0.3265$, $A_2 = -1.0700$, $A_3 = -0.5339$, $A_4 = 0.01569$, $A_5 = -0.05165$, $A_6 = 0.5475$, $A_7 = -0.7361$, $A_8 = 0.1844$, $A_9 = 0.1056$, $A_{10} = 0.6134$, $A_{11} = 0.7210$. The range of applicability is $0.2 \leq p_{pr} < 30$ for $1.0 < T_{pr} \leq 3.0$

and

$$p_{pr} < 1.0 \text{ for } 0.7 < T_{pr} < 1.0$$

Figure 3–10, page 116, Pseudocritical Properties of Heptanes Plus

Sutton and Whitson evaluated several methods for estimating the pseudocritical properties of heptanes plus.[2,5] The Kessler Lee equations are recommended.[6]

$$p_{pc} = EXP\Big[8.3634 - 0.0566/\gamma_{C7+}$$

$$- (0.24244 + 2.2898/\gamma_{C7+} + 0.11857/\gamma_{C7+}{}^2)10^{-3}T_B$$

$$+ (1.4685 + 3.648/\gamma_{C7+} + 0.47227/\gamma_{C7+}{}^2)10^{-7}T_B{}^2$$

$$- (0.42019 + 1.6977/\gamma_{C7+}{}^2)10^{-10}T_B{}^3\Big] \qquad (B-12)$$

and

$$T_{pc} = 341.7 + 811\gamma_{C7+} + (0.4244 + 0.1174\gamma_{C7+})T_B$$

$$+ (0.4669 - 3.2623\gamma_{C7+})10^5/T_B \qquad (B-13)$$

If boiling point is unknown, it can be estimated from the equation developed by Whitson.[5]

$$T_B = \Big[4.5579M_{C7+}{}^{0.15178} \gamma_{C7+}{}^{0.15427}\Big]^3 \qquad (B-14)$$

Figure 3–10 was prepared using these three equations. Pressures and temperatures are psia and °R.

Figure 3–11, page 119, Pseudocritical Properties of Natural Gases

Sutton derived equations for pseudocritical properties of natural gases based on measured z-factors for 264 natural gas samples.[2] He used the Dranchuk Abou-Kassem equation for z-factors and the Wichert Aziz adjustment for nonhydrocarbon components.[4,7,8] Sutton's equations are

$$p_{pc} = 756.8 - 131.0\gamma_g - 3.6\gamma_g^2 \qquad (B-15)$$

and

$$T_{pc} = 169.2 + 349.5\gamma_g - 74.0\gamma_g^2. \qquad (B-16)$$

Figure 3–11 was prepared using these two equations. Pressures and temperatures are psia and °R.

Figure 3–12, page 122, Nonhydrocarbon Adjustment Factors for Pseudocritical Properties of Natural Gases

The equation underlying Figure 3–12 is[7]

$$\epsilon = 120(A^{0.9} - A^{1.6}) + 15(B^{1/2} - B^4), \qquad (B-17)$$

where A is the sum of the mole fractions of carbon dioxide and hydrogen sulfide and B is the mole fraction of hydrogen sulfide. ϵ has units of °R. This equation is based on a set of data with carbon dioxide contents to 55 mole percent and hydrogen sulfide contents to 74 mole percent.

Figure 6–4, page 177, Pseudoreduced Compressibilities of Natural Gases

The equation for pseudoreduced compressibility of gas is

$$c_{pr} = \frac{1}{p_{pr}} - \frac{1}{z}\left(\frac{\partial z}{\partial p_{pr}}\right)_{T_{pr}}. \qquad (6-13)$$

The equation for pseudoreduced density is

$$\rho_{pr} = \frac{\rho_g}{\rho_{gc}}, \qquad (B-18)$$

where ρ_{gc} is the density of the gas at its critical point. Mattar, et al. combined Equation 6–13 with[9]

$$\rho_{pr} = 0.27\left[p_{pr}/(zT_{pr})\right]. \qquad (B-11)$$

where the value of z-factor of the gas at its critical point is assumed to be 0.27, to arrive at

$$c_{pr} = \frac{1}{p_{pr}} - \frac{0.27}{z^2 T_{pr}} \left[\frac{\left(\frac{\partial z}{\partial \rho_{pr}}\right)_{T_{pr}}}{1 + \left(\frac{\rho_{pr}}{z}\right)\left(\frac{\partial z}{\partial \rho_{pr}}\right)_{T_{pr}}} \right]. \quad (B-19)$$

An expression for $(\partial z/\partial \rho_{pr})_{T_{pr}}$ can be derived from Equation B–10.

$$\left(\frac{\partial z}{\partial \rho_{pr}}\right)_{T_{pr}} = A_1 + A_2/T_{pr} + A_3/T_{pr}^3 + A_4/T_{pr}^4 + A_5/T_{pr}^5$$

$$+ 2\rho_{pr}(A_6 + A_7/T_{pr} + A_8/T_{pr}^2) - 5\rho_{pr}^4 A_9(A_7/T_{pr} + A_8/T_{pr}^2)$$

$$+ \frac{2A_{10}\rho_{pr}}{T_{pr}^3} (1 + A_{11}\rho_{pr}^2 - A_{11}^2\rho_{pr}^4) \text{ EXP } (-A_{11}\rho_{pr}^2), \quad (B-20)$$

where values of the constants are given with Equation B–10. Equations B–19 and B–20 are the basis of Figure 6–4.

Notice that the minimum points of the z-factor isotherms of Figure 3–7 are rather sharp at values of pseudoreduced temperature below about 1.4. The slopes of the lines change from negative to positive rather abruptly. Equation B–20 does not predict accurately these slopes near these minima. Thus Equations B–19 and B–20 should not be used at T_{pr} less than 1.4 at p_{pr} between 0.4 and 3.0

This difficulty was recognized in the preparation of Figure 6–4. At the minima, $(\partial z/\partial p_{pr})_{T_{pr}}$ are zero, thus Equation 6–6 applies. The isotherms of Figure 6–4 were adjusted to pass through these points.

Equation 6–16, page 180, Gas Viscosities

Lohrenz, et al. developed a technique for calculating liquid viscosity given composition.[12] The procedure often is used to compute gas viscosity. It is too lengthy for enclosure here.

Figure 6–8, page 182, Viscosities of Natural Gases at Atmospheric Pressure

Standing developed equations representing Figure 6–8 in the ranges $0.55 < \gamma_g < 1.55$ and $100 < T < 300°F.$[10]

$$\mu_{g1} = (\mu_{g1} \text{ uncorrected}) + (N_2 \text{ correction})$$
$$+ (CO_2 \text{ correction}) + (H_2S \text{ correction}), \quad (B-21)$$

where

$$(\mu_{g1} \text{ uncorrected}) = \left[1.709(10^{-5}) - 2.062(10^{-6})\gamma_g\right]T$$
$$+ 8.188(10^{-3}) - 6.15(10^{-3})\log \gamma_g, (B-22)$$

where T is °F,

$$(N_2 \text{ correction}) = y_{N2}\left[8.48(10^{-3})\log \gamma_g + 9.59(10^{-3})\right], (B-23)$$

$$(CO_2 \text{ correction}) = y_{CO2}\left[9.08(10^{-3})\log \gamma_g + 6.24(10^{-3})\right], (B-24)$$

$$(H_2S \text{ correction}) = y_{H2S}\left[8.49(10^{-3})\log \gamma_g + 3.73(10^{-3})\right].(B-25)$$

μ_1 and the units of the corrections are all centipoise.
Equation B–22 fits Figure 6–8 to within less than ½ percent.

Figures 6–9 through 6–12, pages 184-187, Gas Viscosities

The underlying equations are[11]

$$\mu_g = A(10^{-4}) \text{ EXP} (B\rho_g^{\ C}), \qquad (B-26)$$

where

$$A = \frac{(9.379 + 0.01607M_a)T^{1.5}}{209.2 + 19.26M_a + T}, \qquad (B-27)$$

$$B = 3.448 + \frac{986.4}{T} + 0.01009M_a, \qquad (B-28)$$

$$C = 2.447 - 0.2224B. \tag{B-29}$$

ρ_g is gas density in gm/cc, M_a is apparent molecular weight, T is temperature in °R , and μ_g is gas viscosity in centipoise.

Figure 7–1, page 204, Ratio of Reservoir Gas Gravity to Surface Gas Specific Gravity

This figure is represented by[10]

$$\frac{\gamma_{gR}}{\gamma_g} = \left(\frac{240 - 2.22API}{28.97\gamma_g} \right)$$

$$\left(\frac{28.97\gamma_g(131.5 + API) + 18.76Y}{31,560 - 52.2API - 2.22API^2 + 18.76Y} \right), \tag{B-30}$$

where Y is condensate volume in STB/MMscf, API is the stock-tank liquid gravity in °API, γ_g is surface gas specific gravity, and γ_{gR} is the specific gravity of the reservoir gas.

Figure 7–2, page 205, VEQ Correlation for Three-Stage Separation

Gold et al. developed the following equation from three-stage separation of 237 retrograde gas samples.[13] The average absolute error is 5.8%.

$$VEQ = B_0 + B_1(p_{SP1})^{B_2}(\gamma_{SP1})^{B_3} (API)^{B_4}(T_{SP1})^{B_5}(T_{SP2})^{B_6}, \tag{B-31}$$

where

$$B_0 = 535.916, B_1 = 2.62310, B_2 = 0.793183, B_3 = 4.66120,$$

$$B_4 = 1.20940, B_5 = -0.849115, B_6 = 0.269870.$$

T_{SP1} is primary separator temperature in °F, T_{SP2} is second separator temperature in °F, p_{SP1} is primary separator pressure in psia, API is gravity of the stock-tank liquid in °API, and γ_{SP1} is specific gravity of the primary separator gas.

The pressure of the second-stage separator did not affect significantly the correlation and was held at 70 psia. The reservoir samples contained at least four mole percent heptanes plus and less than five mole percent nonhydrocarbons.

Utilizing this equation and the AGP equation which follows in Equation 7–9 results in calculated values of reservoir gas specific gravity which agree with laboratory measurements to within one percent.

Figure 7–3, page 207, AGP Correlation for Three-Stage Separation

Gold et al. developed this correlation for use in calculating the specific gravities of reservoir gases.[13] Nomenclature and comments given above in connection with Equation B–31 are applicable.

$$AGP = B_1(p_{SP1} - 14.65)^{B_2}(\gamma_{SP1})^{B_3}(API)^{B_4}(T_{SP1})^{B_5}(T_{SP2})^{B_6}, \quad (B-32)$$

where

$$B_1 = 2.99222, \quad B_2 = 0.970497, \quad B_3 = 6.80491,$$

$$B_4 = 1.07916, \quad B_5 = -1.19605, \quad B_6 = 0.553670.$$

The ranges of data on which Equations B–31 and B–32 are based are

$$p_{SP1} = 100 \text{ to } 1500 \text{ psia}$$

$$\gamma_{SP1} = 0.60 \text{ to } 0.80$$

$$\gamma_{gr} = 0.80 \text{ to } 1.55$$

$$API = 40° \text{ to } 70°API$$

$$T_{SP1} = 60° \text{ to } 120°F$$

$$T_{SP2} = 60° \text{ to } 120°F$$

This correlation was tested with gases with nonhydrocarbon contents up to 43 percent. Reservoir gas specific gravities calculated with Equation 7–9, using Equations B–31 and B–32 for VEQ and AGP, were within one percent of laboratory measurements when nonhydrocarbon content was below five percent and within six percent of laboratory measurements when nonhydrocarbon content was between five percent and 20 percent. These equations are not recommended for use for gases with nonhydrocarbon contents above 20 percent.

Figure 7–4, page 208, VEQ Correlation for Two-Stage Separation

Gold et al. developed the following equation from two-stage separation of 237 retrograde gas samples.[13] The average absolute error is 6.3 percent.

$$VEQ \ = \ B_0 \ + \ B_1(p_{SP1})^{B_2}(\gamma_{SP1})^{B_3}(API)^{B_4}(T_{SP1})^{B_5}, \qquad (B-33)$$

where

$$B_0 \ = \ 635.530, \ B_1 \ = \ 0.361821, \ B_2 \ = \ 1.05435,$$

$$B_3 \ = \ 5.08305, \ B_4 \ = \ 1.58124, \ B_5 \ = \ -0.791301.$$

Nomenclature and comments in connection with Equation B–31 apply.

Figure 7–5, page 209, AGP Correlation for Two-Stage Separation

Gold et al. developed this equation for use in calculating the specific gravities of reservoir gases.[13] Nomenclature and comments given in connection with Equation B–31 apply.

$$AGP \ = \ B_1(p_{SP1} \ - \ 14.65)^{B_2}(\gamma_{SP1})^{B_3}(API)^{B_4}(T_{SP1})^{B_5}, \qquad (B-34)$$

where

$$B_1 \ = \ 1.45993, \ B_2 \ = \ 1.33940, \ B_3 \ = \ 7.09434,$$

$$B_4 \ = \ 1.14356, \ B_5 \ = \ -0.934460.$$

The ranges of data and comments given in the discussion of Equations B–31 and B–32 apply to Equation B–33 and B–34 except that the range of data for p_{SP1} is 100-700 psia.

Dew-Point Pressure of Retrograde Gases, page 217

The Nemeth Kennedy equation reproduced the 579 data points used in its development with an absolute average deviation of 7.4 percent.[14]

$$\ln p_d = A_1 \Big[y_{C2} + y_{CO2} + y_{H2S} + y_{C6} + 2(y_{C3} + y_{C4})$$
$$+ y_{C5} + 0.4 y_{C1} + 0.2 y_{N2} \Big]$$
$$+ A_2 \gamma_{C7+} + A_3 \Big[y_{C1}/(y_{C7+} + 0.002) \Big]$$
$$+ A_4 T + A_5 (y_{C7+} M_{C7+}) + A_6 (y_{C7+} M_{C7+})^2$$
$$+ A_7 (y_{C7+} M_{C7+})^3 + A_8 \Big[M_{C7+}/(\gamma_{C7+} + 0.0001) \Big]$$
$$+ A_9 \Big[M_{C7+}/(\gamma_{C7+} + 0.0001) \Big]^2$$
$$+ A_{10} \Big[M_{C7+}/(\gamma_{C7+} + 0.0001) \Big]^3 + A_{11}, \qquad (B-35)$$

where $A_1 = -2.0623054$, $A_2 = 6.6259728$, $A_3 = -4.4670559 \times 10^{-3}$, $A_4 = 1.0448346 \times 10^{-4}$, $A_5 = 3.2673714 \times 10^{-2}$, $A_6 = -3.6453277 \times 10^3$, $A_7 = 7.4299951 \times 10^5$, $A_8 = -1.1381195 \times 10^{-1}$, $A_9 = 6.2476497 \times 10^{-4}$, $A_{10} = -1.0716866 \times 10^{-6}$, $A_{11} = 1.0746622 \times 10$.

T is temperature in °R, the y's are compositions in mole fraction, M_{C7+} and γ_{C7+} are the molecular weight and specific gravity of the heptanes plus, and p_d is dew-point pressure in psia. The range of data used in development of this correlation was $106 < M_{C7+} < 235$, $0.7330 < \gamma_{C7+} < 0.8681$ and $T < 320°F$.

Figure 9-3, page 252, Stock-Tank Gas-Oil Ratio Correlation

Figure 9-3 is based on the following equation developed by Rollins, et al.[15]

$$\log R_{ST} = A_1 + A_2 \log \gamma_{STO} + A_3 \log \gamma_{gSP} + A_4 \log p_{SP} + A_5 \log T_{SP}, (B-36)$$

$$A_1 = 0.3818, \quad A_2 = -5.506, \quad A_3 = 2.902, \quad A_4 = 1.327,$$
$$A_5 = -0.7355,$$

where γ_{STO} is the specific gravity of the stock-tank liquid, γ_{gSP} is the specific gravity of the separator gas, p_{SP} is separator pressure in psia, and T_{SP} is separator temperature in °F.

Estimates of R_{ST} from Equation B-36 are accurate to within 12 percent; however, estimates of R_{sb} obtained by adding R_{ST} to producing gas-oil ratio, R_{SP}, are accurate to three percent.

Figure 11-1, page 297, Bubble-Point Pressure–Solution Gas-Oil Ratio Correlation

The following equations are the basis of Figure 11-1.[10]

$$p_b = 18.2\left[(CN)_{pb} - 1.4\right], \qquad (B-37)$$

where

$$(CN)_{pb} = \left(\frac{R_s}{\gamma_g}\right)^{0.83} 10^{(0.00091T - 0.0125API)}. \qquad (B-38)$$

T is reservoir temperature in °F, R_s is solution gas-oil ratio in scf/STB, and p_b is bubble-point pressure in psia. Rs should include separator gas and stock-tank gas. The equations are accurate to within 15 percent for temperatures to 325°F.

Another equation which is slightly more accurate than Equations 37 and 38 is[16]

$$p_b = A_1 R_{SP}^{A_2} \gamma_g^{A_3} \gamma_{STO}^{A_4} T^{A_5}, \qquad (B-39)$$

where

$A_1 = 5.38088 \times 10^{-3}$, $A_2 = 0.715082$, $A_3 = -1.877840$, $A_4 = 3.143700$, $A_5 = 1.326570$.

Temperature, T, is °R and γ_{STO} is the specific gravity of the stock-tank oil. Note that this equation gives better results when the separator gas-oil ratio is used rather than total gas-oil ratio.

Figure 11-2, page 302, Apparent Liquid Densities of Methane and Ethane

The equations

$$\rho_{a,C1} = 0.312 + 0.450\rho_{po} \qquad (B-40)$$

and

$$\rho_{a,C2} = 15.3 + 0.3167\rho_{po} \qquad (B-41)$$

can be deduced from Standing.[10] $\rho_{a,C1}$ is the apparent liquid density of methane, $\rho_{a,C2}$ is the apparent liquid density of ethane, and ρ_{po} is the pseudoliquid density, all in lb/cu ft.

Figure 11–3, page 303, Density Adjustment for Isothermal Compressibility of Reservoir Liquids

The following Equation agrees with the figure to within three percent at pseudoliquid densities above 40 lb/cu ft.[10] Do not use the equation above 9000 psia for pseudoliquid densities below 40 lb/cu ft.

$$\Delta\rho_p = \left(0.167 + 16.181(10^{-0.0425\rho_{po}})\right)\left(\frac{p}{1000}\right)$$

$$- 0.01\left(0.299 + 263(10^{-0.0603\rho_{po}})\right)\left(\frac{p}{1000}\right)^2, \qquad (B-42)$$

where p is pressure in psia, ρ_{po} is pseudoliquid density, and $\Delta\rho_p$ is the adjustment for pressure, both in lb/cu ft.

Figure 11–4, page 304, Density Adjustment for Isobaric Thermal Expansion of Reservoir Liquids

Figure 11–4 was developed from[17]

$$\Delta\rho_T = (0.00302 + 1.505\rho_{bs}^{-0.951}) (T - 60)^{0.938}$$

$$- \left(0.0216 - 0.0233(10^{-0.0161\rho_{bs}})\right) (T - 60)^{0.475}, \qquad (B-43)$$

where T is temperature in °F, ρ_{bs} is liquid density at pressure and 60°F, $\Delta\rho_T$ is the adjustment for temperature, both in lb/cu ft.

Figure 11–6, page 308, Density Ratios at Standard Conditions for Pseudoliquids Containing Methane and Ethane

Figure 11–6 was developed from[17]

$$\rho_{po} = \rho_{C3+}/\left(1.000 + 2.138156W1^{1.1027205} + 0.453717W2^{1.092823}\right), \quad (B-44)$$

where ρ_{po} is pseudoliquid density and ρ_{C3+} is density of the propane and heavier (including hydrogen sulfide) fraction of the liquid at 60°F and atmospheric pressure, both in lb/cu ft. W1 and W2 are defined by Equations 11–1 and 11–2.

Figure 11–7, page 310, Density Adjustment for Hydrogen Sulfide Content of Reservoir Liquids

Figure 11–7 was developed from[17]

$$\Delta\rho_{H2S} = 6.7473\left(\frac{w_{H2S}}{w_{mix}}\right) + 50.2437\left(\frac{w_{H2S}}{w_{mix}}\right)^2 \quad (B-45)$$

where w_{H2S}/w_{mix} is weight fraction of hydrogen sulfide in liquid and $\Delta\rho_{H2S}$ is the density adjustment in lb/cu ft.

The data upon which this equation was based had samples with H_2S content up to 35 mole %. However, little data were above 20%. Use the equation with confidence at H_2S content up to 20 mole %. Use with care at H_2S content up to 35%. Do not use at H_2S content greater than 35 mole %.

Figure 11–8, page 314, Apparent Liquid Densities of Natural Gases

The following equation agrees with Figure 11–8 to within one percent.[10]

$$\rho_a = 38.52(10^{-0.00326API}) + (94.75 - 33.93\log API)\log\gamma_g, \quad (B-46)$$

where API is the gravity of the liquid in °API and ρ_a is the apparent liquid density of the gas in lb/cu ft at 60°F and atmospheric pressure.

Figure 11–9, page 320, Formation Volume Factors of Saturated Black Oils

The following two equations are the basis of Figure 11–9.[10]

$$B_{ob} = 0.9759 + 12(10^{-5}) (CN)_{Bob}^{1.2}, \qquad (B-47)$$

where

$$(CN)_{Bob} = R_s \left(\frac{\gamma_g}{\gamma_{STO}} \right)^{0.5} + 1.25T , \qquad (B-48)$$

where B_o is res bbl/STB, γ_{STO} is the specific gravity of the stock-tank oil, and T is reservoir temperature in °F. Solution gas-oil ratio, scf/STB, at the bubble point is used for R_s to calculate B_{ob}; solution gas-oil ratio at any pressure below bubble point is used for R_s to calculate B_o at that pressure. These equations are accurate to within five percent for temperatures to 325°F.

Figure 11–11, page 327, Coefficients of Isothermal Compressibility of Undersaturated Black Oils

Figure 11–11 was developed from[18]

$$c_o = (A_1 + A_2R_{sb} + A_3T + A_4\gamma_g + A_5 API)/A_6\rho, \qquad (B-49)$$

where

$$A_1 = -1433.0, \; A_2 = 5.0, \; A_3 = 17.2,$$

$$A_4 = -1180.0, \; A_5 = 12.61, \; A_6 = 10^5$$

where c_o is in psi^{-1}, T is reservoir temperature in °F, γ_g is specific gravity of separator gas at separator pressure of 100 psig, API is gravity of stock-tank oil in °API, p is reservoir pressure in psia, and R_{sb} is solution gas-oil ratio, scf/STB, at bubble-point pressure.

Figure 11–12, page 329, Coefficients of Isothermal Compressibility of Saturated Black Oils

Figure 11–12 was developed from[19]

$$\ln(c_o) = -7.633 - 1.497 \ln(p) + 1.115 \ln(T) + 0.533 \ln(API) + 0.184 \ln(R_{sb}), \quad (B-50)$$

where c_o is in psi^{-1}, T is reservoir temperature in °R, p is reservoir pressure in psia, API is stock-tank oil gravity in °API, γ_g is the weighted average of separator gas and stock-tank gas specific gravities, and R_{sb} is solution gas-oil ratio at the bubble point in scf/STB.

When bubble-point pressure is known, a more accurate equation, within 10%, is

$$\ln(c_o) = -7.573 - 1.450 \ln(p) - 0.383 \ln(p_b) + 1.402 \ln(T) + 0.256 \ln(API) + 0.449 \ln(R_{sb}) .(B-51)$$

When neither bubble-point pressure nor solution gas-oil ratio at the bubble-point is known, a less accurate equation, within 15%, is

$$\ln(c_o) = -7.114 - 1.394 \ln(p) + 0.981 \ln(T) + 0.770 \ln(API) + 0.446 \ln(\gamma_g). \quad (B-52)$$

Equations B–50, B–51, and B–52 are valid to 5300 psia and 330°F.

Figure 11–13, page 330, Dead Oil Viscosities

Figure 11–13 was developed from[20]

$$\log \log (\mu_{oD} + 1) = 1.8653 - 0.025086 API - 0.5644 \log(T), (B-53)$$

where T is reservoir temperature in °F, API is stock-tank oil gravity in °API, and μ_{oD} is the viscosity of the pseudoliquid in centipoise at atmospheric pressure and T. This equation is based on data with ranges of 5 to 58°API and 60 to 175°F.

Figure 11–14, page 331, Viscosities of Saturated Black Oils

Figure 11–14 was developed from[21]

$$\mu_o = A\mu_{oD}{}^B, \quad (B-54)$$

where

$$A = 10.715(R_s + 100)^{-0.515}, \qquad (B-55)$$

and

$$B = 5.44(R_s + 150)^{-0.338}, \qquad (B-56)$$

where R_s is the solution gas-oil ratio of the liquid at p and T, μ_{oD} is the viscosity in centipoise of the pseudoliquid at T and atmospheric pressure, and μ_o is the viscosity in centipoise of the liquid at p and T. These equations are based on data with temperatures to 295°F and pressures to 5250 psig.

Figure 11–15, page 332, Viscosities of Undersaturated Black Oils

Figure 11–15 was developed from[18]

$$\mu_o = \mu_{ob}(p/p_b)^B, \qquad (B-57)$$

and

$$B = C_1 p^{C_2} EXP(C_3 + C_4 p), \qquad (B-58)$$

where $C_1 = 2.6$, $C_2 = 1.187$,
$C_3 = -11.513$, $C_4 = -8.98 \times 10^{-5}$,

and μ_{ob} is the viscosity of the reservoir liquid at the bubble-point, cp, p is any pressure above bubble-point pressure and p_b is bubble-point pressure, both in psia; and μ_o is the viscosity of the reservoir liquid at p. These equations were developed from a data base with pressures to 9500 psig; the applicable temperatures were not given.[18]

Lohrenz et al., presented a method of calculating the viscosities of reservoir liquids given composition.[12] The procedure is lengthy and is not reproduced here.

Figure 11–17, page 335, Parachors for Computing Interfacial Tension for Normal Paraffin Hydrocarbons

The data in Figure 11–17 are from Katz et al.[22] The equation of the correlation line is

$$P_j = 2.841M_j + 26.4. \qquad (B-59)$$

Figure 11-18, page 336, Parachors of Heavy Fractions for Computing Interfacial Tension of Reservoir Fluids

The data on the figure is from Firoozabadi et al.[23] The equation of the correlation line is

$$P_{C+} = -12.43 + 3.226M_{C+} - 0.002145M_{C+}^2 . \quad (B-60)$$

Chapter 14, K-factor Correlations

The following equation is not as accurate as the K-factor graphs of Appendix A[24]. However, certain solution techniques for gas-liquid equilibrium calculations with equations-of-state require initial trial values of K-factors. Equation B-61 is useful for this purpose.

$$K_j = \frac{EXP[5.37(1 + \omega_j) (1 - 1/T_{rj})]}{p_{rj}} \quad (B-61)$$

Figures 16-6 and 16-7, pages 447-448, Formation Volume Factors of Water

Figures 16-6 and 16-7 are represented by

$$\Delta V_{wT} = -1.0001(10^{-2}) + 1.33391(10^{-4})T + 5.50654(10^{-7})T^2 \quad (B-62)$$

and

$$\Delta V_{wp} = -1.95301(10^{-9})pT - 1.72834(10^{-13})p^2T - 3.58922(10^{-7})p - 2.25341(10^{-10})p^2 , \quad (B-63)$$

where T is in °F and p in psia.

Figure 16-8 page 449, Effect of Salinity on the Density of Brine

$$\rho_w = 62.368 + 0.438603S + 1.60074(10^{-3})S^2 \quad (B-64)$$

fits Figure 16-8 to within 0.1%. S is salinity in weight percent solids and Pw is the density of the water at standard conditions.

Figure 16-10, page 451, Solubility of Methane in Pure Water

The following set of equations fits Figure 16-10 to within five percent for the full range of temperature and for pressures from 1000 to 10,000 psia. Do not use these equations at pressures below 1000 psia.

$$R_{sw} = A + Bp + Cp^2, \qquad (B-65)$$

where

$$A = A_0 + A_1T + A_2T^2 + A_3T^3, \qquad (B-66)$$

where $A_0 = 8.15839$, $A_1 = -6.12265 \times 10^{-2}$, $A_2 = 1.91663 \times 10^{-4}$, $A_3 = -2.1654 \times 10^{-7}$,

$$B = B_0 + B_1T + B_2T^2 + B_3T^3, \qquad (B-67)$$

where $B_0 = 1.01021 \times 10^{-2}$, $B_1 = -7.44241 \times 10^{-5}$, $B_2 = 3.05553 \times 10^{-7}$, $B_3 = -2.94883 \times 10^{-10}$,

$$C = (C_0 + C_1T + C_2T^2 + C_3T^3 + C_4T^4)(10^{-7}), \qquad (B-68)$$

where $C_0 = -9.02505$, $C_1 = 0.130237$, $C_2 = -8.53425 \times 10^{-4}$, $C_3 = 2.34122 \times 10^{-6}$, $C_4 = -2.37049 \times 10^{-9}$. T in Equations B–66 through B–68 is °F and p is psia.

Figure 16–11, page 452, Effect of Salinity on Solubility of Natural Gas in Water

The following equation fits Figure 16–11 to within three percent.

$$\log \left(\frac{R_{sw} \text{ brine}}{R_{sw} \text{ pure water}} \right) = -0.0840655ST^{-0.285854}, \qquad (B-69)$$

where T is °F and S is weight percent solids.

Figures 16–12 and 16–13, pages 453-454, The Coefficient of Isothermal Compressibility of Brine

The high-temperature, high-pressure part of Figure 16–12 and Figure 16–13 were derived from the equation of Osif.[25]

$$c_w = 1/(A_1p + A_2S + A_3T + A_4), \qquad (B-70)$$

where $A_1 = 7.033$, $A_2 = 0.5415$, $A_3 = -537.0$, $A_4 = 403,300$, and p is pressure in psia, S is salinity in mg/l and T is temperature in °F.

This equation is valid only for temperatures between 200 and 270°F, pressures between 1000 and 20,000 psia, and salinities from 0 to

200,000 mg/l. Osif specifically warns that the equation should not be used outside of these ranges. Osif did not give an estimate of the accuracy of his experimental results or of Equation B–70.

Figure 16–14, page 455, Derivative of Solution Gas-Water Ratio with Respect to Pressure

The equation

$$\left(\frac{\partial R_{sw}}{\partial p}\right)_T = B + 2Cp \tag{B–71}$$

was derived from Equation B–65. Values of B and C are from Equations B–67 and B–68. The results of Equation B–71 do not fit Figure 16–14 very well.

Figure 16–16, page 457, Water Viscosities at Atmospheric Pressure

$$\mu_{w1} = AT^B, \tag{B–72}$$

where T is in °F,

$$A = A_0 + A_1S + A_2S^2 + A_3S^3, \tag{B–73}$$

where $A_0 = 109.574$, $A_1 = -8.40564$, $A_2 = 0.313314$, $A_3 = 8.72213 \times 10^{-3}$, and

$$B = B_0 + B_1S + B_2S^2 + B_3S^3 + B_4S^4, \tag{B–74}$$

where $B_0 = -1.12166$, $B_1 = 2.63951 \times 10^{-2}$, $B_2 = -6.79461 \times 10^{-4}$, $B_3 = -5.47119 \times 10^{-5}$, $B_4 = 1.55586 \times 10^{-6}$, and S is salinity in weight percent solids. The equations fit Figure 16–16 to within five percent at temperatures between 100 and 400°F and salinities to 26 percent.

A more complete study of the viscosity of sodium chloride solutions resulted in correlation equations involving 32 parameters for pressure, temperature, and salinity.[26] Also, a number of tables are given. These tables will not be repeated here.

Figure 16–17, page 458, Effect of Pressure on the Viscosity of Water

Figure 16–17 was plotted from

$$\frac{\mu_w}{\mu_{w1}} = 0.9994 + 4.0295(10^{-5})p + 3.1062(10^{-9})p^2. \quad (B-75)$$

Equation B–75 was developed from data which have a temperature range of 86.5 to 167°F and pressures to 14,000 psia.[27] The equation fits the data to within four percent for pressures below 10,000 psia and seven percent at pressures between 10,000 and 14,000 psia. Pressure is in psia.

Figures 16–18 and 16–19, pages 459-460, Moisture Contents of Natural Gases

Figures 16–18 and 16–19 were developed from[28]

$$W = A/p + B, \quad (B-76)$$

where W is moisture content in lb/MMscf, p is pressure in psia, and

$$A = p_{v,H2O} \frac{18(10^6)p_{sc}}{10.73(459.6 + T_{sc})z_{sc}}, \quad (B-77)$$

where $p_{v,H2O}$ is the vapor pressure of water in psia at the temperature of interest, p_{sc} and T_{sc} are standard conditions in psia and °F, and z_{sc} is the z-factor of the gas at standard conditions.

Also,

$$\log B = -3083.87(1/T) + 6.69449, \quad (B-78)$$

where T is temperature in °R. Bukacek gives a table of A and B versus temperature.[28]

Figure 16–20, page 462, Effects of Salinity on the Moisture Contents of Natural Gases

$$\frac{\text{water content from brine}}{\text{water content from pure water}} = 1 - 4.920(10^{-3})S - 1.7672(10^{-4})S^2 \quad (B-79)$$

fits Figure 16–20 to within one percent. S is salinity in weight percent solids.

Figure 16–23, page 466, Gas-Water Interfacial Tension

The data of Figure 16–23 were fit to within two percent with

$$\sigma_{gw} = A + Bp + Cp^2, \qquad\qquad (B-80)$$

where

$$A = 79.1618 - 0.118978T, \qquad\qquad\qquad (B-81)$$

$$B = -5.28473 \ (10^{-3}) + 9.87913(10^{-6})T, \qquad\qquad (B-82)$$

and

$$C = [2.33814 - 4.57194(10^{-4})T - 7.52678(10^{-6})T^2](10^{-7}) \quad (B-83)$$

T is in °F. Do not use these equations at pressures above 8000 psia or temperatures above 350 °F.

Figure 17–6, page 480, Hydrate-forming Conditions for Natural Gases

Equation B–84 was fit to Figure 16–7.[29]

$$T = 1/[C_1 + C_2 (\ln \gamma_g) + C_3 (\ln p) + C_4 (\ln \gamma_g)^2$$

$$+ C_5(\ln \gamma_g)(\ln p) + C_6(\ln p)^2 + C_7(\ln \gamma_g)^3 + C_8 (\ln \gamma_g)^2(\ln p)$$

$$+ C_9 (\ln \gamma_g) (\ln p)^2 + C_{10} (\ln p)^3 + C_{11} (\ln \gamma_g)^4$$

$$+ C_{12} (\ln \gamma_g)^3 (\ln p) + C_{13} (\ln \gamma_g)^2 (\ln p)^2 + C_{14} (\ln \gamma_g) (\ln p)^3$$

$$+ C_{15} (\ln p)^4], \qquad\qquad\qquad (B-84)$$

where $C_1 = 2.7707715 \times 10^{-3}$, $C_2 = -2.782238 \times 10^{-3}$, $C_3 = -5.649288 \times 10^{-4}$, $C_4 = -1.298593 \times 10^{-3}$, $C_5 = 1.407119 \times 10^{-3}$, $C_6 = 1.785744 \times 10^{-4}$, $C_7 = 1.130284 \times 10^{-3}$, $C_8 = 5.9728235 \times 10^{-4}$, $C_9 = -2.3279181 \times 10^{-4}$, $C_{10} = -2.6840758 \times 10^{-5}$, $C_{11} = 4.6610555 \times 10^{-3}$, $C_{12} = 5.5542412 \times 10^{-4}$, $C_{13} = -1.4727765 \times 10^{-5}$, $C_{14} = 1.3938082 \times 10^{-5}$, $C_{15} = 1.4885010 \times 10^{-6}$.

The authors of Equation B–84 gave no indication of the accuracy of the fit.[29] Do not use this equation at temperatures above 62°F, pressures above 1500 psia, or gas specific gravities above 0.9.

Figure 17–7, page 481, Depression of Hydrate-formation Temperatures by Inhibitors

$$\Delta T_h = AS + BS^2 + CS^3, \qquad (B-85)$$

where

$$A = 2.20919 - 10.5746\gamma_g + 12.1601\gamma_g^2, \qquad (B-86)$$

$$B = -0.106056 + 0.722692\gamma_g - 0.85093\gamma_g^2, \qquad (B-87)$$

and

$$C = 0.00347221 - 0.0165564\gamma_g + 0.019764\gamma_g^2 \qquad (B-88)$$

may be used to estimate the depression in hydrate-forming temperature, °F, for natural gases with specific gravities below 0.68 and brines with salinities to 20 weight percent. When the results are rounded to the nearest whole degree Fahrenheit, the equations fit the curves of Figure 17–7 exactly. S is salinity in weight percent solids.

Figure 17–8, page 482, Depression of Hydrate-Forming Temperatures with Methanol and Diethylene Glycol

The following equation may be used instead of Figure 17–8. Hammerschmidt developed the equation[30]

$$\Delta T_h = \frac{2335W}{100M - MW}, \qquad (B-89)$$

where ΔT_h is the reduction in hydrate-forming temperature, °F, W is weight percent solute, and M is the molecular weight of the solute. The equation was developed for aqueous solutions of methanol, ethanol, isopropanol, and ammonia in concentrations from 5 to 20 weight percent. Scauzillo verified the application of the equation to diethylene glycol solutions up to 42.5 weight percent.[31] Hydrate-forming temperature depressions predicted with Equation B–89 were well within the scatter of the experimental data of Scauzillo.

References

1. Stewart, W.F., Burkhardt, S.F., and Voo, D.: "Prediction of Pseudocritical Parameters for Mixtures," paper presented at the AIChE Meeting, Kansas City, MO, May 18, 1959.
2. Sutton, R.P.: "Compressibility Factors for High-Molecular-Weight Reservoir Gases," paper SPE 14265 presented at the SPE 60th Annual Technical Conference and Exhibition, Las Vegas, Sept. 22–25, 1985.
3. Takas, G.: "Comparisons Made for Computer z-Factor Calculations," *Oil and Gas J.* (Dec. 20, 1976) 64–66.
4. Dranchuk, P.M. and Abou-Kassem, J.H.: "Calculation of z-Factors for Natural Gases Using Equations of State," *J. Can. Pet. Tech.* (July–Sept. 1975) *14*, 34-36.
5. Whitson, C.H.: "Effect of Physical Properties Estimation on Equation-of-State Predictions," paper SPE 11200 presented at the 57th Annual Fall Technical Conference and Exhibition, New Orleans, Sept. 26–29, 1982.
6. Kesler, M.G. and Lee, B.I.: "Improve Prediction of Enthalpy of Fractions," *Hydrocarbon Processing* (March 1976) *55*, 153-158.
7. Wichert, E. and Aziz, K.: "Calculate z's for Sour Gases," *Hydrocarbon Processing* (May 1972) *51*, 119–122.
8. Wichert, E. and Aziz, K.: "Compressibility Factor of Sour Natural Gases," *Can. J. Chem. Eng.* (Apr. 1971) *49*, 267–273.
9. Mattar, L., Brar, G.S., and Aziz, K.: "Compressibility of Natural Gases," *J. Can. Pet. Tech.* (Oct.–Dec. 1975) *14*, 77–80.
10. Standing, M.B.: *Volumetric and Phase Behavior of Oil Field Hydrocarbon Systems*, SPE, Dallas (1977).
11. Lee, A.L., Gonzales, M.H. and Eakin, B.E.: "The Viscosity of Natural Gases," *Trans.*, AIME (1966) *237*, 997–1000.
12. Lohrenz, J., Bray, B.G., and Clark, C.R.: "Calculating Viscosities of Reservoir Fluids from Their Compositions," *J. Pet. Tech.* (Oct. 1964) 1171–76; *Trans.*, AIME (1964) *231*; Reprint Series, SPE, Dallas (1981) *15*.
13. Gold, D.K., McCain, W.D., Jr., and Jennings, J.W.: "An Improved Method for the Determination of the Reservoir Gas Specific Gravity for Retrograde Gases," *J. Pet. Tech.*, (July 1989) *41*, 747-752.
14. Nemeth, L.K., and Kennedy, H.T.: "A Correlation of Dew-point Pressure with Fluid Composition and Temperature," *Trans.*, AIME (1967) *240*, 99–104.
15. Rollins, J.B., McCain, W.D., Jr., and Creeger, J.T.: "Estimation of Solution GOR of Black Oils," *J. Pet. Tech.*, (Jan. 1990) *42*, 92-94.

16. Al-Marhoun, M.A.: "PVT Correlations for Middle East Crude Oils," *J. Pet. Tech.* (May 1988) *40*, 650–666.
17. Witte, T.W., Jr.: *The Development of a Correlation for Determining Oil Density in High Temperature Reservoirs*, M.S. Thesis, Texas A&M University (Dec. 1987).
18. Vazquez, M. and Beggs, H.D.: "Correlations for Fluid Physical Property Prediction," *J. Pet. Tech.* (June 1980) *32*, 968–970.
19. McCain, W.D., Jr., Rollins, J.B., and Villena, A.J.: "The Coefficient of Isothermal Compressibility of Black Oils at Pressures Below the Bubble Point," *SPE Form. Eval.* (Sept. 1988) *3*, No. 3, 659–662.
20. Ng, J.T.H. and Egbogah, E.O.: "An Improved Temperature-Viscosity Correlation for Crude Oil Systems," paper 83-34-32 presented at the 34th Annual Technical Meeting of the Petroleum Society of CIM, Banff, May 10–13, 1983.
21. Beggs, H.D. and Robinson, J.R.: "Estimating the Viscosity of Crude Oil Systems," *J. Pet. Tech.* (Sept. 1975) *27*, 1140–1141.
22. Katz, D.L., et al.: *Handbook of Natural Gas Engineering,* McGraw-Hill Book Co., Inc., New York City (1959)129.
23. Firoozabadi, A., Katz, D.L., Soroosh, H., and Sajjadian, V.A.: "Surface Tension of Reservoir Crude-Oil/Gas Systems Recognizing the Asphalt in the Heavy Fraction," *SPE Res. Eng.* (Feb. 1988) *3*, No. 1, 265–272.
24. Bruno, J.A., Yanosik, J.L., and Tierney, J.W.: "Distillation Calculations with Nonideal Mixtures," *Extractive and Aziotropic Distillation,* Advances in Chemistry Series *115*, ACS, Washington, DC (1972).
25. Osif, T.L.: "The Effects of Salt, Gas, Temperature, and Pressure on the Compressibility of Water," *SPE Res. Eng.* (Feb. 1988) *3*, No. 1, 175–181.
26. Kestin, J., Khalifa, H.E., and Correia, R.J.: "Tables of the Dynamic and Kinematic Viscosity of Aqueous NaCl Solutions in the Temperature Range 20–150°C and the Pressure Range 0.1–35 MPa," *J. Phys. Chem. Ref. Data* (1981) *10*, No. 1, 71–87.
27. Collins, A.G.: "Properties of Produced Waters," *Petroleum Engineering Handbook*, H.B. Bradley et al. (eds.), SPE, Dallas (1987) 24-17.
28. Bukacek, R.F.: *Equilibrium Moisture Contact of Natural Gases,* Research Bulletin, IGT, Chicago (1955) *8*.
29. Kobayoshi, R., Kyoo, Y.S., and Sloan, D.E.: "Phase Behavior of Water/Hydrocarbon Systems," *Petroleum Engineering Handbook*, H.B. Bradley et al. (eds.), SPE, Dallas (1987) 25-13.

30. Hammerschmidt, E.G.: "Formation of Gas Hydrates in Natural Gas Transmission Lines," *Ind. Eng. Chem.* (1934) *26*, 851–855.
31. Scauzillo, F.R.: "Inhibiting Hydrate Formations in Hydrocarbon Gases," *Chem. Eng. Prog.* (Aug. 1956) *52*, 324-328.

INDEX

effect of, on gas hydrate formation, 480; properties of 492-494

Carboxylic acid: functional group of, 34; examples of, 34, 38; on naphthenic ring, 38; relationship of, to acid number, 38-39; in crude oil, 38

Cation: common, in oilfield water, 438; measurement of, 439; as cause of differences between brines, 439

Charles' equation: equation of, 91; use of, to yield ideal gas EOS, 91-94; derived from kinetic theory, 99

Chemical bonding: carbon-to-carbon, 4; types of, 4; in organic compounds, 6; stability of, 21; in cycloalkanes, 26; in benzene, 29; in gas hydrates, 474

Chemical classification of petroleum: according to structure, 1, 40, 41-42

Chemical potential: defined for pure fluid, 415; van der Waals' loop in, 415-416; in gas-liquid equilibria calculations, 417-418, 425; equation of, for ideal fluid, 417; equation of, for real fluid, 417; relationship of, to fugacity, 417; defined for a mixture, 425

Cis-trans isomerism. See Geometric isomerism

Clausius-Clapeyron equation: use of, in vapor pressure calculation, 53-54

Cloud point: defined, 41

Coefficient of isobaric thermal expansion: defined, 237; relationship of, to thermal expansion, 238; in liquid density calculations, 238, 302, 304, 520; mentioned, 301

Coefficient of isothermal compressibility of gas: defined, 170; effect on, of pressure, 170; units of, 170, 172; not equivalent to compressibility factor, 171; of ideal gas, equation, 172-173; of real gas, equation, 174; pseudoreduced, 175-176; estimate values of, 176,177; equivalence of, to oil compressibility, 231

Coefficient of isothermal compressibility of oil: defined, pressure above bubble point, 231; in logarithmic form, 231; values of, above bubble point, 231; in terms of formation volume factor, 232; in terms of density, 234; defined, pressure below bubble point, 234-235; effect on, of pressure, 235-236; of volatile oils, 241; estimate values of, 288-290, 326, 328, 522-523; use of, in oil density calculations, 302-303, 316; use of, in oil formation volume factor calculations, 321; mentioned, 224, 257, 296

Coefficient of isothermal compressibility of

water: effect on, of pressure, 451-452; relationship of, to oil compressibility, 452; defined, pressure above bubble point, 452; estimate values of, 452, 454-455, 526-527; effect of salinity on, 452; defined, pressure below bubble point, 454; mentioned, 438

Coefficient of viscosity of gas: as part of definition of gas, 90; defined, 178; units of, 178; effect of pressure on, 178-179; values of, experimental, 178, 280; of pure gases, 179, 180; relationship of, to volumetric behavior, 180; of gas mixtures, 180; estimate values of, 183-185, 513-514; as factor in gas flow in reservoir, 248

Coefficient of viscosity of oil: units of, 178, 236; defined, 236, 328; effect on, of pressure, 236-237, 328; of volatile oils, 241; values of, experimental, 280; related to ideal solution, 300; estimate values of, 330, 333, 523-524; mentioned, 1, 257, 258, 296

Coefficient of viscosity of water: units of, 178, 456; defined, 456; effect of pressure on, 456, 458; estimate values of, 458, 527-528

Composition of oilfield water. See Brine concentration

Composition of produced gas: values of, 1-2; of dry gas, 167; use of, heating value calculation, 188; of wet gas, 195; use of, recombination calculation, 195-196, 198, 270; use of, wet gas formation volume factor calculation, 211; use of, plant products calculation, 215-216; in sampling, 258, 270; use of, reservoir liquid density calculations, 311; estimate values of, separator calculation, 375

Compositional material balance: for volatile oils, 241

Compressibility. See Coefficient of isothermal compressibility

Compressibility equation. See compressibility equation of state

Compressibility equation of state: defined, 105; accuracy of, 118; limitation of, 129; in gas formation factor equation, 168; in gas compressibility equation, 173; in Joule-Thomson effect, 190; in wet gas formation volume factor calculation, 211, 214; use of, in EOS gas-liquid equilibria calculations, 420, 424; mentioned, 106

Compressibility factor. See z-Factor

Condensate: legal definition, 149; from retrograde gas, 156; from wet gas, 156, 195; estimate production rate of, 204; in